GROWTH
of Farm Animals

To our wives, Elizabeth Ray Lawrence and Janet Fowler, for their help and forbearance in the writing of this book.

GROWTH
of Farm Animals

T.L.J. LAWRENCE

Division of Animal Husbandry
Faculty of Veterinary Science
University of Liverpool
UK

and

V.R. FOWLER

formerly of the
Scottish Agricultural College
and of the Rowett Research Institute
Aberdeen
UK

CAB INTERNATIONAL

CAB INTERNATIONAL
Wallingford
Oxon OX10 8DE
UK

CAB INTERNATIONAL
198 Madison Avenue
New York, NY 10016-4341
USA

Tel: +44 (0) 1491 832111
Fax: +44 (0) 1491 833508
E-mail: cabi@cabi.org

Tel: + 1 212 726 6490
Fax: + 1 212 686 7993

A catalogue record for this book is available from the British Library, London, UK

ISBN 0 85199 143 2

Library of Congress Cataloging-in-Publication Data
Lawrence, T.L.J. (Tony Leonard John)
  Growth of farm animals / T.L.J. Lawrence and V.R. Fowler.
    p. cm.
  Includes bibliographical references and index.
  ISBN 0-85198-849-0 (alk. paper)
  1. Livestock—Growth. 2. Veterinary physiology. I. Fowler,
V.R. II. Title.
SF768.L39 1997   96-48880
636.089 '26—dc21   CIP

Typeset in Stempel Garamond by Solidus (Bristol) Limited
Printed and bound in the UK at the University Press, Cambridge

Contents

Introduction

An understanding of the processes which change the size, shape and composition of farm animals is fundamental to all aspects of production which seeks to meet the dietary and other needs of human populations. This book attempts, within the limits of a basic undergraduate text, to give a comprehensive picture of how animals grow, change in shape and in composition, and to describe those factors which affect growth processes and which dictate the extent and direction of changes within the animal.

The overall scene we have attempted to present is a progression from cell to tissue to entire animal, as well as a description of those factors within the animal, particularly hormones, genes and gender, which fashion this progression. In addition, an attempt has been made to give perspective to the manner in which such a complexity of changes affects the approach to, and understanding of, concepts of efficiency. However, it was felt that the framework would not be complete without first setting the scene in the context of some of the principles which govern size and shape in all animals and, second, describing the methods which may be used to measure the overall results of the processes which have been described. Deliberately, the effect on growth which exogenous factors such as hormonal implants, antibiotics and other intestinal tract manipulators may have has not been considered. Although the use of such manipulators may be transitory and dependent on political and other pressures, basically they were not considered to be appropriate to the general thesis. Whilst it is acknowledged that poultry are farm animals of great importance in producing food for the human, largely through a dearth of appropriate detailed information and because of the approaches to production which are used in practice, this book concentrates on cattle, sheep and pigs.

In our student days we were endlessly fascinated by the seminal studies of D'Arcy Thompson, by the profound work of Brody at Missouri and by the inspired and inspiring work and writing of Hammond and his associates at Cambridge. To no less an extent we found the sheer variety of animal cells and the immensely complex tissues that they form, and which give ultimately the overall growth responses, both amazing and mind gripping. These fascinations have remained with us ever since and in our working lives we have been fortunate to have had the chance to make some contributions to some of those areas which have so consistently intrigued us and occupied our thoughts. As students the writings of those great men above were those to which we turned. Today much of their basic thinking still holds good and leads the enquiring mind in the right direction when considering the problems of how best farm animals may be grown to yield products for the human. However, there is no book which as a complete entity has updated their basic thinking and which takes the reader from cell to complete animal based on original classical anatomical studies and the quantitative and other studies which have followed subsequently. It is hoped that this book will fill this important gap for students and will be of use to all who are interested, whether professionally or otherwise, in animal, particularly farm animal, growth. At a time when man is at the point of being able to manipulate growth, shape and composition of animals by genetic engineering and by other biotechnological processes, we hope that the book will be a timely, and indeed perhaps even a salutory, reminder of the principles which nature has endowed and which man has grappled with for a long time.

Tony Lawrence
Liverpool

Vernon Fowler
Aberdeen

General Aspects of Growth

1

1.1. Introduction

Growth is one of the main attributes of living things and is such an obvious process that it hardly seems to justify any particular formal definition. The simple concept of growth meaning getting bigger, is perhaps rather better than many of the complicated attempts to formalize something of such extraordinary complexity. In general, it is most helpful to use a descriptive word or phrase to qualify growth to identify the broad aspect with which one is concerned. For example, within one individual animal, one may speak about cell growth, organ growth, fetal growth, prepubertal growth, bone growth, chemical growth or negative growth and so on. Indeed it is possible to consider almost every aspect of the expression of genes as an aspect of growth. This discussion will be mainly restricted to the physical aspects of growth but it should not be forgotten that the more abstract expressions of genes in the phenotype, such as immunity and the growth of the mental capabilities and behaviour, are intrinsically related to the physical development of critical cells.

In farm animals the main interest lies in the growth of specific parts of the animal such as bone, muscle, fat or the development of the mammary gland. These aspects of growth are readily appreciated and can be easily subjected to quantification either by weighing or by linear measurement. In this age of dramatic advances in the high technology of biochemistry and genetic engineering, it is helpful to remind ourselves of the biological significance of the size and physical form of animals.

This chapter has a twofold purpose. The first aim is to introduce the subject of growth by a review of the broad perspectives of the biological background of growth as readily perceived by the eye and without the use of sophisticated instrumentation or abstruse algebra. The second objective is to show that many of the principles which apply to growth, at all levels of understanding, can be derived from a reflective consideration of the natural world. Examples are taken across the spectrum of animal life in the belief that it is only after considering animals as a whole that one can grasp some of the essential issues which apply to farm livestock.

1.2. Being the Right Size

There is a tendency during the process of selection of domestic animals for what is regarded as genetic improvement to produce livestock which are larger at a given age and also larger as adults than the ancestral types. The very small breeds of cattle such as the Kerry and Dexter, the primitive Soay sheep and the small pork-type Berkshire and Middle White pigs are virtually obsolete in European production systems. Sadly, for those for whom nostalgia is not a weakness, such breeds have become the stuff of genetic zoos and preservation societies. In their place we have the comparatively gigantic Holsteins and Charolais cattle, large and sturdy Suffolk sheep and the modern bacon-type Large White and Landrace breeds of pigs. There are a number of economic and pragmatic reasons why this has arisen and these will be discussed later (see chapter 7).

In wild animals no such clear advantage exists for size in its own right. Land mammals differ in adult size by a colossal factor. There is some dispute about the smallest which could be either the tiny Etruscan shrew (*Suncus etruscus*) weighing only 2g or the tiny 'bumble bee' bat known also as Kitti's hog-nosed bat (*Craseonycteris thonglongyai*). The largest land mammal today is the African elephant (*Loxodonta africana*) weighing about 5 tonnes with the occasional exceptionally large bull weighing up to 10 tonnes. The factor of difference is a staggering 5×10^6. Even within the ruminants, with their similarity of digestive function, there are very large differences. The small antelope known as the Suni (*Neostragus moschatus*) weighs only about 5kg, whilst the giraffe (*Giraffa camelopardalis*) can weigh up to about 1900kg.

In prehistoric times there were many giant forms of the mammals with which we are familiar today, such as the giant rhinoceros (*Paraceratherium*) thought to stand about 5 metres high and weigh possibly about 15 tonnes. Further back in the era of the dinosaurs, about 120 to 65 million years ago, the scale was extraordinary, with the vegetarian Brachiosaurus estimated to weigh in the order of 80 metric tonnes. The formidable carnivore Tyrannosaurus, which was only capable of locomotion on its two well-developed hind legs, stood an astonishing 6 metres high and had an estimated length of 8 metres.

Surprisingly, these prodigies of size are exceeded by several marine mammals. The largest animal ever known to have lived is the extant blue whale (*Balaenoptera musculus*) which is estimated to have attained weights in excess of 170 tonnes particularly in the days when the species was not exploited by man. Elephants do not dominate on land and nor do the whales rule the seas. The fact is that optimum size depends on a whole range of subtle interacting factors. In a general sense, as Charles Darwin was among the first to point out, this can be described as fitness in relation to the environment, depending on the exact ecological niche of the species or strain.

1.3. Why Do Animals Change in Form as They Grow?

There are two basic reasons why animals change their form during growth. The first is relatively obvious, which is that the animal changes in its physiological needs as it matures. An extreme example is the life history of the common frog or toad which transforms its physical appearance over a very short period to adapt to the change from aquatic to mainly terrestrial living. A less extreme example relating to domesticated livestock is the change which takes place in the calf at weaning. At birth and during the suckling period, there is little use for the rumen which remains small and undeveloped, whilst at this stage it has a relatively large abomasum. However, as soon as roughage feeds feature in its diet, there is a reversal of the relative sizes and soon the rumen is the largest organ in the digestive tract.

The second constraint upon land animals is to respond to the physical consequences of growing bigger. Land animals must contend with gravity, and this poses increasing problems for animals which are very large. The problems were noted by Galileo when he stated (not in English) 'nor can Nature grow a tree nor construct an animal beyond a certain size whilst retaining the proportions and employing the materials which suffice in the case of the smaller structure'. This theme has been taken up by several authors, notably Professor D'Arcy Thompson in his classic book *On Growth and Form* (1942) and Brody (1945).

The physical principles involved in changing scale have been aptly summarized by Brody as:

1. Weight which tends to crush the land animal's limbs and which has to be moved by muscles varies with the cube of linear size.
2. Tensile strength of the muscle and bones which move and support the animal varies with the square of linear size.
3. Surfaces through which diffusion, nutrition and excretion take place vary with the square of linear size.

The organism changes geometrically so as to remain the same physiologically.

What Brody is here amplifying is a principle much respected by biologists in the late nineteenth century, namely that of physiological homeostasis. In other words, it is important for the survival of the animal to keep its critical cells protected from variation in temperature, pressure, nutrition, oxygen supply and so on, and this buffering is achieved by major and minor modifications to its whole strategy for growing and living.

There are a host of interesting structural examples which illustrate this point from the point of view of growth and form. Although these examples are not taken specifically from farm animals, the same general principles apply. Many interesting examples are given in the seminal studies of D'Arcy Thompson (1942), and the reader is urged to consult these highly original studies which have many beautiful illustrations. Here, just three examples will be concentrated on which show some aspects of the problems of scaling and which also show, incidentally, how endlessly fascinating, in terms of the natural world and also in the field of engineering, is the relationship between form and function.

1.3.1. Growth of the eye

The eye is one of the earliest organs to reach its mature size and when it is finally differentiated it has usually already attained a high proportion of its final dimensions. Young children and young animals are characterized by the apparently large relative size of the eye in comparison with the rest of the head or face. The eye is an optically precise 'instrument' and it is obvious from the human analogy that very exact proportions are necessary for its proper function. The retina at the back of the eye forms a part of a virtual sphere, and this shape is maintained partly by the rigidity of the tough sclerotic membrane and also by the pressure generated internally within the aqueous humour. It is easy to understand how this principle operates for relatively small eyes, that is those which have a diameter between 5 and 100 mm, but if eyes were for example as large as a football it would be increasingly difficult to maintain the spherical shape during the normal movements of a land animal. This is because the pressures needed to maintain stability and rigidity would exceed the physical and physiological limitations of the cells and tissues. The principle is easily demonstrated by considering the perfect sphere which is characteristic of small individual soap bubbles and the distortion which occurs as they increase in size.

In land mammals, the principle can be caricatured by taking two extreme examples. The tarsier family (Tarsiidae) is used as a representative of small animals and the hippopotamus as a representative of the large animals. Tarsiers are tiny primates weighing about 100 g, and, weight for weight, have the largest eyes of any mammal. They are a predatory animal and do most of their hunting at night. Their eyes are exceptionally critical for their survival. The weight of their eyes exceeds that of the brain and they occupy about two-thirds of the facial diameter. The hippopotamus weighs about 4000 kg and to have a similar ratio of eye diameter to facial width would require a diameter of eye of about 300 mm, the size of a football. In fact the diameter of the eye of the hippopotamus is about 100 mm and that of other very large mammals, for example whales and elephants, is about the same and suggests that 100 mm is close to the largest size which can be adequately functional in the case of land animals. Even the eye of the massive blue whale which under normal circumstances is supported by water has been reported as only being about 120 mm in diameter in the largest recorded specimen. However, in the case of that other monster of the seas the giant squid (*Architeuthis* sp.), which again is supported by the surrounding water of its habitat, the diameter of the eye is claimed to be the largest of any living animal at 380 mm, that is just over one-third of a metre (*Guinness Book of Records*, 1994).

1.3.2. Growth of wings

The power of flight is characteristic of many thousands of different animals ranging in size from the tiny fairy flies such as the 'battledore winged fairy fly' (Mymaridae) which is a mere 0.2mm long and an infinitesimal 5×10^{-6} g in weight to, at the other extreme, the rarely seen frantic efforts to fly of the adult domestic turkey weighing up to about 20kg. It can sustain flight for barely a few seconds. The power to weight ratio is absolutely critical and as a general rule no wild bird exceeding 15kg in weight can sustain flight for any lengthy period. The only flighted birds which exceed this weight and then only in rare instances are the mute swan (*Cygnus olor*) and the Kori bustard (*Otis kori*).

The wing systems that can operate for any given size vary considerably. The agile flight of insects and dragonflies in particular can be achieved by the rapidly beating membranous wings. The smallest birds such as the humming birds also have relatively small wings which beat rapidly and give them great agility so that they can hold their station whilst sucking nectar even in gusting winds. As birds get larger the problem of achieving sufficient lift demands another type of specialization and, increasingly, the wing takes on the aerodynamic features seen in a less sophisticated form in aircraft and gliders. Large birds like the swan (*Cygnus* sp.), albatross (*Diomedea*) and the condor of the Andes (*Vultur gryphus*) all have a wing span of about 3m but still have enormous difficulty in becoming airborne because of the problems of generating sufficient power for the initial lift. They tend to augment their own muscular effort by running into the prevailing wind or by launching themselves from elevated positions. Ultimately, birds which consistently attain weights of greater than 15kg as adults have settled for a mode of existence which does not require being airborne. The largest birds of the present era are the North African ostriches (*Struthio camelus camelus*) which weigh about 150kg. These and most lesser sized flightless birds have no 'keel' on the sternum to support the enormous pectoral muscles which are necessary for flight.

1.3.3. The pinna of the ear

An easily observed feature of animals is the pinna of the ear. In small animals these function to collect and focus sound waves on to the eardrum. The relatively huge ears of the long-tailed Mongolian gerboa and the jack rabbit of the Central Southern States in the USA are probably used to dissipate heat. In the Arctic hare and rabbit the animal has an awkward compromise between the need for acuity of hearing and the potential hazard of frostbite, and in these high latitude species the ears are relatively smaller. The elephant too uses its ears to focus sound and to dissipate heat, but the problem of an ear weighing 100kg necessitates a different anatomical arrangement. Instead of being positioned on the top of the head, the ears are attached down the side and swing to and fro like a well-hinged and well-oiled door.

1.4. Shape and Mass

Those land mammals which attain a very great size, that is in excess of about 100 kg, must forfeit some of the options available to small animals. In general the bodies become progressively shorter in relative terms so that they do not have such a great problem in supporting a lengthy arch. The problems are not dissimilar to those of the bridge builder attempting to span a wide river. A structure which would be suitable for transport over a small stream could effectively be a mere plank of wood with hand rails. This simple structure is not at all suitable for longer spans. The engineering glories of the great cantilevered arches or suspension bridges of the world, such as the rail and road bridges over the Forth estuary near Edinburgh in Scotland, illustrate one particular engineering approach to the problem. Indeed one can liken the skeletal structure of the elephant to that of a Roman arch, the structure of the large dinosaurs to the Forth Railway Bridge, and the structure of giraffes and for that matter Giant Sequoias to that of the Eiffel Tower in Paris. In essence, much that is deemed essential in the engineering of large structures has already been pioneered in the architecture of animals. Even in the construction of nuclear submarines the profile owes much to the shape and streamlining of the large whales.

Animals which live in water do not have gravity to contend with because of the buoyancy of the medium in which they live. In fact, very often young fish look very much like the mature fish of the same species several orders larger than themselves, and baby whales look almost like miniatures identical to their parents except for size. Many aquatic animals have developed a very streamlined appearance, particularly if they need speed through the water to escape from predators or, on the other hand, if they are themselves predators which capture their prey by superior speed. However, where aquatic and marine animals can achieve defence or sustenance by camouflaging themselves, then very exotic forms are produced such as those of the angler fish and octopus.

1.5. Domestication and Size of Animal

Rather surprisingly the range of animals domesticated by man for food is only about a score out of about 4500 mammalian species. The determinants appear to have been very complex.

In general, early man excluded the very largest which could break out of enclosures and which, unless allowed to roam and graze, were demanding in terms of their food requirement. The smallest were difficult to confine due to their mobility, many of them being able to dig or gnaw their way to freedom, although the Romans were happy to use the edible dormouse (*Glis glis*) which they housed in earthenware jars.

It is interesting to note that in his account entitled 'The taming of the few', Davis comments that in certain Natufian sites in the Middle East there is evidence that attempts were made to domesticate antelope of various kinds but that these were eventually replaced by sheep and goats (Davis, 1982). This was possibly because of their dual- and even triple-purpose attributes milk, wool and meat, but also because the extreme agility of the antelopes and their flighty nature, mitigated against domestication. There were advantages in choosing those with herding or flocking instincts, those which were not expert at climbing and those which did not create immediate danger either by their ferocity or by their physical strength. Depending on the main function of the animal the priorities lay with docility, the ability to breed regularly in the unnatural and often stressful environment of captivity and the ability to fatten, that is to store energy in the adipose depots since, above all, prior to the advent of the refrigerator, man has relied on animals as a feed store. Even in the sophisticated ambience of the end of the twentieth century many peasant communities are forced in the extremity of famine to slaughter their highly prized breeding livestock.

The species which fulfilled the requirements for domestication better than most were certain members of the Artiodactyla (the even-toed ungulates), such as pigs, sheep, goats and cattle, and the Camelidae, such as the Arabian and Bactrian Camels (*Camelus dromedarius* and *C.bactrianus*) and the llamas and alpaca of South America (*Lama peruana* and *L. pacas*).

Unfortunately there appear to be no unmodified ancestral types of modern cattle in the wild although wild sheep, such as the Mouflon in Europe and the Bighorn in North America, give a good indication of the form of ancestral sheep. The origins of the domestic pig are clearly seen in the European wild boar (*Sus scrofa*) and in the wild pigs of South-East Asia such as the bearded pig (*Sus barbatus*). There is no longer any wild equivalent of the single-humped camel of Africa, but the analogy is provided by the two-humped Bactrian camel of the Gobi desert which remains wild in one or two reserved areas.

1.6. Growth and Form

Small species of land animals are rarely miniatures of large ones nor are young land mammals tiny replicas of the adult. The changes from the egg to the adult are not merely ones of scale but also of form. The process of new structures and organs being formed is called differentiation, whilst the remodelling of these structures and the changing proportion which they constitute of the whole body can be described as differential growth.

1.7. Domestication and Growth

Many of the general principles explored above have some application to our domestic livestock and it is extremely helpful to carry over some of the ideas into this new domaine. Juliet Clutton-Brock (1987) has given a very full account of some of the relevant factors in her *Natural History of Domesticated Mammals*. However, one must be extremely careful of extrapolating principles of growth from the natural world to domesticated species because there has been a profound change in the rules. Farmers provide food and shelter for their livestock and protection from predators and disease. Within this environment, man pursues breeding programmes with livestock which are far removed from the principles of natural selection. Over a relatively short period of time, man has substantially modified the form of domesticated animals and transformed their efficiency in terms of converting a food resource into meat.

New techniques have brought with them the power to introduce new genetic material into the genome. It is now feasible to consider that new genes could be derived from other species, without the long process of either natural or artificial selection. Also, it is now possible to use modern biochemistry to manufacture the actual molecules which control growth and introduce these substances directly into animals, so that their growth patterns no longer conform to the general rules but take the whole subject of growth into totally uncharted combinations of high growth rate and abnormal body proportions.

These developments have huge implications for science, technology and indeed the whole role of animals in relation to man. It is perhaps more important now than it ever was, to understand the underlying biology of growth, the inextricable relationship between form and function and the implications for the future of farm livestock. The basic biology of growth at the cellular and tissue level will now be considered in the next few chapters.

References

Brody, S. (1945) *Bioenergetics and Growth*. Reinhold Publishing Company, New York.

Clutton-Brock, J. (1987) *A Natural History of Domesticated Mammals*. Cambridge University Press, Cambridge and British Museum (Natural History), London.

Davis, S. (1982) The taming of the few. *New Scientist*, 95: 697.

Guinness (1994) In: P. Matthews (ed.) *The New Guinness Book of Records 1995*. Guinness Publishing Ltd, Enfield, UK.

Thompson, D'Arcy W. (1942) *On Growth and Form*. Cambridge University Press, Cambridge.

Cells

2

2.1. Introduction

Cells are the basic building blocks of all animal tissues. In multicellular animals there are enormous ranges in size, type and function with cells specializing in certain functions to give a maximum overall efficiency. The mammal is the most advanced type of animal and in the higher mammals the degree of cell differentiation and interaction reaches a very high level, with the motor neuron cell perhaps representing the ultimate in cell differentiation and specialization.

With the exception of the process of wound healing and repair, where cellular growth reflects a specific response to injury, the normal course of events is one in which individual cells die and have to be replaced throughout the life of the organism. The replacement rate depends on the type of cell and on the stage of growth of the animal. When the animal is mature a balance has to be struck between depletion and repletion. When the animal is young and growing quickly the repletion rate must exceed the death rate of old cells. In the central nervous system the life of cells is long and the turnover rate is slow. In contrast, the white phagocytes of the blood have a life span of 1–3 days whilst red blood cells have a much slower turnover rate with an average life span of 120 days. Overall the processes of growth and replacement have to be balanced to maintain the structure and coordinated function of the organism as a whole but sometimes the balance is upset when cells grow and divide at abnormal rates, as in for example malignant diseases.

All cells are capable of moving but the cells most specialized for effecting movement are the muscle cells. Muscle tissue is discussed in more detail later (see chapter 3), but in terms of cell movement the point to note is that either under voluntary (striated muscle) or involuntary (smooth muscle) control, muscle cell movement is responsible for, respectively, limb movement, where a very quick action is mediated, and movement in sites such as the gut, in the arteries and in the uterus where the speed of action is much slower. Therefore in the former case the cell is responsible for physically moving the entire organism, whereas in the latter cases the cell helps in moving within the animal the digesta in the gut, the blood in the vascular system and the fetus in the uterus.

2.2. Cell Structure

2.2.1. General

A generalized diagram of an animal cell is given in Fig. 2.1. The division into nucleus and cytoplasm for the purpose of description is both conventional and convenient but in reality is very artificial because the cell functions as a complete entity.

The micron or micrometre (μm) is the unit used to describe the size of cells, whilst the angstrom unit (Å) is used to describe the size of the atoms and molecules of which cells are composed. One μm is equal to 10^{-6} metres (m) and one Å is equal to $10^{-4}\,\mu m$ or $10^{-10}\,m$. Sometimes, the nanometer (nm), which is

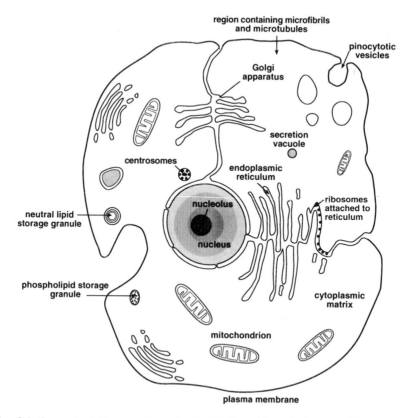

Fig. 2.1. Generalized diagram of an animal cell (adapted from Ambrose and Easty, 1977).

equal to 10^{-9}m or 10 Å is used to measure size. Examples of cell and other structure sizes are: the hydrogen atom, 1 Å; a protein molecule such as haemoglobin, 80–90 Å; bacteria, 1–15 μm; most cells in higher organisms that are multiplying, 20–30 μm; human ovum, about 200 μm and the largest cell of any animal body, the yolk of the ostrich egg, about 50^3 μm in diameter.

2.2.2. The nucleus

The nucleus itself is very much a functional unit in its own right. Its functions are twofold. First, it preserves genetic material intact and passes it from one generation to the next. Second, it is responsible for the direct synthesis of ribonucleic acid (RNA) and other cellular components. In the mitotic interphase of growing and dividing cells (see section 2.5) it assumes variable shapes, different to those in the phases when the cell is actually dividing, but invariably it increases in actual size. Within the nucleus of the resting cell the nucleolus is the body that is always clearly visible and is the site of synthesis of ribosomal RNA. It is a dense regular-shaped body consisting mostly of protein (proportionately up to 0.80 on a dry-weight basis) and most of the nucleus RNA. Thread-like strands of deoxyribonucleic acid (DNA) are only visible in cells actually engaged in dividing. It has no limiting membrane separating it from the rest of the nucleus and it may act in transferring genetic information from the nucleus to the cytoplasm. In contrast, the nucleus has a limiting membrane separating it from the cytoplasm. This membrane is comprised of two unit membranes, each of three layers and 40–60 Å thick, separated by a clear space of variable thickness but usually approximately 200 Å in width. Proportionately about 0.10 of the surface area of the membrane is covered with small holes which are about 500 Å in diameter and which possibly assist in the transfer of materials such as RNA into the cytoplasm. The membrane is not immortal and during cell division disappears. During metaphase it splits into fragments which then form rounded vesicles in the cytoplasm. The vesicles move into both daughter cells, aggregate around the chromosomal material and then become flattened to form new nuclear membranes. In rapidly dividing embryonic cells, but rarely in cells of differentiated tissue, the membrane is continuous with the endoplasmic reticulum and sometimes ribosomes are attached to it.

The chromosomes are often clearly visible in the nucleus and are considered later, relative to protein content, genetic coding and mitosis.

2.2.3. The cytoplasm

Collectively the various bodies found in the cytoplasm are often referred to as organelles. The cell membrane itself contains mostly lipid and protein with a small amount of carbohydrate. The proportion of lipid to protein varies greatly, but in most cells a phospholipid, lecithin and a steroid molecule, cholesterol, account for a large part of the lipid component. However, not only do different species and different cells within species have different compositions, but also the same cell type can change the composition of its membrane in response to changes in its environment, such as the diet received by the organism as a whole. Therefore some cells have dynamic membranes whilst others, such as the cells of myelin sheaths of nerve fibres, have more rigid, static membranes.

The endoplasmic reticulum consists of membrane-bound cavities which are linked together to form a complex branching network. The networks thus formed tend to be concentrated more in the inner, endoplasmic region of the cell than in the peripheral or ectoplasmic region. The form of the endoplasmic reticulum varies. It may consist of varying proportions of tubules, vesicles and large flattened sacs or cisternae and it may have a rough granular appearance on the outside, where ribosomes have become attached, compared with a smooth appearance in other areas. The rough appearance seems to be accentuated when cells are actively producing proteins and the Golgi apparatus is thought to be an extension of the smooth part of the reticulum. The amount of endoplasmic reticulum is not constant and varies with the age and with the function of the cell. Sometimes the endoplasmic reticulum is disrupted to form separate small spherical vesicles known as microsomes.

Ribosomes consist of RNA and protein and have an important function, to be elaborated upon later, in orientating RNA molecules towards appropriate sites on amino acids to build up polypeptide chains. The proteins synthesized on the ribosomes penetrate cavities in the endoplasmic reticulum and are stored there in a segregated manner from the other cytoplasmic proteins. This contrasts to the situation when the proteins synthesized are incorporated into the cell itself as happens, for example, in the case of the precursors of red blood cells. Here there is little or no endoplasmic reticulum and the proteins are synthesized on free ribosomes and stored in the cell matrix. The chief function of the endoplasmic reticulum, with its associated Golgi apparatus, is therefore one of storing, segregating and transporting substances synthesized by the cell, particularly proteins, for extracellular use. It is probable that proteins synthesized on the rough endoplasmic reticulum are first stored in cisternae and that subsequently they become enclosed in vesicles, formed by the budding of the smooth endoplasmic reticulum, before migrating to the cell surface in the vesicles to be released. If this is so it appears likely that most budding takes place in the Golgi apparatus.

The storage body for digestive enzymes used inside the cell is a vacuole known as the lysosome. The digestive enzymes can break down materials such as proteins, nucleic acids and polysaccharides and include various proteases, nucleases and glycosidases. They are believed to play a part in the fertilization of ova and in ageing processes and if the cell is starved of nutrients the lysosome can effect an autodigestion in which some of the contents of the cell itself are engulfed by the lysosome and are degraded. This catabolic potential implies that lysosomes fulfil the extremely important function of enabling the cell to adopt metabolically to conditions in which food supply, and perhaps other environmental factors, change rapidly. Also, in the case of the cell that is dying, the lysosome enzymes are released to destroy the cell as it becomes increasingly useless. Linked with the lysosomal function are the pinocytotic vesicles.

Pinocytotic vesicles are formed when cell membranes invaginate and then bud-off internally. By this process of pinocytosis cells can absorb extraneous material into the cytoplasmic medium – a process known as phagocytosis – and in so doing appear to fuse the absorbed and engulfed extraneous material with the lysosome membrane and its enzymes.

The mitochondria found in the cytoplasm are thread-like or round compartments surrounded by a membrane containing a high proportion of phospholipid. Internally they are divided by the inner membrane forming projections known as cristae. These give a large surface area for the attachment of enzymes which can in consequence be packed in tightly. Mitochondria contain their own RNA and DNA and can therefore reproduce themselves. They are the powerhouses of cells in that they are the sites for energy production. The enzymes they contain assist in the extraction of energy from the breakdown products of glucose and of other food and in releasing it for cellular use.

The energy released from the breakdown of sugars, lipids and proteins is utilized by a few energy-rich compounds, present in the mitochondria, for cellular needs. The most important compounds are the tri-, di- and mono-phosphate esters of adenine ribonucleoside, particularly the former. Adenosine triphosphate (ATP), adenosine diphosphate (ADP) and adenosine mono-phosphate (AMP) store the energy released from the breakdown of foodstuffs and then act as donors of that energy for a wide variety of biochemical reactions inside cells. Hydrolysis of ATP yields about 33,600 joules (J) of energy, hydrolysis of ADP yields about 27,300 J of energy, whilst hydrolysis of AMP yields about 9250 J of energy. Other compounds are involved in cellular energy storage and transfer but are of less importance. An example is phospho-enolpyruvic acid which on oxidation yields about 53,750 J energy and pyruvic acid.

The dynamic functions of the structure described above are shown in Fig. 2.2.

2.3. Chemical Composition of Cells

2.3.1. General

The four principle classes of cellular molecules are nucleic acids, proteins, lipids and carbohydrates.

Nucleic acids are long-chained molecules built of repeating units of comparatively small molecules known as nucleotides joined end to end in the chain. Each nucleotide consists of three subunits: a basic nitrogen-containing ring compound, a pentose sugar and a phosphate group. The pentose sugar in nucleic acids is always ribose and there are two types of nucleic acid, DNA and

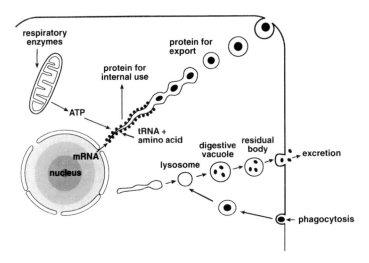

Fig. 2.2. Schematic representation of the functions of cytoplasmic structures (adapted from Ambrose and Easty, 1977).

RNA. The former plays a supremely important role in cellular inheritance, in cell division and in the synthesis of protein molecules and other cell constituents, the latter plays an essential role in protein synthesis. A fuller discussion of these two vitally important nucleic acids follows in the next section of this chapter.

Proteins are also long-chain molecules but are composed of different building units, namely amino acids. Twenty amino acids are found commonly in cells and in solution they act as buffers and resist change in pH in their environment. Proteins have both structural and non-structural parts to play in cellular function. They complex with lipids to form nuclear and plasma membranes with unique properties and, through being enzymes, act as catalysts and facilitate many chemical reactions within the cell. The site of protein synthesis is the ribosome and the synthesis itself is assisted by the RNA molecule. Ribosomes may be found scattered randomly in the cytoplasm but it is likely that those which are attached to the endoplasmic reticulum will be the greatest secretors of protein into the surrounding medium.

Although lipids are also long-chained compounds, their chain lengths are shorter than those of proteins and nucleic acids. They are insoluble in water and this imparts unique properties of diffusion and transport to the membranes in which they are mixed with proteins. Reserves of lipid molecules are found in the form of lipid granules in most cells.

The polysaccharide which predominates in cells is glycogen but in most cells glucose is the ultimate source from which energy is derived. In connective tissue there is a wide distribution of a group of polysaccharides known as the mucopolysaccharides which contain a repeating disaccharide unit of an amino acid and a uronic acid. The most abundant of these is hyaluronic acid and this occurs in subcutaneous tissues as a cementing substance and in the synovial fluids of joints as a lubricant. Other important polysaccharides are chondroitin, which is found in cartilage and tendons (see chapter 3), and heparin which is found mostly in the lungs, in the liver and in the walls of large arteries.

2.3.2. Deoxyribonucleic acid (DNA) and ribonucleic acid (RNA)

Deoxyribonucleic acid is the genetic material of life and in the nuclei of animal cells this fundamental nucleic acid is present in the chromosomes in association with protein molecules. Four different bases form the building units of DNA, two purines – adenine and guanine – and two pyrimidines – thymine and cytosine. A combination of one of these bases with deoxyribose forms a nucleoside and the phosphate ester of a nucleoside is known as a nucleotide. Units of nucleotides are joined together to form long polynucleotide chains. In DNA there are equal proportions of the large purine bases and the smaller pyrimidines with adenine and thymine, and cytosine and guanine, present in equimolar proportions within pairs. There is regularity in the placing of the purine and pyrimidine bases along the molecule and in the placing of the nucleotide units. The former are placed regularly at distances of 3.4 Å and the molecule itself is twisted one complete turn every 34 Å or 10 nucleotide units, therefore indicating a non-linear but helical-type structure. In fact, and as elucidated by the Nobel Prize-winning work of Watson and Crick in 1953, DNA is a double helix consisting of two twisted but complementary poly-nucleotide chains. In each of these chains the deoxyribose sugar units on adjacent nucleotides are linked by phosphate groups to form an outer sugar–phosphate backbone (Fig. 2.3). The structures of the nucleotide units are such that the purine and pyrimidine bases are turned inwards and are linked by hydrogen bonds with each base on one chain being paired with a base on the other chain. This pairing is very specific and is only between adenine and thymine and between cytosine and guanine.

Ribonucleic acid, like DNA, is a long-chain molecule built of repeating nucleotide units linked by phosphate diester bonds. Compared with DNA there are two differences: the sugar component is ribose and not deoxyribose and the fourth base of DNA, thymine, is replaced by uracil which has one methyl group less. The RNA molecule is built from nucleotide triphosphate units by a copying process from one strand of the DNA molecule. The DNA strand, as for DNA replication, must unwind for RNA to be copied from it and a specific enzyme, RNA polymerase, is required to link the ribonucleotide units by means of ester linkages. Therefore RNA chains do not pair in a complementary way along their length as do DNA chains and RNA exists as a single-stranded molecule with the strands being synthesized on a DNA template (Fig. 2.4). By this process a number of RNA molecules, all smaller than the DNA template, can be synthesized on one DNA molecule.

Cells containing a lot of protein contain large amounts of RNA and three

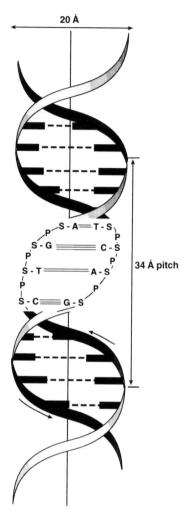

Fig. 2.3. The double helix of DNA as proposed by Watson and Crick (1953), with the hydrogen bonds between the bases adenine (A), thymine (T), guanine (G), cytosine (C), phosphate (P) and deoxyribose sugar (S) holding the two chains together.

types of RNA, all synthesized on the DNA template, are involved in protein synthesis: messenger (mRNA), transfer (tRNA) and ribosomal (rRNA).

2.4. Protein Synthesis and the Genetic Code

Chromosomes are composed of a single strand of double helix DNA together with RNA and two types of protein, the histones or basic protein, which contain high proportions of lysine and arginine, and acidic protein which is linked to DNA and which contains a high proportion of decarboxylic amino

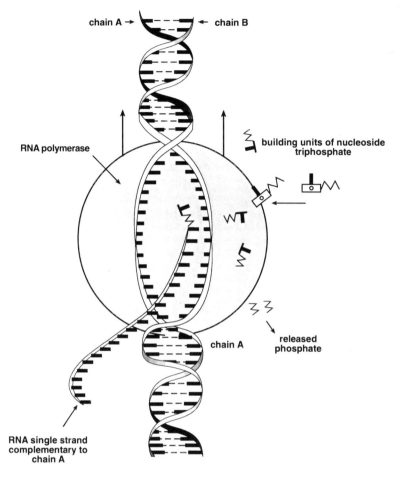

chain A → ← chain B

RNA polymerase

building units of nucleoside triphosphate

chain A

released phosphate

RNA single strand complementary to chain A

Fig. 2.4. Diagrammatic representation of the synthesis of RNA on a DNA template. As in DNA synthesis, the double helix again unwinds and on one of the strands of DNA a ribonucleotide building chain is synthesized (adapted from Ambrose and Easty, 1977).

acids. Histones act as a type of chromosomal glue, which binds the genetic units of DNA, but, more importantly, they repress the genetic activity of cells. In contrast the acidic nuclear proteins probably act as derepressors by making specific regions on the DNA available for RNA synthesis.

The sequence of nucleotide bases in the DNA molecule determines the structure of proteins and the relationship between the nucleotide sequence in DNA and the amino acid sequence in proteins is known as the genetic code. The message of the genetic code is carried from the nucleus, through its membrane, to the site of protein synthesis in the ribosomes by mRNA. In other words it acts as a template for the translation of the DNA code into a specific protein. The length of mRNA which carries the information necessary to determine the complete polypeptide chain of a protein molecule is called a cistron and is the present concept of a gene. Each cistron codes for a complete polypeptide chain. Following this transcription stage there is a stage of translation which involves a change from the nucleotide language of mRNA to the amino acid language of the proteins. During this process the tRNA acts as an adaptor or selector molecule between a particular amino acid in the cytoplasm and the triplet combination of the bases adenine, cytosine, guanine and uracil that code for it on the mRNA molecule. It is possible that the unpaired bases in the rRNA molecule bind mRNA and tRNA to ribosomes.

The way in which genetic information is carried in the DNA molecule depends on the sequence in which the four bases adenine, thymine, cytosine and guanine are arranged along the DNA chain. The way in which the same sequence of nucleotide bases is transmitted exactly from one generation to the next is by the DNA molecule unwinding so that each strand may serve as a template for a new complementary strand (Fig. 2.5). The addition of each of the new nucleotide units to the template strand eventually forms a continuous complementary strand.

The genetic control of growth is discussed in chapter 4. In chapter 11 some of the most recent advances in molecular biology, and their application in practice, for example the use of recombinantly derived somatotropins in controlling growth, are considered. Such considerations make it clear that a thorough understanding of the structure of the proteins of DNA and RNA is necessary in the first instance.

2.5. Mitosis

The mitotic process is that by which single cells, having reached a certain size, divide to form two more or less equal daughter cells. Identical replication of DNA genetic material, and duplication of RNA, protein, lipid and carbohydrate molecules, is effected. The chromosomal material is doubled and divided precisely between the two daughter cells. The period between the end of one cell division and the next division is known as a cell cycle and the period between two divisions is known as an interphase. Therefore interphase follows the previous mitosis and precedes the onset of DNA synthesis (Figs. 2.6 and 2.7). Details of the cycle are presented in Table 2.1, but it is important to remember that cell division involves not only the division of the nucleus between two daughter cells but also the division of the cytoplasm as well. Most cytoplasmic organelles are present in large numbers, for example there are several hundred mitochondria, and DNA is localized in these and controls protein synthesis within the organelle.

2.6. Cellular Proliferation

The proliferation of cells may follow, although not absolutely in all organs and tissues, one of three basic patterns (Fig. 2.8). Reference to this figure indicates

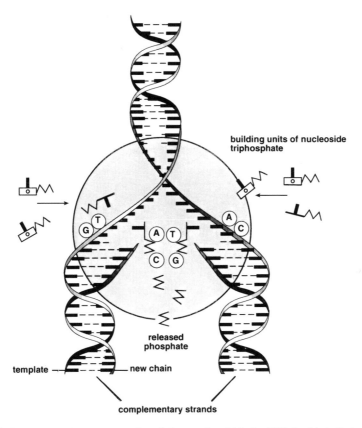

complementary strands

building units of nucleoside triphosphate

released phosphate

template — new chain

complementary strands

Fig. 2.5. Diagrammatic representation of the way in which the DNA double helix is replicated. After the unwinding of the strands the nucleotide triphosphate building units on each of the old chains are used to build a new polynucleotide chain. As a result one old and one new chain is continued within each of the two identical double helices (adapted from Ambrose and Easty, 1977).

the failure of smooth muscle cells to fit neatly into any one pattern whilst some characteristics of renewing and static tissues are combined in some tissues such as cartilage and bone.

In renewing tissue an important characteristic of the differentiated cells is that they have limited life spans, in fact considerably shorter than those of the organisms of which they are a part. Some renewing tissues, for example the epidermis and blood cells, are in a lifelong state of renewal at cellular level.

Expanding organs differ from renewing organs in three main ways. First, they have potentially indefinite longevity, secondly, they do not possess a growth zone and thirdly, differentiation is not incompatible with mitosis, virtually every cell having the potential, and using that potential, to divide frequently during the course of development. Although achievement of the

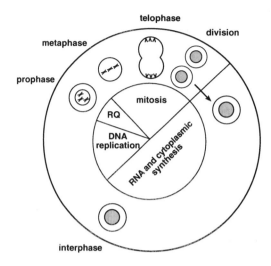

Fig. 2.6. The life cycle of a cell where RQ = period of relative quiescence (adapted from Ambrose and Easty, 1977).

mature state negates the necessity for cell division because the organ does not have to keep pace with the overall growth of the body, a potential must be retained in the fully differentiated cells to contend with events such as injury or reductions in cell mass. Many of the exocrine organs fall into this category.

Muscle and nerve tissues fall into the category of tissues that are mitotically static. In these cases the fully differentiated cell is so specialized that it has abandoned its capacity to divide in the early stages of its development. Such cells are characterized by having no growth zones and by their longevity. In other respects they are similar to the cells of renewing tissue in being fully differentiated and in being incapable of mitotic activity. Therefore, differentiated cells in renewing and static tissue, but not in expanding tissue, have lost their mitotic potential.

There are further distinct differences between the various classes of cells related to physiological activity, composition and structure. For example, differentiated cells which are no longer capable of dividing retain specific end products such as keratin and haemoglobin in their cytoplasm, whilst their activity tends to be of a physical nature, such as contraction or mechanical protection. Expanding organ cells, by comparison, export their synthesized products and therefore have a more chemical function. Examples of the latter are exocrine secretions such as milk, bile or digestive enzymes. It has been suggested that the disposition of specific cell end products has an effect on mitotic potential by means of what may be termed a 'negative feedback' (Goss, 1978). Using this hypothesis the dissipation of end products leaves the mitotic potential of the cell uninhibited whereas intracellular retention mitigates against cell division. Such a principle may be observed in cartilage and bone where the absence of mitotic activity in chondrocytes is perhaps attributable to the close

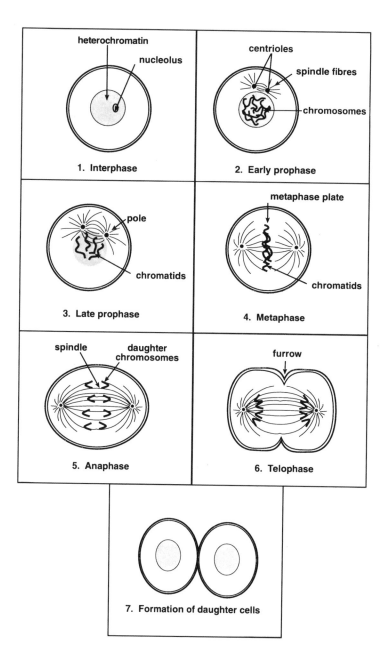

Fig. 2.7. The stages of mitosis: 1, interphase nucleus; 2, early prophase; 3, late prophase; 4, metaphase; 5, (mid-)anaphase; 6, telophase; 7, two daughter cells.

Table 2.1. The phases of mitosis.

Phase	Nucleus	Nucleolus	Cytoplasm
Interphase	Little structure apparent	Distinct appearance; contains RNA and protein and is centre for synthesis of ribosomal material	Protein synthesis
Prophase Early	Chromosomes: (1) condense and RNA deposited on chromosomal strands (2) appear double stranded, spiralize independently and lie beside one another	Shrinks and finally disappears	Centrioles (two in each dividing cell) lie adjacent to nuclear membranes as hollow cylinder-like structures. Spindles formed around and radiate from the centrioles and form asters
Late	Sap mixes with cytoplasm - important in development of spindle material		Spindles develop, asters move apart and nuclear membrane disintegrates
Metaphase			Spindle and chromosomes interact to form a metaphase plate Chromosomes attach to centromere Centromeres separate
Anaphase			Pairs of centromeres move apart and carry daughter chromosomes of each pair to opposite poles. Spindles increase in size
Telophase	Two daughter nuclei reform Nuclear membranes appear and nucleus is re-established	Nucleoli re-established	Chromosomes drawn towards the poles and become shortened and thickened Division furrow appears on cell surface and cell division occurs

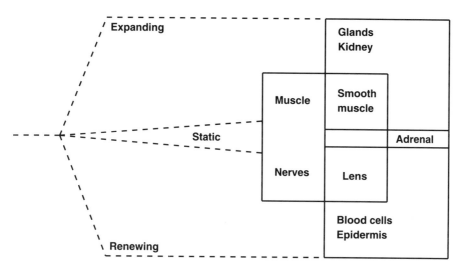

Fig. 2.8. Pathways of the development of three types of tissue. Mitotic proliferation is represented by dashed lines. As maturity is approached cells cease to divide (top). When mitotic competence is lost the cells of static tissue hypertrophy (middle) but renewing tissue (bottom) retains a line of proliferative stem cells from which cells with no mitotic potential differentiate. The three categories of adult tissues, with some that have overlapping attributes, are shown on the right (adapted from Goss, 1966).

proximity to their end products in the surrounding matrix. However this is not an absolute case because although their cells are not being continually turned over, there may be phases of resorption and deposition of osteons in solid bone, akin more to a recycling rather than a replacement process.

2.7. Cell Hyperplasia and Hypertrophy

Goss (1978), in an extremely eloquent and perceptive discourse, indicates the shortcomings of the cell proliferation hypothesis discussed above. He points out the limitations of accepting, important though they undoubtedly are, mitotic concepts alone because they are limited to the cellular level of organization and not necessarily to the most meaningful physiological mechanisms of tissue growth. If this shortcoming is accepted a more appropriate classification may be based on the relative abilities of different tissues to increase their functional capabilities as they grow in mass. In this context, it is important to realize that each organ or tissue is constructed of subunits upon which specific physiological activity depends. The definition of a functional unit is 'the smallest irreducible structure still capable of performing physiological characteristics of the organ of which it is a part' (Goss, 1978). Such units may in many, but not in all, cases be cells. Therefore, Goss proceeds to point out 'if the growth of an organ or a tissue is to be more than an increase simply in mass, it must involve a multiplication of the functional units (hyperplasia) rather than an enlargement of pre-existing ones (hypertrophy)'.

The modes of growth from organ to organ are extremely unequal in the animal body. Goss (1978) has philosophized about this, suggesting that the paradox of vitally important organs, such as the brain, the kidneys, the heart and the lungs, having no regenerative abilities, compared with some of the glands which have, has evolved to limit the stature of mammals and birds, or is itself caused by, the attrition of irreplaceable functional units in vital organs.

Nevertheless, it is important to appreciate that even in mitotically static tissues growth does involve some hypertrophy and, also, that hyperplasia can be reinstated in some organs to increase functional ability. An example here is the liver which retains the capacity, after the initial burst of hyperplasia has ceased, to grow new tissue units in the form of lobules and hepatic cords (Table 2.2). In all tissues in animal bodies it is extremely difficult to know the relative contributions of hyperplasia and hypertrophy to the overall increases in size which take place at any point in time, apart from the very first stages of fetal growth where obviously hyperplasia must, by definition, precede hypertrophy. In the next chapter this difficulty is highlighted where adipose tissue growth and cellularity are considered in cattle, sheep and pigs.

Table 2.2. Possible durations of hyperplastic phases in different types of human tissue. Cell hyperplasia (proliferation) may remain active throughout life in indeterminate tissues but may be lost before maturity in the determinate tissues (adapted from Goss, 1966).

Tissue	Type	Period	Approximate possible hyperplastic duration
Neurons	Determinate	Prenatal	2.5 months
Skeletal muscle fibres	Determinate	Prenatal	3.5 months
Seminiferous tubules	Determinate	Prenatal	6 months
Renal nephrons	Determinate	Prenatal	8 months
Heart muscle fibres	Determinate	Prenatal	9 months (birth)
Pulmonary alveoli	Determinate	Postnatal	9 years
Intestinal villi	Determinate	Postnatal	10 years
Ovarian follicles	Indeterminate	Postnatal	40 years
Thyroid follicles	Indeterminate	Postnatal	90 years
Hepatic cords	Indeterminate	Postnatal	90 years
Exocrine acini	Indeterminate	Postnatal	90 years
Osteone	Indeterminate	Postnatal	90 years
Endocrine cells	Indeterminate	Postnatal	90 years
Blood cells	Indeterminate	Postnatal	90 years

References

Ambrose, E.J. and Easty, D.M. (1977) *Cell Biology*, 2nd edn. Nelson, London.

Goss, R.J. (1966) *Science*, 153, 1615–1620.

Goss, R.J. (1978) *The Physiology of Growth*. Academic Press, New York.

Watson, J.F. and Crick, F.H.C. (1953) *Nature, London* 171, 737–738.

3 Tissues

3.1. Introduction

If cells are the basic building blocks of the animal body a further, larger type of building block is the tissue. All parts of the animal body are constructed from tissues and the four basic types are nervous, epithelial, connective and muscle. In this chapter, to conform with the main overall aims of the book, connective and muscle tissues will receive most attention.

The tissues are not only vital to the living, growing animal, but are also of fundamental importance to considerations of the quantitative and qualitative yields of products for human consumption both during the lifetime of the animal and after it has been slaughtered. In Table 3.1 the proportions of live weight which the various components of the animal body account for in cattle, sheep and pigs are presented. In Table 3.2 the proportions of the major tissues in cattle, sheep and pig carcasses are tabulated. The overall picture presented by these tables is, in the meat animal, the major but not sole goal that is aimed at in the descriptions of the various tissues which follows.

3.2. Nervous Tissue

3.2.1. Introduction

Nervous tissue is organized in an unique way to form what is known as a system. A system may be defined as a group of organs or structures that work together to carry out special functions for the body. Thus it provides a means of instantaneous communication between different cells and tissues. The primordial importance of the system to the animal in terms of coordinating, both on a voluntary and an involuntary basis, all activities essential to both maintaining existing tissues, and in growing new tissues, is obvious, but it has an important part to play too in terms of meat quality by virtue of its function immediately prior to stunning and exsanguination during the slaughtering

Table 3.1. Components of the live animal expressed as g kg^{-1} of live body weight. Typical figures for average pigs, cattle and sheep in Great Britain, based on Meat and Livestock Commission information (adapted from Kempster *et al.*, 1982).

	Pigs	Cattle	Sheep
Gut fill	100	170	120
Skin (hide/fleece)	—	70	135
Empty gut	30	45	65
Intestinal and caul fat	15	45	35
Heart, lungs and trachea	15	15	15
Liver, gall bladder, pancreas and spleen	25	15	15
Head	—	30	40
Feet	—	20	20
Blood	40	30	40
Other components	10	10	15
Hot carcass including KKCF	765	550	500
Total	1000	1000	1000
Weight loss on cooling	15	10	15
Cold carcass including KKCF	750	540	485
KKCF	—	20	—
Cold carcass excluding KKCF	—	50	—
Weight loss during dissection	10	10	15
Sum of dissected parts*	740	510	470

*Including KKCF for pigs and sheep; excluding KKCF for beef (KKCF = kidney + perinephric + retroperitoneal fat).

process. On the other hand, and although some parts are regarded as a delicacy for the human palate, it forms a small proportion, proportionately less than 0.01, of the meat of the carcass.

3.2.2. Structure of basic tissue: cells and fibres

The essential cell making up nervous tissue is the neuron. The neuron is of polyhedral shape and contains a single nucleus. Emanating from the cell itself are two or more nerve processes which are called axons if they conduct impulses away from the cell body and dendrites if they transmit impulses towards the cell body. The neuron is referred to as unipolar if it has a single process extending from it, bipolar if it has two processes extending from it and multipolar if there are more than two processes. The bipolar cell usually has one dendrite and one axon and the multipolar cell a number of dendrites but usually only one axon (Fig. 3.1).

Table 3.2. Composition (g kg^{-1}) of average beef, sheep and pig carcasses* and fat and lean to bone ratios in Britain and the typical range (adapted from Kempster et al. 1982).

	Beef			Sheep			Pigs		
	Lean	Average	Fat	Lean	Average	Fat	Lean	Average	Fat
Lean meat	660	590	500	640	570	480	670	590	530
Total fat	160	250	370	140	240	380	220	310	380
Subcutaneous	30	80	150	50	110	200	150	220	280
Intermuscular	100	130	170	70	100	130	50	60	70
KKCF/flare	30	40	50	20	30	50	20	30	30
Bone (including small waste component)	180	160	130	220	190	140	110	100	90
Total	1000	1000	1000	1000	1000	1000	1000	1000	1000
Subcutaneous/intermuscular fat ratio	0.3	0.6	0.9	0.7	1.1	1.5	3.0	3.7	4.0
Lean/bone ratio	3.7	3.7	3.8	2.9	3.0	3.4	6.1	5.9	5.9

*All carcasses excluding head, feet and skin (KKCF = kidney + perinephric + retroperitoneal fat).

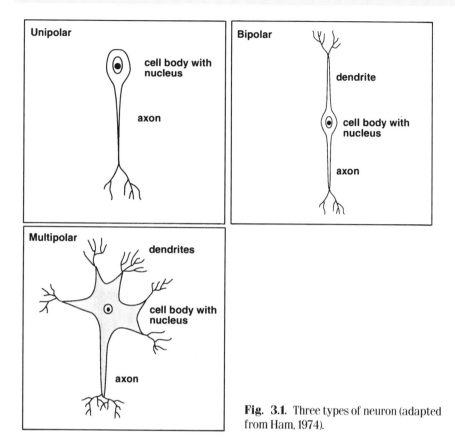

Fig. 3.1. Three types of neuron (adapted from Ham, 1974).

Neurons are individual units but connect with each other, the junction of the axon of one neuron with another being called a synapse. The cytoplasm of the neuron is often referred to as neuroplasm and the dendrites are short protoplasmic processes that branch repeatedly. The neurons themselves are carried in a supporting tissue.

With the exception of some eggs there are no other cells which surpass in cytoplasmic volume nerve cells with their associated axons and dendrites. However, compared with other cells that rival them in size, for example the multinucleated muscle fibres, they are puny in containing but a single nucleus. The outside dimensions may be the result of a long period of development in which continuous growth to keep pace with the rest of the body, following a premature loss of mitotic competence in the prenatal stages, was essential (Goss, 1978).

Nerve fibres are composed of groups of neural axons and are either myelinated or unmyelinated. In the former the fibre is surrounded by a white sheath of fatty material, the sheath itself being formed from many layers of a cell membrane of a Schwann cell wrapped around the nerve fibre. In the latter the fibres are invaginated into the cell membrane of a Schwann cell. Nerve fibres appear to control the proliferation of their associated Schwann cells. Also there appears to be a critical axon diameter of about 1 µm before myelenation is initiated. Before this the Schwann cells themselves multiply and a number of axons become embedded in the cytoplasm of each cell. The thickness of the myelin sheath increases less rapidly than does the diameter of the neuron axon and because of this the relative thickness of the myelin sheath decreases with the growth of the nerve fibre. Mitotic potential is restored in Schwann cells if myelin is lost: otherwise there is no proliferation of Schwann cells when a myelin sheath is present.

3.2.3. *Major divisions and development of the nervous system*

Although the nervous system is an integrated unit it is convenient to regard it as composed of two divisions. The central nervous system (CNS) is surrounded and protected by bone. The brain, enclosed in the cranial part of the skull, and the spinal cord, which is continuous with it and enclosed in the vertebral canal, form the CNS. The second division is known as the peripheral nervous system (PNS) and consists of cranial and spinal nerves which lead off from the brain: the cranial nerves emerge through the cranial foramina of the skull and the spinal nerves emerge through intervertebral foramina. The nerves are in pairs, one pair going to one side of the body and one pair to the other side. A further subdivision of the PNS is the autonomic nervous system (ANS) consisting of the sympathetic nervous system – the thoracolumbar portion – and the parasympathetic nervous system – the craniosacral portion. The ANS innervates smooth muscle, cardiac muscle and the glands of the body, that is the visceral structures. This is in contrast to the rest of the PNS which is associated with somatic structures.

In the earliest stages of embryonic life the ectoderm of the dorsal midline thickens to form the neural plate. The events subsequent to this in the development of the entire nervous system are shown in Fig. 3.2. As can be seen, the entire nervous system originates from the neural tube, the cells of which, together with the neural crests, constitute a proliferative neuroectoderm.

The growth of the spinal cord is characterized by the walls becoming increasingly thickened, by the cavity initially remaining more or less constant in size and then decreasing in size markedly and by the cord becoming flattened in cross-section. These changes result from cell proliferation in the wall of the tube. The wall itself consists of three concentric layers: an inner ependymal layer, a middle mantle layer and a superficial marginal layer. The mantle layer and the superficial marginal layer contain, respectively, the so-called grey and white matters of the cord.

The immensely complex brain, incredibly, grows from the simple neural tube. It is postulated that its development results from two basic mechanisms which give differential growth rates in the wall of the tube and a vast amount of longitudinal growth in the site where the brain develops. The vesicles, or swellings (Fig 3.2f), are separated by two constrictions and form the prosence-

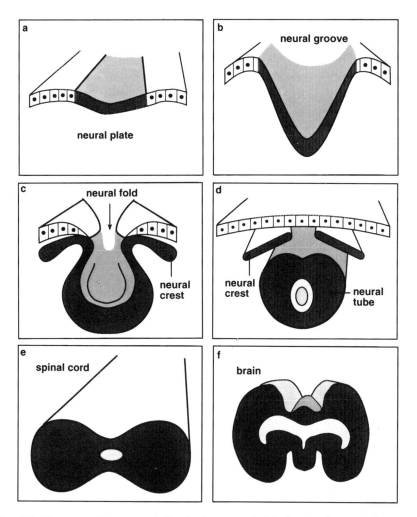

Fig. 3.2. Diagrammatic representation to show neural plate forming from ectoderm (a), the neural groove forming as a result of faster growth of the neural plate along the lateral margins compared with the centre (b), the neural groove developing into the neural fold and crest (c), the edges of the fold joining to form the neural tube and the persistence of the neural tube lumen to give the central canal of the spinal cord (d), or the two (e) lateral ventricles and the 3rd and 4th ventricles of the brain (f) (a–f adapted from Ham, 1974).

phalon or fore-brain, the mesencephalon or mid-brain and the rhombencephalon or hind-brain.

The major components of the three sections of the brain are shown in Table 3.3 and the reader is referred to other texts such as Ham (1974) and Frandson (1981) for further details. Relative to the overall structure depicted in Table 3.3, it would be wrong to emphasize the importance to the animal's integrity of any

Table 3.3. Major divisions of the brain.

Rhombencephalon or hind-brain		Mesencephlon or mid-brain	Prosencephalon or fore-brain	
Major divisions	Major components	Divisions	Major divisions	Major components
Metencephalon	Cerebellum pons	Cerebral peduncles	Telencephalon	Cerebral cortex Corpora striata Rhinencephalon (Olfactory brain)
Myencephalon	Medulla oblongata	Quadrigeminal bodies	Diencephalon	Thalamus Epithalamus (including pineal body)
Fourth ventricle				Hypothalamus (including pituitary gland)

one particular part relative to any other part or parts: all have a vital part to play in maintaining overall the ability of the animal to grow from conception to maturity. Nevertheless, the main thrust of this book towards an overall understanding of growth would be less complete if it were not pointed out that the hypothalamus may play a fundamental role in controlling growth whilst the pituitary gland and the pineal gland also play important roles in controlling growth processes (see chapter 4).

3.3. Connective Tissue

3.3.1. Structure and classification

The tissues of the body other than connective tissue are composed mostly of cells and are soft. Connective tissue connects and holds these tissues together and gives the body a coherent form. It differs from the other major tissues in consisting of a few cells, mostly fibroblasts, mast cells and macrophages, resting in a large intercellular matrix of both inorganic (in the case of bone) and organic substances which are non-living. These impart strength to the tissue and to the body as a whole but at the same time are media for the transport of nutrients to cells within the intercellular matrix. The principle components of the matrix are mucopolysaccharides, chondroitin sulphates and hyaluronic acid set in a framework of elastin and collagen fibres.

The strength of connective tissue is derived from its collagen fibres. Resiliance depends on the elastic fibres within the intercellular matrix. Goss

(1978) portrayed the many physical characteristics in a picturesque manner: 'connective tissue can be as tough as leather, as sinewy as gristle, as soft as adipose tissue, as transparent as the cornea, or as liquid as the fluids filling the body cavities'.

The basic component of the collagen fibres which form the chief structural element of connective tissue is the tropocollagen molecule. In fact there is no single tropocollagen molecule but a family of closely related molecules (Simms and Bailey, 1981). The molecule itself is a triple helix of polypeptide chains about 15 Å in width and about 3000 Å in length, with each chain containing about 1050 amino acids. The polymerization of these macromolecules and their alignment side by side forms collagen fibrils which in turn form the fibres. Fibroblast cells from the mesenchyme are the specialized cells responsible for the synthesis of collagen molecules (Priest and Davies, 1969) but it is still a considerable puzzle as to how such cells, which show virtually no differentiation, can instigate the growth of connective tissues which emerge in so many different forms and patterns, from the criss-cross laminae of the dermis to the parallel structures of tendons.

The fibres of collagen are straight, inextensible, non-branching and white in colour, and vary in diameter from about 16 nm in some fetal tissues to about 250 nm in some tendons of adult animals. In contrast elastic fibres are elastic, branching and yellow in colour. The collagen fibril, built from aggregates of tropocollagen molecules as illustrated in Fig. 3.3, contains larger quantities of hydroxyproline (proportionately about 0.14 more) than other proteins, with the polypeptide chains of its primary structure having the repeating sequence – glycine–proline–hydroxyproline–glycine–another amino acid. One chain in three has a different composition from the other two. The chain types are referred to as α_1 and α_2 and currently five different forms of α_1, which are designated commonly as α_1, I–V are recognized. In total there are thus five chain types (Simms and Bailey, 1981) and the collagen fibril in its secondary structure is arranged first as a left-handed helix and then, with three of these intertwining, as a right-handed super helix. The types of collagen found in different tissues are characterized by the nature of their constituent polypeptide chains (Table 3.4). Types I, II and III form by far the largest proportion of the extracellular fabric of the major connective tissues.

Intermolecular cross-links are largely responsible for giving collagen its strength. Intramolecular cross-linkage also occurs but its functional significance is not clear. The various cross-linkages found vary in stability during the growth of the animal but become stable in non-reducible forms when the animal is mature.

The structure and arrangement of elastin fibres differ according to their origin. They are embedded in a mucopolysaccharide ground substance of which the chondroitin sulphates are most important. Essentially, elastin has a rubber-like consistency and consists of randomly coiled peptide chains cross-linked at intervals by stable chemical bonds and the extent and stability of the cross-links, even in immature tissue, is much greater than in collagen. In some tissues (e.g. arteries and veins) elastin fibres are associated with collagen fibres and smooth

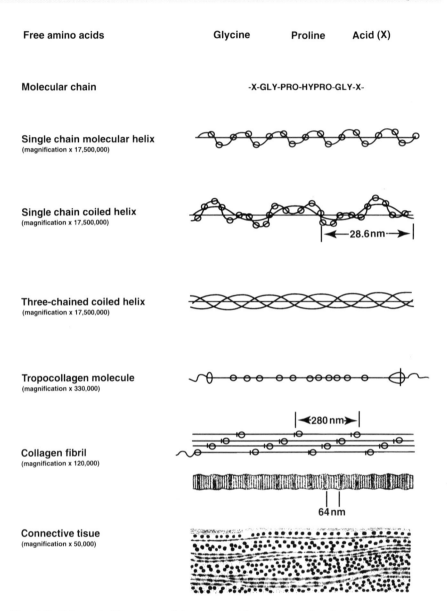

Fig. 3.3. Schematic illustration of the amino acid sequence and molecular structure for collagen and tropocollagen; and collagen fibril formation (based on original in Gross, 1961).

muscle, in others (e.g. elastic cartilage) they form a network of fibrils. Proportionately elastin accounts for only about 0.05 of the overall connective tissue in muscle and in essence the importance of this fibrous protein is in fibres which require a high degree of elasticity (e.g. blood vessels and ligaments).

Table 3.4. Collagen types and their location (based on Dutson (1976) and Simms and Bailey (1981)).

Type	Location and proportion	Molar composition	Main features
I	By far the most abundant type of collagen. Very dominant in muscle, proportionately over 0.90 of the collagen in bone and in tendon and about 0.80 of the collagen in adult skin	$[\alpha_1\,(I)]_2\alpha_2$	Two identical polypeptide α_1 (I) chains hydrogen bonded to each other and to a third chain with a different amino acid sequence and designated α_2. Six to eight hydroxylysine residues in each α_2 chain, (note difference to types II, III and IV in containing α_1 and α_2 chains)
II	Prominent in cartilage and in invertebrate discs	$[\alpha_1\,(II)]_3$	Three identical polypeptide chains having a different amino acid sequence to either of the α chains in type I collagen. Between 20 and 25 hydroxylysine residues in each α_1 chain
III	Occurs in many tissues (except in cartilage and bone): placenta, blood vessels, spleen, liver, muscle and skin (proportionately about 0.60 of the collagen in fetal skin but decreasing to the adult proportion of about 0.10 just before birth)	$[\alpha_1\,(III)]_3$	Although similar to type II in molecular composition the α chains differ in both amino acid composition and sequence. The α chains contain the same number of polypeptide residues as the α_2 chains in type I but are linked via disulphide cystine bonds
IV	Unique connective tissue found in basement membranes which have specialized functions within the body such as providing elastic support in the lens capsule or in acting as a basis for filtration in the glomerular basement membrane. Also found in the placenta	$[\alpha_1\,(IV)]_3$	Probably three α_1 peptide chains originating from two different molecules with the α_1 chains containing 60–70 hydroxylysine residues in each chain
V	Small proportion in a number of tissues including muscle, placenta, aorta, skin, lung, nerve, synovial membrane, bowel and liver	?	Suggestions include single and double molecular species

There are several ways in which connective tissue may be classified. One possibility is that shown in Table 3.5. In this book, where one of the major aims is that of elucidating the principles of overall growth relative to the development of tissues in meat animals, bone (and cartilage) and adipose tissue will receive most attention. However, it is important to appreciate that collagen has a fundamental role to play in determining the texture of muscle, and therefore of meat, and this role will be considered later (see section 3.4.6).

3.3.2. Supportive connective tissue

Dense regular and irregular ordinary connective tissue

There is not always a distinct line of demarcation between on the one hand loose connective tissue and on the other hand dense supportive connective tissue which is composed mostly of collagen fibres. In the few places where elastin fibres are found they occur in dense concentrations and the few cells found are mostly concerned with producing intercellular substance. Because collagen is a non-living material this type of connective tissue requires a small blood supply and therefore contains few capillaries.

There are two main types of dense connective tissue: regular and irregular. In regular tissue the collagen fibres are arranged more or less in the same plane or in the same direction. It follows that structures based on this tissue have great tensile strength and can withstand tremendous pulls exerted in this plane, and in the direction of its fibres, without stretching. The tissue is therefore ideal for tendons and ligaments which join muscles to bones and bones to bones and for

Table 3.5. Classification of connective tissue.

Type	Occurrence and subtype

Loose
  Mixture of cells and intercellular
  substances
	Distributed widely throughout body, e.g. to provide a substrate on which epithelial tissues lie and in which glands rest. One type of its cells can synthesize and store fat

Haemopoietic Almost entirely cells	Blood cells Myeloid tissues – bone marrow Lymphatic tissues – thymus, lymph nodes and spleen

Supportive
  Mostly connective tissue (intercellular
  substances)
  Strong
	Dense regular and irregular arranged ordinary connective tissue and cartilage
	Bone and joints
	Teeth

where a pull is exerted in one direction. The cells present are nearly all fibrocytes which are located between parallel bundles of collagen fibres.

In irregular connective tissue the collagen fibres run either in different directions but in the same plane or in every direction. In various types of sheaths composed of tissues the fibres are more or less in the same plane but may run in different directions. In these cases resistance to stretching lies in the direction in which the fibres run. In other sites in the body, such as the recticular layer of the dermis of the skin, the collagen fibres run in different directions and in different planes. In consequence the dermis can stretch in any direction.

In many organs such as the spleen and in lymph nodes, the encapsulating tissue is based on irregularly arranged connective tissue and this often extends into the organs themselves as septa or trabeculae. In addition, the external wrappings of various tubes in the body, of muscles and of nerves and of the sheath enclosing the CNS (the brain and spinal cord) are all based on this type of connective tissue.

The tendon will be the only type of dense connective tissue to be examined in detail. In the embryo, tendons first appear as dense bundles of fibroblasts which are orientated in the same plane and which are packed together tightly. Growth in the tendon proceeds through the fibroblasts arranging themselves in rows and secreting ever increasing quantities of collagen between the rows. This process changes the character of the tissue from one in which cells predominate to one in which intercellular substance predominates. The diameters of collagen fibres in tendons increase as the animal grows to match the tensions to which they are likely to be subjected. Increase in length is achieved by internal expansion, mostly at the junction with the muscle. At the opposite end the tendon reorganizes itself to allow alterations in site of attachment to the bone commensurate with the animal's growth. In certain sites where tendons might rub against each other or against friction-generating surfaces, they are enclosed in sheaths.

Cartilage and bone

Introduction
Cartilage and bone are specialized types of connective tissue. Bone has many functions in the animal body: it assists in maintaining mineral homeostasis, it gives to the body a certain rigidity whilst at the same time allowing some flexibility in development to allow for growth, it provides 'levers' to facilitate movement, it gives protection to certain organs, it stores some energy in the form of lipids and it stores minerals and key elements of the immune system.

The significance of allowing flexibility in development, relative to the evolutionary pattern of the vertebrates, has been discussed by Goss (1978). He argues that whilst cartilage may be responsible largely for the versatile characteristics of skeletal tissues, it has the important limitation, because it has no vascular system, of being stagnant and unable to turn over its population of cells once they are trapped in their matrix. This limitation, together with the limitation that growth is restricted to the perichondrium (outer membrane), equips cartilage poorly to remodel itself and restricts its capacity for repair and regeneration. He proposes that the evolution of bone has provided a solution to these shortcomings, although the vertebrate still makes good use of cartilage through its properties of toughness and resilience for articulating surfaces and through the cartilaginous plates in the long bones which allow for growth in length during the process of maturation.

Cartilaginous plates first appeared in the reptiles. Their appearance signified the first steps in evolution of a skeletal structure, which was capable of elongating for a finite period of time, by separating the articular and growth components of the cartilaginous epiphysis. At the same time the plate could be disposed of when growth ceased. This was of immense evolutionary significance in limiting the growth of terrestrial vertebrates to the total mass capable of being supported by the skeletal structure.

Therefore, in the higher terrestrial vertebrates, evolution produced animals with determinant body sizes because they could terminate their own growth around the point of sexual maturity by switching off their cartilaginous plates. The weight of the skeleton therefore limits the size to which the land vertebrate can grow. The immense size of the blue whale is only possible because its weight does not have to be supported wholly by its skeleton, the sea in which it lives being the major supportive element for its soft tissue and organs (Widdowson, 1980).

Similarities and differences

Although cartilage and bone are, according to Table 3.5, different subtypes of tissue, they are nevertheless two closely interlinked types of special dense connective tissue and have basic similarities and differences (Table 3.6). It is wholly appropriate that they be considered together as the development of the cartilaginous models of bones in the developing embryo, and the development and function of cartilaginous plates in the long bones, are so central to skeletal growth to maturity in postnatal life.

Cartilage structure

Compared with the other tissues there are small quantities of cartilage in the body of the non-fetal animal. Nevertheless it is a vitally important tissue, with its unique properties allowing the free movement of joints (e.g. knees and elbows) and with the growth of long bones being totally dependent on its existence in the first instance.

In physical characteristics the intercellular substance of cartilage differs from that of tendon. It will not stretch but it will bend easily because the cartilage fibres are embedded in a mucopolysaccharide. This has similar physical attributes to a plastic, giving sufficient firmness to bear a certain amount of weight.

There are three types of cartilage: hyaline, elastic and fibrocartilage. Hyaline is the most common and the type that will be considered in detail. Elastic cartilage is found in sites which require a tissue that is both stiff and yet to some degree requires some give in it. Examples of sites are the external ear

Table 3.6. Similarities and differences between cartilage and bone.

Similarities	Differences
1. Both tissues are composed mostly of intercellular substances	1. Cartilage can grow by interstitial (growth within the tissue) as well as by appositional mechanisms. Bone cannot and the reasons relate to:
2. Cells within the intercellular substances, chondroblasts in cartilage and osteoblasts in bone, lie in little hollows known as lacunae, where they are known as chondrocytes and osteocytes in cartilage and bone respectively	2. There is no calcification of the intercellular matrix of cartilage. In bone there is calcification and this makes interstitial growth impossible (one exception is when chondrocytes hypertrophy and begin to secrete alkaline phosphates)
3. With the exception of articular cartilage, the outer surface is covered with a membrane, the perichondrium in cartilage and the periosteum in bone. Each layer contains an outer fibrous layer and an inner layer contains either chondrogenic (in cartilage) or osteogenic (in bone) cells	3. Resultant from 2 the intercellular matrix of bone contains both organic and inorganic components: that of cartilage consists of organic components only
4. In both tissues growth can take place by appositional mechanisms, i.e. by adding new layers of tissue to old on the outside	

and the epiglottis. As the name suggests elastic cartilage contains considerable numbers of elastic fibres in its intercellular substance. Fibrocartilage is found in tendons.

The name hyaline is derived from the fact that to the naked eye this type of cartilage has a pearly white, glassy, transluscent appearance due entirely to the special character of the intercellular substance. In adult life it persists in the articular surfaces of the joints and in parts of the ear and gives support to the nose, the larynx, the trachea and the bronchi and the walls of the upper respiratory tract. In fetal tissue an abundance of hyaline cartilage is found in the cartilaginous models of the future bones and some of this persists in postnatal life, until growth ceases at maturity, in the epiphyseal plates.

Hyaline cartilage develops from mesenchyme cells. Condensates of these cells, which lose the cytoplasmic threads that had previously joined them together, develop. The cells then differentiate into chondroblasts and separate from each other due to the initiation of the deposition of the intercellular matrix (Fig. 3.4.b). Next the chondroblasts hypertrophy, stretch the intercellular substance and lie in their lacunae as chondrocytes. They continue to secrete the intercellular substance, largely composed of the mucopolysaccharide peculiar to cartilage, and as a result they are pushed further apart (Fig. 3.4.c). The

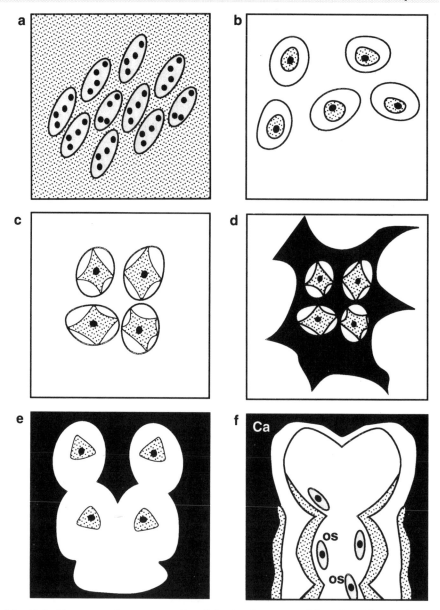

Fig. 3.4. Diagrammatic representation of the development and fate of cartilage in the body. a, Condensate of mesenchymal cells forming. b, Mesenchymal cells differentiated into chondroblasts and the laying down of intercellular substance commencing. c, Hypertrophy of chondroblasts into chondrocytes lying in lacunae with a stretching of the intercellular substance. d, Secretion of phosphatase calcifying the intercellular substance. e, Chondrocytes becoming shut off from nutrient sources and dying and the intercellular substance starting to disintegrate. f, Osteoblasts with capillaries forming bone on the cartilage remnants: Ca = intercellular cartilage substance; os = osteoblasts; stippled areas = bone intercellular substance (a–f adapted from Ham, 1974).

intercellular substance itself is a firm gel and contains collagen fibres immersed in the mucopolysaccharide chondroitin sulphuric acid.

Cartilage growth

Growth can proceed by two different methods. Interstitial growth is caused by the chondrocytes, until they become mature, retaining their ability to divide. The new cells formed by this method can give rise to new and more intercellular substance. For this type of growth from within, the intercellular substance must be sufficiently malleable to allow the necessary expansion to take place. It follows that most interstitial growth occurs in young cartilage.

The second mechanism of growth is by apposition, that is by adding new layers of cartilage on top of old rather than by growth occurring from within. In the developing embryo, in the cartilaginous models of future bone, those mesenchyme cells proximal to its sides form a surrounding membrane known as the perichondrium (Vaughan, 1980). The outer cells of this develop into fibroblasts which form collagen and thus the outer layer of the perichondrium becomes a connective tissue sheath. The inner cells remain unchanged and constitute what is known as the chondrogenic layer of the perichondrium. Growth in length of the model occurs near its ends. This leaves the chondrocytes in the middle of the model time in which to mature. Growth in width is effected by proliferation and differentiation of the cells of the chondrogenic layer of the perichondrium.

The intercellular matrix in which the chondrocytes in the middle of the model remain thins out, and when a certain degree of cell hypertrophy has taken place (Fig. 3.4.c) the chondrocytes start to secrete the enzyme alkaline phosphatase (Fig. 3.4.d). This is a characteristic of the mature osteoblast. Thus the intercellular substance becomes increasingly calcified and because the nutrient supply to the hypertrophied chondrocytes becomes severed, they die (Fig. 3.4.e). As a result of these changes in the intercellular substance the mid-part of the model starts to break up leaving cavities.

At about the same time that the above changes are taking place capillaries invade the perichondrium, which becomes increasingly thickened due to new layers being added to the side (appositional growth). The appearance of capillaries, bringing more oxygen to the cells, signifies a change in the differentiation of the cells in the chondrogenic layer: the chondrocytes commence a transformation into osteoblasts and osteocytes. As a result of these changes a very thin layer is deposited around the cartilaginous model and the perichondrium becomes the periosteum (Fig. 3.4.f). However the cells of the inner layer of the periosteum retain an ability to differentiate into chondroblasts and to form cartilage even in adult life.

A feature of this stage of development is that the cartilage which has become calcified in the mid-part of the model begins to disintegrate. The disintegrating tissue is invaded by osteoblasts and capillaries from the inner layer of the periosteum. In so doing they form a diaphyseal (diaphyseal = shaft) centre of ossification from which bone formation will start and spread to replace most of the cartilaginous model (Fig. 3.5). The first bone formed on the remnants of the cartilage is known as cancellous bone and will be described in more detail later.

Interstitial growth at both ends of this cartilage model continues and further lengthens the model overall. However the proportion of cartilage within the model progressively declines because cartilage is destroyed at the edge of the ossification front by the invasive ossification of the intercellular matrix. Increases in the width of the model are caused by the osteogenic cells of the periosteum adding further bone to the sides of the model. Because of this the periphery of the model becomes stronger and the need for cancellous bone as a supportive element in the centre diminishes. This leads to the resorption of the cancellous bone with the consequent development of a cavity which becomes the marrow cavity of the future bone. However, the marrow is always separated from the cartilaginous ends by longitudinally arranged trabeculae of bone and these will be described in more detail later.

Bone structure

Two types of bone structure are found in the mature skeleton. Hard, compact or cortical bone is found largely in the shafts of long bones surrounding the marrow cavities. Spongy, cancellous or trabecular bone, constituted from a network of fine interlacing partitions or trabeculae enclosing cavities containing red or fatty marrow, is found in the vertebrae, in the majority of flat bones and in the ends of long bones.

In both types of bone both appositional growth and resorption (which is to be discussed later) take place throughout life. In young bone these processes increase both bone length and diameter. In old bone the modelling process of resorption occurs internally with no significant alteration in bone shape.

The previous sections dealing with intercellular growth in the cartilage model were terminated where activity of osteoblasts was just commencing and where the ultimate structure of bone tissue was beginning to emerge in a very rudimentary form. It was convenient to stop at that point in order that true bone growth could be considered in this section. However, such an approach is artificially divisive, as in reality there is but one progressive process in the animal.

It has been pointed out already that the formation of new bone is the function of a specialized cell known as an osteoblast and that a vascularized environment enhances its growth and activity. The osteoblast is responsible for depositing the intercellular organic matrix known as the osteoid and the differentiation of the osteoblast and the deposition procedures are referred to collectively as ossification. Each osteoblast has a number of cytoplasmic threads projecting from it and these connect with each other or with adjacent threads from other osteoblasts. The osteoblasts secrete the intercellular organic matrix of bone around themselves and around the cytoplasmic threads, which act at this time as moulds for future minute passageways known as canaliculi. These passageways remain to provide communication between adjacent osteoblasts and the surface on which the bone is forming. They allow the permeation of tissue fluids from the capillaries at the surface to the cells entrapped within the

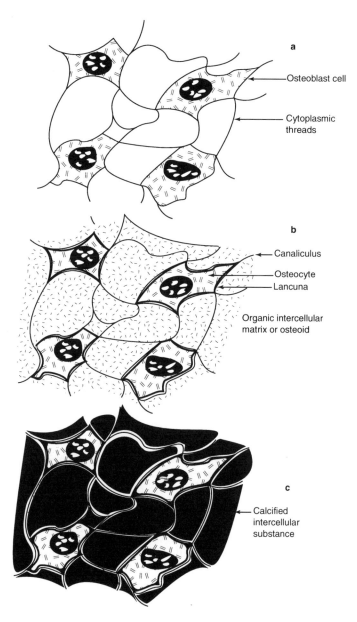

Fig. 3.5. Diagrammatic representation of stages in the process of ossification and calcification of bone tissue from the point where the osteoblasts have differentiated into groups and become joined by cytoplasm threads (a), through the intermediate stage of the deposition of the organic intercellular matrix, or osteoid, with the retention of passageways by the formation of canalicula and the development of the osteocyte lying in its lacuna (b) to the final entombment of the osteocytes in their lacunae by the calcification of the organic matrix (c) (adapted from Ham, 1974).

intercellular matrix and are concerned with the exchange of nutrients between matrix, bone fluid and extracellular fluid. Eventually the osteoblasts become completely entombed within the intercellular organic matrix they have secreted. They are then known as osteocytes and lie in their own individual cavities or lacunae (Fig. 3.5). Not all osteoblasts behave in this way and a few continue to form osteoid tissue to surround capillaries which allow the transport of the haemopoetic elements of the marrow.

The chemistry and mineralized matrix of bone are discussed fully by Vaughan (1975). The organic intercellular matrix contains collagen, mucopolysaccharide and glycoprotein. The mineral which impregnates this organic matrix is mostly in the form of crystals of hydroxyapatite $(Ca_{10}(PO_4)_6(OH)_2)$. In mature bone the proportion of the dry weight in an inorganic form is about 0.75 whilst proportionately about 0.88 of the organic form is collagen. The crystals of hydroxyapatite are needle or rod shaped, 30–50 Å in diameter and up to 600 Å in length. The consequence of these various changes to the original organic matrix is that spicules of bone are formed which radiate out from the ossification centre. These are known as trabeculae and are a common feature of spongy bone.

Before considering bone growth further it is important to appreciate that ossification can occur under conditions of abnormal calcium metabolism without any accompanying calcification. Thus osteoid tissue is uncalcified bone. This is different to decalcified bone which can only be produced in the laboratory by using decalcifying agents. A decalcified bone will not differ in gross appearance from a calcified bone but it will not be capable of carrying weight: literally it can be tied in a knot if it is of sufficient length.

Bone growth and modelling
Introduction   In the previous section the initial processes of ossification and calcification in the formation of bone were considered. After these initial stages further growth proceeds by means of two different processes: endochondral and intramembranous ossification. At the same time the bones are modelled, by structural changes which take place in the adult and in the developing skeleton, by resorption of existing bone. These processes of apposition (growth) and resorption continue throughout adult life and are responsible for a continuous remodelling and turnover of bone tissue.

The terms 'endochondral' and 'intramembranous' refer to the sites or environments in which formation and ossification occur. The term 'endochondral' infers that growth is taking place 'in cartilage' and the term 'intramembranous' infers that growth is taking place 'in membrane'. The fundamental process of endochondral ossification is responsible for growing most bones in the skeleton and those at the base of the skull. It is characterized by bone tissue being deposited on a strong network of calcified cartilage and in normal circumstances can occur only when an epiphyseal plate is present. Basically the process is responsible for the development of length and bulk in the growing skeleton. Intramembranous ossification deposits bone on the surface of pre-existing bony structures. Modelling and remodelling deposit bone similarly after specialized cells known as osteoclasts have assisted in the resorption of existing bone.

Before considering the two processes of ossification in detail it is important to bear in mind that bones increase in length by growing from their ends. This has been clearly demonstrated by many workers who, by inserting two metal pins in a growing bone, have shown that the distance between the pins remains constant even though the bone increases in length greatly.

Endochondral ossification The centres of ossification responsible ultimately for increases in length of bone appear in the cartilaginous ends of the model as seen in Fig. 3.4f. However not all cartilage is replaced and at each end of the model sufficient remains to form both the articulating cartilage of the future bone and the transverse disc or plate of cartilage, which traverses the bone from one side to the other, and which separates the bone of the epiphysis from that of the shaft (metaphysis + diaphysis). This epiphyseal plate persists until the longitudinal growth of bone is completed.

The epiphyseal plate thickens but the increases are limited by the effects of calcification and death taking place on the diaphyseal side of the plate which give a continual appositional growth of bone. This is the major influence in increasing bone length and the epiphyseal plate and the diaphysis adjacent to it constitute what is often referred to as the growing zone of the long bone.

The region directly adjacent to the epiphyseal plate on the diaphyseal side is known as the metaphysis of the bone. This is composed of bony trabeculae and eventually bones funnel out and increase in diameter as they approach their epiphyses (Fig. 3.6). To maintain appropriate proportions and shape this flared section must continually be reabsorbed from the outside. To balance this, apposition of bone has to occur internally. The trabeculae at this point form tunnel-like structures and have cartilage linings. Very active osteoblasts in these linings deposit bone on the lining surfaces and this counterbalances the resorption and enhances strength and rigidity. This development is the basis of the formation of the Haversian or osteon canal system (Fig. 3.6). Each canal has one or two blood vessels and thus provides the tissue fluid to nourish the surrounding osteocytes.

Increase in bone width is achieved by new layers being added to the outer aspect of the shaft, beneath the periosteum, by the osteogenic layer of the membrane. At the same time bone is dissolved away from the inner aspects of the shaft. These simultaneous processes widen the shaft, without materially thickening the walls, and therefore increase the size of the marrow cavity. This means that the bone of an older or of an adult animal will not be the same bone that was present when the animal was younger. A stylistic representation of progressive growth and ossification in a long bone is given in Fig. 3.7. Therefore, growth in length is a consequence of endochondral ossification which finishes relatively early in life when the hyaline cartilage of the epiphyseal plate has been eliminated. In contrast growth in thickness has no such definite end point and although it decreases with increasing age, it is a later developing characteristic, positively correlated with increasing live weight

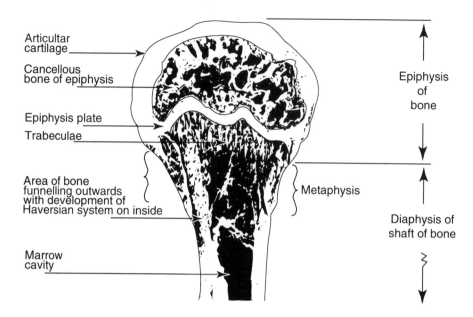

Articultar cartilage

Cancellous bone of epiphysis

Epiphysis plate

Trabeculae

Area of bone funnelling outwards with development of Haversian system on inside

Marrow cavity

Epiphysis of bone

Metaphysis

Diaphysis of shaft of bone

Fig. 3.6. Diagrammatic representation of longitudinal section of a long bone in a growing animal (adapted from Ham, 1974).

and dependent on a different process, namely periosteal or intramembranous ossification.

Intramembranous ossification   The skull bones are formed by this process and merit a more detailed consideration. However, first a little more detail on the process in general is required. As described above no cartilage model is formed in the first place and no epiphyseal plate is involved. Osteoblasts and hence osteocytes develop directly from mesenchyme cells and form spicules of bone. New osteoblasts arise from a self-maintaining population of stem cells which are perpetuated from the mesenchyme. The spicules of bone formed are covered with osteogenic cells and osteoblasts. Those spicules of bone that radiate out from the ossification centre are also termed trabeculae and bone that consists of a network of trabeculae joined together forms, as was pointed out previously, spongy or cancellous bone. However, the continued deposition of fresh bone on trabeculae soon changes the bone to a structure containing few spaces, that is to compact or dense bone. Therefore spongy or cancellous bone may occupy the centre of the mass and layers of compact bone may be formed on the surface by the activity of the osteoblasts. The skull is the obvious bone in which to consider this process in a little more detail.

At birth the bones of the skull have grown to the point where only small spaces or sutures separate them. At points where more than two bones meet the sutures are wide and are known as fontanelles. Fontanelles are less prominent

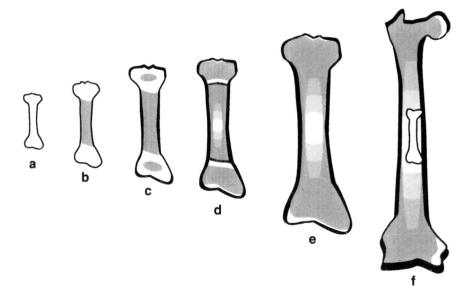

Fig. 3.7. Progressive ossification and growth of a long bone. a, The cartilaginous stage. b, Deposition of spongy, endochondral bone (stipple), and the compact, perichondral bone (black). c, An epiphysis appears at each end. d, The marrow cavity (sparse stipple) appears as endochondral bone is resorbed. e, Each epiphysis ossifies, leaving articular cartilage at both ends. Notice that the enlargement of the marrow cavity continues by the resorption of bone centrally as deposition continues on the periphery. f, A bone at birth superimposed over the same bone of an adult to show their relative sizes, and the amount of deposition and internal resorption that occurs during growth (adapted from Forrest *et al.*, 1975).

in farm animals than in the human. In the horse fontanelles are virtually absent whilst their presence in the calf is very much more marked than in the young pig or lamb. Interestingly, fontanelles exist throughout the life of the Chihuahua dog. In postnatal life the vault of the skull increases by appositional growth. As the cranium enlarges the curvature of its bones must decrease. This necessitates continuous remodelling by the deposition of bone from other surfaces. The single plates of bone over the surface of the skull at birth thus become double plates of compact bone with cancellous bone and marrow between them.

Remodelling   One kind of remodelling during growth in length and width of long bones, involving resorption and apposition, has already been described in the previous sections. This is often termed structural remodelling. Another type is internal remodelling. This is necessary because the Haversian systems in compact bone and the trabeculae of cancellous bone do not last throughout adult life. Lacroix (1971) discusses the process in detail.

Parts of the Haversian systems which have died are resorbed by osteoclasts where the capillary and osteogenic cell supplies are good. In this process elongated tubular resorption cavities are formed and become lined with the osteogenic cell precursors of osteoblasts. The osteoblasts then secrete the matrix of new bone which becomes calcified. These new layers of bone are thought to be of metabolic significance in that the calcium remains more or less in equilibrium with the ionized calcium of the blood and therefore may act as a pool of available calcium if sudden emergencies occur.

Factors affecting bone growth   Vaughan (1980) lists ten variables that can affect bone growth and modelling (Table 3.7). Clearly these may be divided into exogenous (dietary) factors on the one hand and endogenous factors (mostly hormonal) on the other. Enhancement and retardation of the various processes involved in bone growth and modelling may be induced and in many cases the variables are interactive. For example, the action of parathyroid hormone is very complex. It reacts with certain vitamin D metabolites, it may be catabolic or anabolic depending on its concentration, it may act upon osteoblasts, osteo-clasts and osteocytes and it may have an important role to play in maintaining calcium homeostasis. Calcitonin also is complex in its action and is capable of inhibiting osteoclastic activity and resorption. The effects of the various hormones are discussed relative to general growth elsewhere (see chapter 4) but the effects of both oestrogen and testosterone on the seasonal growth of the antlers of deer are considered appropriately here.

Antlers are cast and regrown each year and represent a unique growth sequence. The casting is precipitated by the activation of osteoclasts, in response to falling levels of testosterone, at the base of the pedicle of the antler where the living bone of the skull joins the dead bone of the antler (Short, 1980). As the testosterone activity varies throughout the year it is usually the regression of the testes that precipitates the casting. Re-growth is stimulated by small quantities of testosterone but exogenous oestrogen can induce a similar effect (Fletcher and Short, 1974). Eventually, however, the rising level of testosterone cuts off the blood supply to the antler and arrests its growth.

Table 3.7. Some factors affecting normal bone growth and modelling (Vaughan, 1980).

1. Parathyroid hormone
2. Calcitonin
3. Vitamin D:
   25-hydroxycholecalciferol (25-(OH) D_3)
   1,25-dihydroxycholecalciferol (1,25-(OH)$_2D_3$)
4. Vitamin A
5. Vitamin C
6. Thyroid hormones
7. Corticosteroids
8. Testosterone
9. Oestrogen
10. Growth hormone

If antler growth is unique because of its seasonality, tooth growth is unique because of the substances that characterize the calcification process. Teeth are held in their sockets or alveoli by bundles of connective tissue fibres containing mostly collagen. The main calcified connective tissue is known as dentine and the part of the tooth which projects through the gums into the mouth is further covered with a cap of very hard calcified epithelial-derived tissue known as enamel. As in other bone tissues, mesenchyme cells are the progenitors of odontoblasts which form successive layers of dentine to support enamel and some internal remodelling takes place.

3.3.3. Haemopoietic connective tissue

Important though the various blood cells and the lymphatic tissues are to growth and development in animals, in a book of this type a detailed account is not warranted. Only a brief mention of the structure of myeloid or bone marrow tissue is to be made to complete the picture of bone and cartilage growth and structure which has been dealt with in some depth already.

In postnatal life myeloid tissue is confined mostly to the cavities of bones. There are two types, red and yellow. The red type derives its name from the vast number of red cells contained within it and which it is actively engaged in multiplying. The yellow type derives its name from the large quantity of lipid it contains but it is capable of manufacturing red cells as well.

The two main components are the connective tissue elements known as stroma and the unattached, or free, blood cells in various stages of formation. Blood vessels constitute the basic framework of the stroma and fibroblasts form collagen around the larger of these blood vessels. The blood vessels supported in this way provide the main skeleton for the bone marrow. Therefore the stroma of bone marrow has a backbone of arteries supported by small amounts of connective tissue. Wide tubular channels known as sinusoids connect the arterial and venous sides of the circulation, blood from arterial vessels being delivered into sinusoids and flowing along these channels to reach a vein. There are three main types of cells: osteogenic cells which are capable of forming bone but which are few in number, reticular cells which produce the delicate reticular fibres found in the marrow and fat cells.

This type of connective tissue, vitally important as a source of blood cells, and therefore of basic importance in controlling many aspects of body function, is clearly very different in both structure and in function to the hard and rigid connective tissue of bone in which it is enclosed.

3.3.4. Loose connective tissue

General structure

As pointed out in the section dealing with the structure and classification of connective tissue, mesenchyme cells of similar morphology have the potential to differentiate along different pathways to give cells of very different appearance. In a group of mesenchyme cells which eventually forms loose connective tissue some may remain undifferentiated, some may form fibroblasts, which in turn synthesize the intercellular substance and fibres, and some may form mast cells with phagocytic properties. Others may differentiate to form the cells of the endothelial lining of blood vessels and yet others may form adipose tissue cells and store lipid. The next section is concerned with this latter type of cell and the tissue surrounding it. In chapter 4, the importance of endocrine (hormonal) influence on adipose tissue growth and on lipolytic and lipogenic activities will be discussed.

Adipose tissue

Structure

Fat cells filled with lipid are known as adipocytes. Single or small groups of fat cells are normal constituents of loose connective tissue. If the tissue consists almost entirely of fat cells that are organized into lobules, the tissue is known as adipose tissue. In pigs and in ruminant animals this is the principle site of fatty acid syntheses, the liver participating to a limited extent only.

Lobules of fat cells are separated from each other and supported by partitions of loose connective tissue known as septa. As pointed out in the previous section, the strands of septa form collectively the stroma of the connective tissue and this is responsible for carrying blood vessels and nerves into the adipose tissue. Individual fat cells within a lobule are supported by stroma that consist of nets of delicate reticular and collagenic fibres which are richly endowed with capillaries in their meshes. This brings capillaries into intimate contact with fat cells. The proportion of cells in adipose tissue which are not fat cells but which are cells of the stroma is about 0.05. The lipids found in the tissue are of two basic types: those found as an integral part of cell structure, mostly phospholipids, and those which form reserves of energy in depot fats, which are mostly triglycerides.

Types

Two types of adipose tissue are found in the animal body. They differ in function, colour, vascularity and metabolic activity and are known as white adipose tissue and brown adipose tissue. Their characteristics are presented in Table 3.8.

White adipose tissue acts as a long-term energy store. In common with brown adipose tissue its triglycerides are formed from free fatty acids that are

Table 3.8. Characteristics of white and brown adipose tissue.

Character	White adipose tissue	Brown adipose tissue
Cell shape and size	Spherical and large (up to 120 μm) with small rim of cytoplasm and flattened peripheral nucleus	Polygonal and smaller (25-40 μm in diameter and 8-32 pl in volume) with a greater cytoplasm to lipid ratio, with the cytoplasm not reduced to an outer rim and with the nucleus sometimes eccentric in position but flattened at the cell periphery
Type of lipid	Proportionately 0.98-0.99 of lipid occurs as triglyceride. With the exception of stearic acid, which is less in this tissue, fatty acid make-up of triglycerides is similar to that in brown adipose tissue in several species	Proportionately 0.75-0.90 of lipid as triglyceride with phospholipids forming a high proportion of other lipid
Gross appearance of lipid	Triglycerides form one large amorphous fat vacuole of lipid	Triglycerides form many lipid droplets and cells contain many mitochondria
Fatty acid oxidation	Fatty acids are mobilized and transported via the plasma to the liver and to peripheral tissues for oxidation	Fatty acids are oxidized *in situ* without concomitant stoichiometric ATP synthesis resulting in the energy release appearing as heat
Vascularity	Not prominent	Prominent vascular network with characteristic venous drainage – which is a factor in allowing the quick release of heat for oxidative process
Frequency of occurrence	High. Most abundant adipose tissue occurring at many sites within the body	Low. Relatively small quantities occurring in more specific sites

released from the lipoproteins in the bloodstream by a coupling reaction involving glycerol-phosphate. The fatty acids assimilated through the plasma membrane of the cell are accompanied by other metabolites, including glucose and acetate, which are required with insulin for the synthesis of triglycerides. The new lipid appears first in the form of tiny droplets which are covered with an electron-dense single layer limiting membrane. These are known initially as liposomes and the individual liposomes fuse with each other to give lipid droplets. The plasma membranes themselves contain more protein than lipid but because of differences in molecular weights far more lipid molecules than protein molecules (Garton, 1976).

If the energy supply to the animal is in excess of its total needs for growing non-adipose tissue and for maintenance, lipid will be deposited in adipose tissue in this way. However, should the animal fall on hard times and have a need for energy not available from the food which it can obtain, then it can tap this source to meet its requirements. In this event adrenaline, glucagon and growth hormone can all stimulate lipolysis and the discharge of free fatty acids from cells, although ruminant animals are less responsive generally to lipolytic stimuli than are animals with simple stomachs. The enzyme lipase is active in this process and is aroused from its normally dormant state by lipolytic hormones such as adrenaline. When released from the adipose tissue the free fatty acids are dispatched to other tissues and organs for oxidation.

As will be evident from Table 3.8, brown adipose tissue, with its colour reflecting an extensive vascularity and a high proportion of cytochromes, has, in contrast to white adipose tissue, a metabolic function. The lipolysis process releases fatty acids which are oxidized *in situ*, thus producing a localized heat by a non-shivering thermogenesis. The metabolic function is important in maintaining body temperature in the critical period immediately after birth and in arousal from hibernation in those animals that hibernate. It is therefore not surprising that brown adipose tissue is found in the newborn of many animals but may gradually disappear with increasing age (e.g. sheep), whilst in other animals it may persist into adult life (e.g. rodents, hibernating animals and possibly man). On the other hand, it is perhaps surprising that the environment to which the dam is exposed during late pregnancy may favour the development of, or arrest the normal decline in, brown adipose tissue in her fetus. The findings of Stott and Slee (1985) show this to be the case with the ewe and her lamb and therein indicate an improved capacity for non-shivering thermogenesis in response to an anticipated increased environmental stress.

The ability of brown adipose tissue to generate heat is different in different species and Girardier (1983) calculated the range to be between 350 and 500 watts (W) per kilogram of tissue. Slee *et al.* (1987) calculated that for a lamb of about 4.7 kg birth weight there might be about 70 g brown adipose tissue present and that with a heat output of 400 W per kilogram the total heat output would be 28 W. In this context some work indicates that the ability to generate heat is dependent on the concentration of tissue-specific mitochondrial uncoupling protein and that simple histological appearance is not a satisfactory basis for differentiating between white and brown adipose tissue (Trayhurn, 1989). Interestingly, the young pig appears to have insufficient concentrations of this protein in its brown adipose tissue to support thermogenesis and the functional nature of this tissue in this species must therefore be questionable (Trayhurn *et al.*, 1990).

Histogenesis of the fat cell

There are two important stages in development before the true adipocyte stage is reached. The undifferentiated mesenchyme cells destined to form adipocytes are small, have no lipogenic enzymes, contain no lipid droplets and are known as adipoblasts. They multiply and differentiate to form pre-adipocyte cells which are still small, relative to the true adipocyte, but which by this time have become truly differentiated and contain lipogenic enzymes and a few lipid droplets. The true adipocyte develops from the pre-adipocyte by increasing in size. It has an increased lipogenic enzyme activity and large, single droplets of lipid can be observed. For a fuller account of this developmental process the reader is referred to the paper by Vernon (1986).

The points in time at which mesenchyme cells either differentiate into droplets or proliferate are still unknown. Similarly, the further differentiation into pre-adipocytes is difficult to time with any degree of certainty. In calves the initial detection of pre-adipocyte lobes depends on the region examined. The intermuscular tissue proximal to the sternum will exhibit pre-adipocyte lobes by about 4 months postconception whilst in the subcutaneous and perineal tissues the first appearance is about 1–2 months later. In lambs too, different regions exhibit pre-adipoyte cell accumulation at different times, for example, postconception at about 55 days in the perirenal region and no earlier than about 85 days in the subcutaneous region.

It is probable that there are genuine differences between the development of white and brown adipose tissue in different species (Leat and Cox, 1980). These workers point out that although white adipose tissue appears to develop from brown adipose tissue in some situations, brown adipose tissue is not an obligatory intermediate stage and there is good evidence that both types of adipose tissue can develop from a common or very similar precursor cell. It seems probable, and the probability is supported by the work of Leat and Cox with the fetal lamb, that cells of both types have a common origin initially (up to 80 days of gestation in the fetal lamb), but that there is some divergence subsequently. In this context it is possible that brown adipose tissue is a transitory form in the development of white adipose tissue in the perirenal, but not in the subcutaneous, region and this could be regarded as a special adaptation to provide for non-shivering thermogenesis in the neonate (Vernon, 1986). This capacity for non-shivering thermogenesis is usually lost during the first 7–10 days after birth and this loss is accompanied by a transition of brown to white adipocytes in the perirenal tissue. Nevertheless there is a certain inherent adaptability in this process, because if lambs are exposed to cold after birth the transition can be delayed for several weeks (Gemmell *et al.*, 1972). When the transformation from brown to white adipocyte does occur there is a marked increase in cell volume, usually by a factor of two, resulting from an increased fatty acid deposition, mostly from the blood stream and effected by an enhanced lipoprotein lipase activity (Vernon, 1977). A schematic representation of the whole developmental process is given in Fig. 3.8 and photomicrographs of typical brown and white adipose tissue cells are given in Fig. 3.9.

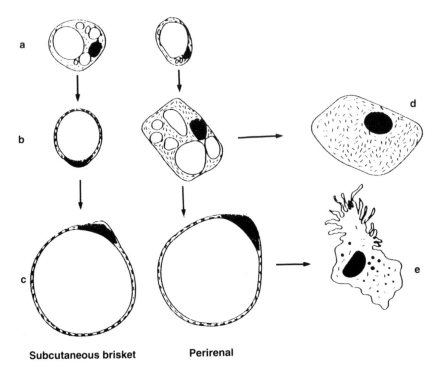

Subcutaneous brisket   **Perirenal**

Fig. 3.8. Scheme for the development of white adipose tissue (WAT) and brown adipose tissue (BAT) in fetal and neonatal sheep. a, 80 days' gestation. The fat cells of subcutaneous brisket (potential WAT) and perirenal adipose tissue (potential BAT) are indistinguishable from one another. They are about 10μm in diameter, either unilocular or multilocular, and contain a moderate number of mitochondria. b, 137 days' gestation. The fat cells of the two tissues are now markedly different in appearance, although approximately of the same size (10–20μm). Subcutaneous brisket fat cells are unilocular, with few mitochondria, and have the typical appearance of WAT. Perirenal adipose tissue has the typical appearance of BAT, with closely packed multilocular cells containing numerous mitochrondria. c, 30 days' postpartum. The fat cells have become much larger and in both tissues have the appearance of WAT. d, Fat-depleted BAT cell showing numerous, closely packed mitochondria. e, Fat-depleted WAT cell. The cell outlines are irregular, with many folds. Mitochondria are few, and there are some locules of fat present (reproduced from Leat and Cox (1980) by kind permission of the copyright holder, W.M.F. Leat).

Cellular aspects of development

General The size of an organ or of a tissue may increase by cell hyperplasia and/or cell hypertrophy, and this has been discussed already (see chapter 2). However, in fatty tissues it is very difficult indeed to disentangle the relative effects of true hyperplasia on the one hand and the expansion of empty cells on the other and there is great uncertainty where, if at all, adipocyte hyperplasia stops and hypertrophy alone becomes responsible for any further adipose tissue

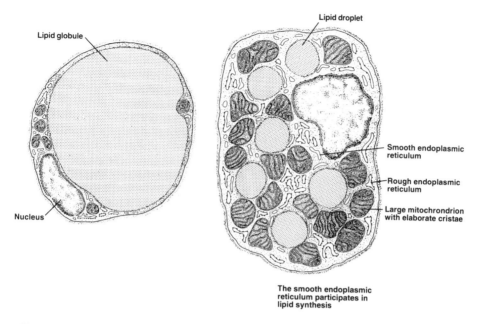

Fig. 3.9. Diagrammatic representation of typical brown (right) and white (left) tissue cells (courtesy of Dr T.P. King, Rowett Research Institute, Aberdeen).

growth. The ultimate size of adipose tissue depots is not necessarily limited by hyperplasia in the early life of the animal because the attainment by adipocytes of a large size can either stimulate adipogenesis and/or fill quiescent pre-adipocytes. If there are great uncertainties with this, the growth which takes place in the adipose tissues of animals carries no uncertainties at all for it is very spectacular and the capacity of an adipocyte to store lipid is very great indeed. A typical average fat cell may have a diameter of about 100 µm. In very fat animals this diameter can increase to 250 µm. As lipid is deposited the cell increases in diameter and volume. It is the relationship between these two dimensions that gives the clue to the enormous propensity for lipid storage, for the volume of a sphere is proportional to the cube of its radius. Therefore a twofold increase in diameter will give an eightfold increase in volume, whilst a tenfold increase in diameter will give a thousandfold increase in volume. The rate of lipid deposition and therefore of hypertrophy depends on the relative rates of esterification and lipolysis and when these two processes are equal hypertrophy ceases.

The actual changes that have been recorded in Friesian bulls give a practical perspective to the generalizations expounded above. The Friesian bull is proportionately about 0.70 of its mature size at a live weight of 700 kg and has reached this size after dramatic changes in adipocyte number and in diameter and in the amount of lipid stored have taken place. Based on the findings of Robelin (1985) the diameter of adipocyte cells in the intermuscular, perinephric and omental depots at birth is between 45 and 50 μm and slightly less (35–40 μm) in the subcutaneous region. At about 700 kg live weight the perinephric and omental cells will have diameters of 140–145 μm, the intermuscular cells diameters of 130–135 μm and the subcutaneous cells diameters of about 110 μm only. Taking the mean diameter of all cells as 45 μm at birth, the mean at 700 kg was found by Robelin (1985) to be 135 μm and during this time period their volume had increased by a factor of 29 and their total number by a factor of 6.7. The lipid stored during this period had increased by a factor of 197, from 0.7 kg to 138 kg.

Cattle   As discussed elsewhere, there are distinct differences in adipose tissue partitioning within the body and in the total adipose tissue content of different breeds of farm animals and between sexes within breeds. These differences are accompanied by variation in the cellular aspects of the adipose tissue itself. Also, in addition to the variation in adipocyte size detailed above, in some species the size may follow a gradient within a depot.

In cattle, and as referred to in other parts of this book but particularly below, there are large differences in the amounts of adipose tissue in animals of traditional beef breeds such as the Hereford compared with animals of dairy breeds such as the Friesian, particularly when compared at young ages (Truscott *et al.*, 1983b; Table 3.9). Reference to this table leaves no doubt that it is impossible to generalize on the cellularity of adipose tissues. The *ad libitum* feeding was deliberately imposed to obtain maximum adipose tissue deposition in the two genotypes and this gave greater empty body weights for the Friesians at the three ages but smaller proportions of adipose tissue. On the whole, during growth the differences in cellularity between breeds were smaller than those between depots within breeds. In particular a striking feature is the lack of a consistent trend between breeds in adipocyte size. The authors argue that if the two breeds had partitioned their body adipose tissue similarly, then the Friesians, because of a heavier final weight than the Herefords (proportionately 0.02), would have been expected to have had more cells in all their depots. However, this expectation was clearly not completely realized when the number of adipocytes was expressed in relation to fat-free body weight in that although the ratio figures were higher for the perirenal, omental and prescapular regions, the three subcutaneous ratios and the brisket ratio indicated a true breed effect independent of size.

When comparisons of different breeds of cattle with different propensities to lay down adipose tissue have been made at similar live weights, differences in hypertrophy have been shown to be more important than differences in hyerplasia. A comparison of Friesian and Charolais bulls grown similarly to 500 kg live weight illustrates this point (Robelin, 1986). In this comparison Robelin found that at this live weight the total weights of adipose tissue deposited in the entire body were 110 kg and 60 kg for the Friesian and for the Charolais respectively. The total numbers of adipocytes associated with these amounts of adipose tissue were 120×10^9 and 100×10^9 respectively and the average adipocyte diameters were 110 μm and 92 μm respectively. Therefore, the

Table 3.9. Cellularity of adipose tissue in Hereford and Friesian castrated males given food *ad libitum* (from Truscott *et al.,* 1983a,b).

	Age (months)					
	6		13		30	
	Hereford	Friesian	Hereford	Friesian	Hereford	Friesian
Live weight (kg)	151	169	353	422	487	568
Proportion of adipose tissue in empty body	0.106	0.073	0.234	0.211	0.309	0.285
Mean adipocyte volume ($\times 10^{-8}$ cm^3) (a) and mean numbers of adipocytes ($\times 10^8$) (b) at various depots						
Rump						
(a)	26	10	170	188	224	257
(b)	3.2	6.5	4.5	3.0	9.6	4.4
Midloin						
(a)	18	10	172	168	242	253
(b)	1.3	2.9	2.2	1.8	4.4	2.4
12th rib						
(a)	16	3	217	156	239	265
(b)	6.0	12.8	5.4	6.5	11.4	7.7
Brisket						
(a)	16	11	45	72	92	104
(b)	5.2	2.3	20.1	9.4	27.0	18.6
Perirenal						
(a)	15	16	183	180	392	446
(b)	66.6	71.6	42.9	60.6	41.2	54.0
Omental						
(a)	24	22	164	164	344	382
(b)	35.8	28.4	39.8	56.2	40.0	53.0
Prescapular						
(a)	17	13	134	145	175	153
(b)	29.3	18.7	13.9	17.7	25.2	30.3
Ratio of number of adipocytes to fat-free body weight (10^6 cells kg^{-1}) at 20 months						
Subcutaneous depot						
Rump	—	—	—	—	3.2	1.2
Midloin	—	—	—	—	1.5	0.7
12th rib	—	—	—	—	3.7	2.2
Brisket	—	—	—	—	9.2	5.2
Perirenal	—	—	—	—	13.6	15.1
Omental	—	—	—	—	13.2	14.9
Prescapular	—	—	—	—	8.3	8.4

major difference was in adipocyte volume (proportionately 0.70) rather than in adipocyte number (proportionately a difference of 0.20).

The effects of castration can influence markedly adipose tissue cellularity. For example, in comparisons at 300 kg empty body weight of similarly grown Charolais bulls and castrated male Herefords, Robelin (1986) found that animals of the latter breed had considerably more fatty tissue in their bodies than did animals of the former breed and that there were big differences in adipocyte cell diameters, although this feature was more marked in the subcutaneous depots, compared with the intermuscular, perinephric and omental depots. In the subutaneous depots the cell diameters of 50 μm for the Charolais and 160 μm for the Hereford imaged to a large extent the effects of both breed and castration in the total adipose tissues present. However, when the comparisons were made at identical fat proportions of empty body weight, the differences, although still apparent, were greatly reduced. For example, when the fat proportion was 0.15 the diameters were 80 μm for the Charolais and 120 μm for the Hereford. The reduction in size differential, from proportionately 2.20 in the comparison at similar live weight to proportionately 0.50 in the comparison at 0.15 total adipose tissue content of the body, reflects largely the removal of the contributory effects of castration. This is supported by comparisons of castrated males with entire males within breeds. For example, Robelin (1981) found that in the Friesian breed, at any given age, the bull had smaller adipocytes than did the castrated male and it may be assumed, therefore, that one of the roles of androgens is in modifying adipocyte hypertrophy.

At this stage in the discussion it would be appropriate if some indication could be given on the age at which hyperplasia ceases to play an important part in adipose tissue growth in cattle. Unfortunately it is impossible to give such an indication other than in the most general terms; namely that it appears that most hyperplastic activity is complete by about 15 months of age.

Pigs In common with cattle, different pig genotypes with different propensities to grow adipose tissue, exhibit differences in the cellularity of that tissue (Table 3.10). The data in this table reveals that the pigs of this experiment selected for low adipose tissue proportions in their bodies had at any given weight more but smaller adipocytes in the tissues studied, thereby indicating that adiposity may be due primarily to cellular hypertrophy, particularly as the pig nears its maximum or mature size, and is under genetic control. It is clear, however, that adipocyte hyperplasia was still very much in evidence in the live weight interval from 54 kg to 109 kg and Hood and Allen (1977) found that most of this activity took place between 54 kg and 83 kg live weight. During this live-weight interval hypertrophy was less pronounced than in the period which preceded it. From this and other work it is difficult to know the age at which hyperplasia ceases and the age from which hypertrophy is solely responsible for increases in adipose tissue mass but there is a large body of evidence to support the contention that in pigs which are grown well and without interruption, in postnatal life hyperplasia is the more important force up to about 2 months of

Table 3.10. Effects of breed (see note below), live weight and anatomical location on the cellularity of porcine adipose tissue in castrated male pigs (Hood and Allen, 1977).

	Live weight (kg)								
	28			54			109		
	H×Y	M3×L	HM	H×Y	M3×L	HM	H×Y	M3×L	HM
Carcass weight (kg)	18.4	17.2	17.8	38.7	37.3	38.8	82.5	79.6	—
Fat-free carcass weight (kg)*	14.1	12.8	12.2	28.1	24.0	22.1	54.7	39.7	—
Age (days)	79.8	70.0	158.8	117.3	99.7	200.7	167.8	159.6	—
Proportions of body fat									
Perirenal	0.007	0.008	0.022	0.013	0.014	0.040	0.023	0.032	—
Extramuscular carcass	0.210	0.227	0.281	0.249	0.317	0.397	0.306	0.466	—
Intramuscular carcass	0.024	0.030	0.032	0.023	0.033	0.029	0.031	0.035	—
Subcutaneous fat thickness (cm)									
Outer	0.41	0.56	0.64	0.71	0.69	0.94	0.89	1.27	—
Middle	0.43	0.58	0.69	0.79	0.91	1.50	1.32	2.10	—
Cell volumes ($\mu m^3 \times 10^4$)									
Subcutaneous fat									
Outer	17.7	19.8	41.0	32.4	41.0	65.5	41.0	72.9	—
Middle	18.3	20.2	43.9	35.9	58.9	88.3	50.4	90.5	—
Perirenal	15.2	16.8	50.3	33.8	41.5	94.5	54.0	87.2	—
Numbers of adipocytes ($\times 10^9$)									
Extramuscular	25.3	23.0	13.8	33.2	29.7	23.7	64.5	52.4	—
Perirenal	1.1	1.0	0.9	1.7	1.5	2.2	4.4	3.5	—

H×Y: pigs selected to have lower proportions of fatty tissue at same carcass weight than M3×L pigs.
HM: pigs with small body size at maturity.
*Regarded as equivalent to true body size.

age whilst hypertrophy is the more important force after about 5 months of age.

In common with the data of Table 3.9 for cattle, the data of Table 3.10 for pigs indicate that there are differences in adipocyte size between adipose tissue depots. Additionally in pigs adipocyte size is smaller immediately beneath the skin than in the middle layers of the subcutaneous depots (Table 3.10). Had subcutaneous adipocytes of the inner layers of that depot been studied, then it is likely that those next to muscle tissues would have been nearer in size to those on the outside, rather than to those in the middle of the tissue.

Sheep   The relative contributions of hyperplasia and hypertrophy to adipose tissue growth in sheep, at any one age, are difficult to quantify, but it appears that *in utero* the first adipocytes to exhibit lipid accumulation are those of the perirenal depot at between 80 and 90 days postconception, to be followed by the adipocytes of the subcutaneous depots about 14 days later. Some 60 days later when the lamb is born (or even as early as 40 days later in some cases (Broad *et al.*, 1980)), most hyperplasia in the perirenal tissue will be complete and the sizes of the pre-adipocytes and the adipocytes will have increased by about 30- and 20-fold respectively. Even at this stage of life hypertrophy clearly has a significant part to play in adipose tissue growth, but compared with the newborn calf the adipocytes of the lamb are smaller (Robelin, 1981).

During the first 50–60 days of the suckling period after birth, when milk fat forms a high but progressively decreasing proportion of total food intake, the perirenal and subcutaneous adipocytes exhibit considerable hypertrophy. Accompanying this hypertrophy, up to about 100 days after birth, hyperplasia increases the numbers of adipocytes in the subcutaneous and intermuscular depots by a factor of between two and three but, in contrast, there is little or no hyperplasia in the perirenal region. Based on the work of Robelin (1985), Vernon (1986) points to a distinct difference here compared with the events which occur in Friesian bull calves where there is little hypertrophy of carcass or abdominal adipocytes, but a substantial increase in the number of adipocytes, in the first 100 days after birth.

The period of rapid adipose tissue growth in lambs is followed by a period in which the growth of the same tissues is minimal. Again this differs from the case with both cattle and pigs where there are no obvious breaks in the growth processes, although the comparison with cattle is not quite as simple as it would at first appear. This is because although in the abdominal and intermuscular depots there is no hyperplasia between 100 and 400 days after birth, in the same period of time hyperplasia does occur in the subcutaneous depot and hyper-trophy in all three depots begins during the first 100 days. Overall the present stage of knowledge suggests that in sheep, hyperplasia is an active force up to about 12 months of age after birth, although during this period hypertrophy will play an important part also in total adipose growth. After 12 months hyperplasia will play little, or at the very most a very small part, and growth will become increasingly dependent on hypertrophy, perhaps particularly in the non-carcass depots compared with the subcutaneous and intermuscular depots of the carcass (Thompson and Butterfield, 1988). At maturity, differences due to the castration of male animals may be apparent, with entire males having smaller adipocytes but in larger numbers, particularly in the subcutaneous and omental depots (Thompson and Butterfield, 1988).

Allometric growth coefficients can be useful for differentiating between changes in adipocyte volume and changes in lipid weight. The standardized allometric equation developed by Taylor (1980) may be used for this purpose and a standardized growth coefficient of less than unity from this equation will indicate that adipocyte volume has increased at a slower rate than the lipid weight in the particular depot under consideration. Thompson and Butterfield

(1988) used this equation to study changes in composition in rams and in castrated males of the Australian Dorset Horn breed. In this case the allometric coefficients for the rates at which adipocyte volume increased to mature values, relative to the rate at which lipid was deposited, indicated that in all depots adipocyte volume increased at a slower rate than total lipid weight.

Differences between breeds in adipocyte volume have also been found and lambs from breeds with differing body compositions have distinct differences in lipid metabolism and in adipocyte characteristics (Sinnett-Smith and Wooliams, 1988). In this work the subcutaneous depots of three large breeds of sheep (East Friesland, Texel and Oxford) were compared with a small feral breed (Soay). Differences in volume were not apparently related to carcass fatness of the breed. Thus the leaner Soay lambs had larger adipocytes than the larger and fatter Texal lambs, whilst the former and the East Friesland lambs had greater rates of incorporation of acetate into fatty acids than did either of the two other breeds.

Distribution of adipose tissue

The distribution of adipose tissue within the mammalian body is basically the same in all species (Pond, 1984). However adipose tissue is most extensively present in the subcutaneous, perinephric, omental and muscular regions. The proportion of adipose tissue in newborn farm animals is between 0.01 and 0.04, and from these low levels there is a massive increase to maximum proportions of about 0.40 at maturity. Within this overall increase in cattle and pigs there is evidence of a higher lipogenic activity in the subcutaneous region compared with other internal sites. The proportion of lipid in mammalian skeletal muscle may vary from about 0.015 to 0.130 on a fresh-weight basis. Neutral lipids and phospholipids may each account for about one-third of this total, with cholesterol and cerebrosides accounting each for about one-sixth. The composition of the lipid in other adipose tissue is highly variable and influenced by a number of factors as discussed in the next sections. The overall growth of adipose tissue receives further attention in chapter 6.

Lipid chemistry

Major components   The lipids of fully developed adipose tissue consist mostly of triglycerides (proportionately 0.90–0.98) with small amounts of diglycerides (proportionately 0.01–0.02), phospholipid (proportionately 0.0025) and cholesterol (proportionately 0.0025). This contrasts strongly to the make-up of immature adipose tissue where the proportion of triglyceride is low. For example, in pigs the triglyceride proportion increases from about 0.07 at birth to about 0.90 at 160 days of age (Leat and Cox, 1980). The fatty acids which make up the triglycerides, and therefore the lipids, are of immense importance in determining the physical characteristics of adipose tissues and man's perception of quality in adipose tissue in the overall context of meat quality. Those that are found most commonly in animal tissues are detailed in Table 3.11, although in addition to some of these the shorter chain fatty acids, butyric ($4:0$), caproic ($6:0$), caprylic ($8:0$) and capric ($10:0$) occur in milk. The

Table 3.11. Some commonly found fatty acids in the adipose tissues of animals (adapted from Enser, 1984).

Number of carbon atoms and double bonds	Systemic name	Common name	Melting point (˚C)
14:0	*n*-Tetradecanoic	Myristic	54.4
16:0	*n*-Hexadecanoic	Palmitic	62.9
-16:1	*cis*-9-Hexadecanoic	Palmitoleic	0.0
17:0 br*	14-Methyl hexadecanoic	†	39.5
18:0	*n*-Octadecanoic	Stearic	69.6
-18:1	*cis*-9-Octadecanoic	Oleic	13.4
-18:2	*cis*-9, 12-Octadecadienoic	Linoleic	-5.0
18:3	*cis*-9, 12, 15-Octadecatrienoic	α-Linolenic	-11.0
20:0	*n*-Eicosanoic	Arachadic	75.4
20:4	*cis*-5, 8, 11, 14-Eicosatetraenoic	Arachadonic	-49.5
22:1	*cis*-13-Docosenoic	Erucic	33.5

*Branched chain.
†No common name.

perception of quality is, however, dependent too on the changes which take place in the relative proportions of lipid, which is the major constituent, relative to water and connective tissue. Adipose tissue quality will be dealt with in the next section and the aim here is to describe the chemical composition of adipose tissues and those factors that can modify it.

Age effects on major components Compared with more mature adipose tissue, young adipose tissue contains higher proportions of both water and of connective tissue and a lower proportion of lipid contained in small adipocytes. As the animal grows and ages the adipose tissues increase in size progressively from inclusions of lipid of dietary origin in the adipocytes. In consequence the adipocytes increase in size, the lipid proportion of the entire tissue increases and the water and connective tissues decrease. These changes are reflected in the appearance and in the feel of the tissue. Compared with older tissue young tissue feels wet, it lacks firmness because the adipocytes containing small quantities of lipid are not packed together tightly, it has a greyish hue partly because of the higher proportion of connective tissue and it will separate from proximal muscle tissue relatively easily. These changes occur in all adipose tissues but different depots contain different proportions of water. In cattle the subcutaneous and intermuscular depots of the carcass have higher proportions than the perirenal and omental depots of the abdomen and in the carcass depots the brisket subcutaneous depot has higher proportions than the midloin, 12th rib and rump depots (Truscott, 1980).

Overall nutrition, sex and genotype effects on major components The proportions of water, lipid and connective tissue in adipose tissue will depend also on the speed of growth of the animal, and therefore on its plane of nutrition, and on its sex and genotype. The better fed, faster growing animal will have greater deposits of adipose tissue in its body and this adipose tissue will contain less water than in the animal which is less well treated. As a result of these effects the adipose tissue which is present in animals that have been poorly fed will have features similar to that found in younger animals. Sex effects have been found in cattle, sheep and pigs. The difference between the castrated male and the bull tends to be greater than the differences between breeds at the same age and Wood (1984) reports, in *ad libitum*-fed animals at 400 days of age, proportionately 0.23 more water and proportionately 0.34 less lipid in the bull compared with the castrated male. In pigs, as in cattle and sheep, the effects of castration of increasing the total adipose tissue content of the body at earlier ages and at lighter live weights are accompanied by that tissue having a floppy texture. This texture partly reflects the fatty acids in the lipid but is also related to the higher proportions of water and connective tissue in the tissues of the boar (Wood and Enser, 1982). Pig genotypes selected for increased muscle content in their carcasses also have differently constituted adipose tissues compared with those having greater amounts of carcass adipose tissue. As will be discussed below, the softer texture of the fat is particularly related to the fatty acid content of the lipid in the adipose tissue, but in addition to this there is a lower proportion of lipid and a higher proportion of water (Wood *et al.*, 1983).

Fatty acid composition: general Lipids are very complexly structured and in the live animal the mixture of fatty acids imparts to the lipid a variable fluidity at body temperature. The degree of fluidity depends on the proportions of certain fatty acids present and the melting points of the resultant triglycerides are determined by the fatty acids of which they are composed. The melting point is probably the most important physical characteristic in determining quality, as it affects the firmness of the tissue at any particular temperature. As the carbon chain lengths of fatty acids increase so too do melting points. In pigs, sheep and cattle, the fatty acid most strongly linked to the melting point of the lipid and its firmness is stearic acid. In sheep and cattle the concentration of stearic acid in adipose tissue is greater than in pigs and because of this the adipose tissues of beef and lamb carcasses tend to be harder and firmer when cold than do the adipose tissues of pig carcasses, although in the case of lamb adipose tissue the stearic acid imparts a sticky feel to the palate. The fatty acid composition of adipose tissue lipids changes with age and with a variety of other factors as discussed below but an idea of the proportions which may be found in the subcutaneous depots of pigs, sheep and cattle and in the phospholipids of the muscle tissue of these species is given in Table 3.12. In cell membranes the proportions of fatty acids in the phospholipids are of no less importance in the context of cell fluidity than are the proportions of fatty acids in the triglycerides in the lipids contained within the cells.

Table 3.12. Major proportions by weight of fatty acids in the triglycerides from the subcutaneous adipose tissue and in the phospholipids of the longissimus dorsi muscle of cattle, sheep and pigs (adapted from Enser, 1984).

Fatty acid: number of carbon atoms and double bonds*	Triglycerides[†]			Phospholipids[‡]		
	Cattle	Sheep	Pigs	Cattle	Sheep	Pigs
14:0	0.037	0.029	0.015	0.004	0.021	0.002
16:0	0.298	0.237	0.276	0.226	0.220	0.189
16:1	0.047	0.035	0.032	0.025	0.023	0.016
18:0	0.171	0.183	0.122	0.078	0.132	0.120
18:1	0.423	0.432	0.451	0.243	0.303	0.188
18:2	0.023	0.038	0.104	0.230	0.180	0.255
18:3	–	–	–	0.020	0.039	0.002
20:4	–	–	–	0.125	–	0.077

*Number of carbon atoms in chain followed by number of double bonds.
[†]Cattle: loin fat from 400-day-old Friesian bulls; sheep: inguinal fat from 224-day-old Hampshire lambs; pigs: outer subcutaneous loin fat from Large White pigs 87 kg in weight.
[‡]See original reference for sources for different species.
–, Information not available.

Species and nutrition effects on fatty acid composition It was pointed out above that the factors of age, growth rate, overall nutrition, sex and genotype affected the proportions of water, lipid and connective tissue in the adipose tissue itself. These same factors are important in determining the fatty acid composition of lipids and in the case of pigs prolonged exposure to cold gives a higher proportion of unsaturated fatty acids which can influence the melting point and physical characteristics of the lipid (Fuller *et al.*, 1974). Also there are distinct differences between the species and in the site of deposition. In general the greatest species differences are found between ruminant animals on the one hand and simple stomached animals on the other. In this particular respect the reasons for the firmer adipose tissues of cattle and sheep have been outlined above and on the whole their tissues are less susceptible to change by dietary factors than are those of pigs. The higher proportion of saturated fatty acids in the lipids of the adipose tissue of cattle and of sheep is in turn a reflection of the events which take place in the rumen, particularly the process of biohydrogenation. Biohydrogenation is the process whereby hydrogen is added to the double bonds of the unsaturated fatty acids of the dietary lipids which enter the rumen and in so doing converts them to their saturated analogues. An important example is the conversion of oleic acid to stearic acid. By this process the highly unsaturated fatty acid content of most ruminant diets is not reflected in the adipose tissues of the animals which eat them. On occasions, for example when there is a big influx into the rumen of unsaturated fatty acids from young herbage, the mechanisms of the hydrogenation process may, in effect, be swamped, with the result that some of these acids escape hydrogenation and are absorbed in the lower parts of the gut in the form that they appeared in the diet.

In consequence, the adipose tissue will contain higher proportions of them and this is, in fact, found in practice where there is a noticeable seasonality in adipose tissue firmness in cattle and sheep.

Other dietary factors can affect the fatty acid content of the lipids in the adipose tissue of cattle and of sheep. If the proportion of concentrates (particularly cereal) to roughage in the diet is increased, more propionic acid (C3 : 0 fatty acid) is produced in the rumen and this results in higher proportions of unsaturated fatty acids. This effect will be most noticeable when each cereal grain has been fragmented, as in for example the rolling process, and may be due to the fermentation process of the rumen producing high proportions of branched chain fatty acids, because if the grain is given whole there is little or no effect. There is, however, a marked difference between cattle and sheep in this respect in that cattle produce fewer branched-chain fatty acids than do sheep when given the same processed cereal. The reasons for this are complex but basically may be a consequence of the difference between the two species in the production and/or metabolism of propionic acid. Excess propionic acid produced in the rumen is incorporated into long-chain fatty acids and its primary metabolite, methylmalonate, is similarly utilized to give numerous branched-chain fatty acids, the presence of which in lipids makes tissues softer than usual.

The deliberate inclusion of concentrated sources of lipid in the diet may or may not affect the fatty acid composition of body lipids. Some unsaturated fatty acids may escape biohydrogenation and therefore have an influence on adipose tissue but this depends on the concentration in the diet and the other food that is given. If for any reason the degree of saturation of the body tissue lipids needs reducing, for example to placate the lobby within the human population which associates high saturated fat intakes with certain types of heart disease, then this can be accomplished relatively easily by enriching the diet with concentrated sources of lipid containing high proportions of unsaturated fatty acids wrapped in protective envelopes, such as casein, to protect them from the effects of the rumen and to allow an uptake of the unchanged acids in the lower gut.

Pigs are often given diets with concentrated sources of lipid included to improve food conversion efficiency and to reduce production costs. Any concentrated source of lipid given in this way must be considered in the first place relative to the effects which it may have on the adipose tissues of the animal. In this particular respect the proportions of linoleic acid in the dietary lipid, and the contribution of linoleic acid energy to total dietary digestible energy, where a critical concentration of 0.035 MJ for each megajoule of digestible energy has been identified (Prescott and Wood, 1988), are of extreme importance. Linoleic acid is not synthesized by mammalian tissues but is highly digestible compared with saturated fatty acids such as those found in animal tallows and also is preferentially deposited compared with other fatty acids. The same is true of linolenic acid. In terms of concentration in the lipid of adipose tissue relative to melting point, Wood and Enser (1982) suggest that melting points are likely to be unaffected if the concentration is less than 150 mg g^{-1} of total fatty acids, a concentration that is not exceeded often in most pigs with average adipose tissue content and given conventional diets containing less than 40 g kg^{-1} lipid per kilogram. If, however, the dietary lipid is of plant origin and the concentration exceeds 40 g g^{-1} with a concentration of linoleic acid of about 16 mg g^{-1}, then concentrations of linoleic acid in adipose tissue lipid greater than 150 mg g^{-1} of total fatty acids will be induced and these will have an increasingly important effect on the melting point of the lipid and on its physical characteristics (Wood, 1984). Therefore a simple relationship exists: the higher the concentration of linoleic acid in the diet, the higher will be the concentration in the adipose tissue.

Another factor that has been shown to affect the physical characteristics of the backfat of pigs is the concentration of copper in the diet. In the UK the use of high concentrations of copper in the diet for growth promotion purposes has been widely practised for many years. This practice has been shown to produce lipids with decreased melting points and to give softer backfat through the copper activating the desaturase enzymes to give an increased ratio of oleic to stearic acid. However European Union legislation which precludes the inclusion of copper in the diet at the commonly used previous concentration of 250 mg g^{-1}, but still allows an inclusion level massively above that required to meet normal physiological requirements, may reduce this effect considerably if not totally eliminating it.

Age and fatty acid composition   Changes in the fatty acid composition of lipids due to age, to speed of growth and to overall plane of nutrition are evident in all species but it is difficult to separate the relative effects of any one of these interlinked factors. In the pig the half-life of lipid fatty acids is about 180 days and it follows that in pigs slaughtered for bacon curing at about 90 kg live weight there is a 50% or better chance that a fatty acid deposited in early life will still be present (Enser, 1984). However, the fatty acids present in lipids do change during growth and fattening. Typically, pigs given conventional diets containing less than 40 g kg^{-1} lipid up to about 6 months of age, have lipids which are very rich in palmitic and palmitoleic acids (about 980 mg g^{-1} of total fatty acids) with the only other fatty acid exceeding a concentration of 10 mg g^{-1} of total fatty acids being myristic, which reaches a concentration of about 17 mg g^{-1} at 184 days of age (Wood, 1984). These terminal figures are reached after the palmitoleic, and to a lesser extent the palmitic acids, have decreased from even higher levels at birth. Compensating these decreases are increases in stearic and oleic acid. Within this overall time period other changes are found. For example linoleic acid increases sharply from birth to about 3 days of age, mirroring the high levels of linoleic acid in sow's colostrum and its preferential deposition, but then progressively decreases to give an increased melting point and firmness to the adipose tissue. In contrast to this sequence of events in the pig, in cattle in particular, and in sheep to a lesser extent, the concentration of saturated fatty acids in the lipids of adipose tissue decreases with age and the unsaturated fatty acids increase. The contrast is of course relative because, and as pointed out previously, cattle and sheep lipids are on the whole more saturated and firmer than pig lipids because of the higher proportion of stearic acid.

Growth rate and fatty acid composition The growth rate, and therefore by implication the overall plane of nutrition, will cause changes in the overall patterns discussed above. In pigs a decrease in the growth rate resultant from a lowering of the plane of nutrition affects the fatty acid and water contents of the adipose tissue (Wood, 1984; Table 3.13). In this particular instance the growth rate reduction of proportionately 0.18 was very marked but it is of interest that in those countries where there has been a deliberate policy of reducing backfat thickness sometimes there have been allegations of poor (soft) adipose tissue. For example, in the UK, softness of backfat has been shown to increase with increasing lean content of the carcass, particularly in carcasses with less than 12 mm backfat (Dransfield and Kempster, 1988). As will be evident below, other factors undoubtedly play a part in this overall effect but nevertheless the effect is of importance in its own right, even though in this set of data the concentrations of linoleic acid were below the concentrations mentioned earlier as representing a watershed in terms of distinct differences in softness becoming important relative to concepts of quality. In cases where restrictions of food and of growth are more severe the effects on lipid composition are correspondingly more severe. Although the total withholding of food from pigs and from sheep for up to 4 days results in the mobilized fatty acids resembling those of the adipose tissue lipids, if for sheep the period is extended, then the mobilization becomes more selective and therefore the body tissue lipids change in their composition. In general, in cattle and sheep, progression from the fat animal to

Table 3.13. Effect of food intake on the chemical composition of the backfat of the pig (adapted from Wood, 1984).*

	Food intake	
	High	Low
Daily growth (g)	667	547
P_2 backfat thickness (mm)[†]	17.0	13.3
Muscle in side (g kg^{-1})	547	598
Composition of backfat		
Outer		
Water (g kg^{-1})	120	196
Lipid (g kg^{-1})	841	750
Linoleic acid (mg g^{-1} fatty acids)	90	117
Fat firmness score[‡]	3.6	3.1
Fat whiteness score[∥]	3.7	3.1

*Female pigs grown between 20 and 68 kg live weight receiving a diet containing 35 g total lipid and 13.0 MJ of digestible energy per kg.
[†]Depth of skin and adipose tissue measured with an intrascope at the level of the last rib 65 mm from the mid-dorsal line.
[‡]Scores: 1 (very soft) to 5 (very hard).
[∥]Scores: 1 (very grey) to 5 (very white).

the emaciated animal is accompanied by increased proportions of stearic acid, and decreased proportions of oleic acid, in the adipose tissues.

Site and fatty acid composition   The proportions of fatty acids in the lipids of different adipose tissues vary in a manner that is only partly related to the amount or rate of lipid that is deposited. A generalization is that there is a progressive increase in saturation from peripheral (subcutaneous) tissues through intermuscular and intramuscular deposits to deep body sites in cattle, sheep and pigs. This trend is apparent in the data of Table 3.14 for cattle. There is much speculation on the reasons behind this distribution and one plausible hypothesis is that temperature differences between the sites are responsible. For example, the lower temperatures proximal to the subcutaneous depots may necessitate a lower melting point in the lipid.

Breed, genotype, sex and fatty acid composition   Lastly, the role of breed, genotype and sex in determining the fatty acid composition of lipids is considered. In cattle differences in total fatness between breeds are responsible for a high proportion of the differences found in fatty acid composition. For example, high positive correlations have been found between the amounts of subcutaneous adipose tissues in the hindquarters and in the concentrations of palmitoleic and stearic acids in the brisket subcutaneous depots (Pyle *et al.*, 1977). However, because double-muscled Charolais cattle are leaner than their normal contemporaries, the concentrations of stearic acid have been shown to be higher, and those of linoleic acid lower. In this particular respect it is important to appreciate that floppy adipose tissue from extremely lean breeds is more likely to be due to the effects of differences in gross chemical composition than to changes in fatty acid composition, because leaner breeds have more saturated lipids. It follows from the hypothesis presented here that if there are small differences in total adipose tissue between breeds, then it is likely that there will be small differences in fatty acid concentrations as well. In pigs, the fact that genetically lean strains have relatively unsaturated and soft

Table 3.14. Fatty acid composition (mg g^{-1}) of lipid from four body sites in Aberdeen Angus and Friesian steers and heifers aged 16–22 months given a hay diet, with equal numbers of each sex from each breed contributing to each mean (Wood, 1984).

Fatty acid	Body site			
	Subcutaneous	Intramuscular	Intermuscular	Perirenal
14:0 Myristic	33	30	35	45
16:0 Palmitic	260	316	312	336
9–16:1 Palmitoleic	94	43	41	20
18:0 Stearic	82	189	224	252
9–18:1 Oleic	447	366	322	282
9, 12–18:2 Linoleic	21	12	11	10

adipose tissues has already been mentioned. There are small differences in cattle between the sexes and those that do exist can be explained by differences in total adipose tissue in the body. Similarly, there are small differences between sexes in sheep but in pigs the differences are more marked. Boars tend to produce carcasses containing lipid which is more unsaturated than is that from castrated males of the same carcass weight. However, boars are leaner at similar carcass weights and the differences are a consequence of higher concentrations of linoleic acid in the lipid. The practical reality is that boars have slightly lower fat quality scores in most situations because they are leaner. If comparisons are made at similar adipose tissue content, differences become smaller but are still detectable. In consequence, the practical implication is that boars from lean strains should be fed to grow as quickly as their genetic make-up will allow on diets that do not contain too much linoleic acid.

Adipose tissue quality
The concept of quality has both subjective and objective components and may have a different meaning for the butcher/meat wholesaler/packer than for the consumer. In the previous section a number of different factors were identified as causing changes in the chemical and physical structures of adipose tissues, but to what extent these have any real bearing on quality depends very much on the criteria that are being used to judge quality and the way in which these criteria are applied by the individual who is doing the judging. Overall the senses of sight, touch, taste and smell can play a part in quality assessments and more attention has been given to the carcass adipose tissues of pigs than to those of cattle and sheep. This is perhaps not surprising in view of the massive production changes and intensification that have occurred in the pig industry compared with the cattle and sheep industries.

The colour and firmness of adipose tissue are important components of the concept of quality to the consumer. Some individuals prefer white fat to yellow fat, others the converse. Similarly some individuals have no strong dislike of soft fat but others object strongly to it. As pointed out in the previous section, adipose tissue assumes a greyish colour if it contains a high proportion of connective tissue. In addition to this the colour is partly determined at any particular temperature by the extent to which the lipids have solidified and the quantity of capillary tissue present, the gradual solidification of the lipids effecting a change from a grey/yellow colour to a creamy white colour. The yellow colour can, however, persist in some cases and be unrelated to the solidification process. In cattle the Channel Island breeds give yellower adipose tissue than do other breeds but sex differences within breeds appear to be virtually non-existent. A yellower adipose tissue colour is found also in cattle that have either been finished at grass or on diets containing high levels of carotenoids, compared with animals finished on diets containing high levels of cereals. In the case of the Channel Island breeds the propensity to lay down yellow adipose tissue reflects an inefficient mechanism for breaking down carotenoids and any consumer dislike of this adipose tissue, or other tissue similar to it, is in many ways irrational because it is of higher nutritive value because of its high content of carotenoids. Pig adipose tissue too can be coloured yellow if the diet given contains high proportions of carotenoids and other retinol precursors such as cryptoxanthis as found in maize.

For reasons already discussed, on the whole problems associated with firmness of fat are found in pig adipose tissue rather than in cattle and sheep adipose tissue. The desire to produce pigs with less and less adipose tissue in their carcasses has given rise to some problems with firmness and colour and to separation of lean but it would be wrong to give the impression that this has assumed alarming proportions because there are no serious deteriorations until very low concentrations of adipose tissue in the carcass are reached. Apart from appearance there is no reason to assume that eating quality is in any way affected and the problem is one more for the butcher than for the consumer. The separation of muscle from adipose tissue and the failure of joints to set correctly and to look collapsed are problems that he has to contend with in presenting his meat to the consumer who in any case may or may not be particularly interested in these aspects. For the manufacturer of vacuum-packed rindless rashers of bacon, soft adipose tissues present problems in that their presence in individual rashers coalesces the rashers to give a product that is not attractive in appearance. For the curer of bacon sides that have little adipose tissue there may be problems with concentrated pockets of brine being deposited between the muscle and adipose tissue layers.

The answer to the question 'can the problem of adipose tissue quality and the leanness of pig carcasses be quantified' is an equivocal 'yes'. A body of evidence is fast accumulating which suggests that adipose tissue quality problems begin at P_2 (depth of fat and skin measured at the level of the last rib 65 mm from the mid-dorsal line) backfat depths of 8–10 mm and that at depths of 8 mm or less the separation of backfat from lean may occur in proportionately 0.50 of all carcasses. In practice obviously any steps which are taken to reduce overall adipose tissue concentrations in carcasses must effect a balance between age, sex and diet if soft adipose tissue is to be avoided. In particular if boars and/or genetically lean pigs are involved, then feeding diets with low concentrations of linoleic acid to grow at maximum rates must be practised.

A further aspect of pig adipose tissue chemistry that may have a bearing on quality in the context of consumer acceptance is the taste and aroma of the meat which contains the adipose tissue. The lipids of adipose tissue are important to the development of flavour in meat which stems from flavour volatiles produced in the cooking process by the interaction between the lipid, the proteins and the other constituents of muscle. The flavouring volatiles increase with age in all animals and it therefore follows that, in common with other species, the older the pig the more likely it is to have meat with a greater flavour. Indeed muscle from pig, cattle and sheep carcasses cooked without its own associated adipose tissue is not easily, if at all, distinguished between by the human palate.

High concentrations of linoleic acid in the lipid of adipose tissue can also have a marked effect on flavour. If, in sheep and cattle, high levels of linoleic acid are present in the lipid, these produce oily, sweet or bland tastes during cooking, whereas high concentrations of oleic acid in the lipid improve the flavour.

Temperature control in storing and in packing pig meat is also important if high concentrations of linoleic acid are present in the lipids. This is because linoleic acid, being unsaturated, is very much more prone than is, for example, oleic acid to combine with atmospheric oxygen to give oxidative rancidity, the products of which can be unpalatable and have an objectionable aroma to the human senses. Aroma can also be markedly affected by the androstenone content of the lipid and this is particularly important in consideration of acceptability to the human of meat from boars compared with castrated males and gilts. In this particular context the general consensus of opinion is that the androstenones will not be of importance in affecting aroma in the meat derived from boars grown quickly to live weights which do not exceed 90–100 kg.

Functions of adipose tissue

In the previous sections the metabolic function of lipid was discussed. Other functions of adipose tissue and the lipid within it are listed in Table 3.15.

The storage capacity of lipid is of immense importance to many types of animal, allowing them to have available a rich source of energy for emergency purposes. In terms of subcutaneous adipose tissue the insulatory qualities imparted are of great significance, particularly to the marine animal, where the tissue may also help to streamline the shape of the body to assist passage through the medium in which it lives. The future need for increased insulation seems to be anticipated in some species. For example, if the rat is raised in a cold environment a greater number of adipocytes develop, although the dimensions of the cells remain smaller than those in rats reared in a warmer environment. Also, in animals which hibernate, exposure to short photoperiods enhances the development of brown fat, presumably in anticipation of hibernation.

The presence of lipid in the gut enhances the absorption of fat-soluble vitamins but is also an important source of these vitamins itself. However, it is with lipid storage that the greatest importance lies and the lipids are in a constant state of physical turnover which allows physiological adaptation in the animal.

Table 3.15. Functions of adipose tissue and lipids.

1. Metabolic
2. Insulatory
3. Source of metabolic water
4. Storage:
   a. lipid
   b. fat-soluble vitamins
5. Intestinal absorption
6. Animal form

3.4. Muscle Tissue

3.4.1. Introduction

The total mass of muscle in the body exceeds that of all other organs and tissues. Unless the animal is excessively fat the skeletal muscle accounts for the bulk (proportionately from about 0.35 to 0.68) of the carcass weight of meat animals and in cattle, sheep and pigs for proportionately between 0.30 and 0.40 of the total live weight. There are a number of features which are unique to muscle tissue (Goss, 1978): first, the concentration of protein in the muscle fibre is surpassed by few cells with the possible exception of erythrocytes and, secondly, the intimacy, both structural and functional, with which the skeletal muscle fibre is associated with nerve fibres, as well as with tendons and bones, is unsurpassed by other cellular associations. In addition to these unique characteristics muscle has the capacity to undergo hypertrophy as well as dystrophy.

In the carcasses of meat animals, excluding the head, there are over 100 different muscles. In beef carcasses 30 of the largest muscles account proportionately for over 0.75 of the total weight of muscle (Brown *et al.*, 1978) and their anatomical distribution and identity are shown in Fig. 3.10. For a guide to the location of muscles in the carcasses of pigs and sheep, the reader is referred to Kauffman and St Clair (1965) and to Kauffman *et al.* (1963) respectively.

It is perhaps all too easy, when considering muscle tissue as the most important and, from the point of view of human food, highly sought after tissue in the animal carcass, to overlook the importance of its functions in the live animal. Thus the various muscles allow the animal to move, to stretch and, generally speaking, to perform a variety of functions essential to life. In this sense considerations of growth and development of the tissue relative to the functioning of the animal in the live state are of no less importance than are considerations relative to the economic worth of the carcass. Without the first the second could not be realized.

3.4.2. Structure

General

The structure unit of skeletal muscle tissue is the highly specialized cell known as the muscle fibre. An entire muscle is usually surrounded by a heavy connective tissue sheath termed the epimysium. Other connective tissue is contiguous with, and may arise from, the epimysium which not only acts as a wrapping agent to muscle bundles but also as a divider within bundles by entering the bundle itself. Thus the connective tissue which is contiguous with the epimysium but which penetrates between groups of fibres and segregates them into muscle bundles (the perimysium), in turn is contiguous with the

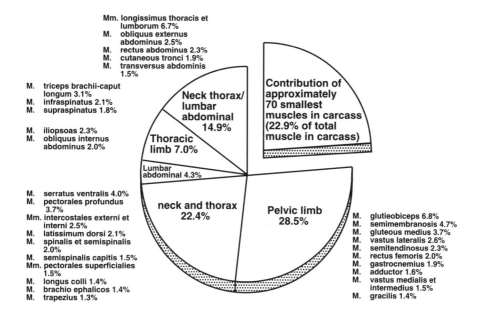

Mm. longissimus thoracis et lumborum 6.7%
M. obliquus externus abdominus 2.5%
M. rectus abdominus 2.3%
M. cutaneous tronci 1.9%
M. transversus abdominis 1.5%

M. triceps brachii-caput longum 3.1%
M. infraspinatus 2.1%
M. supraspinatus 1.8%

M. iliopsoas 2.3%
M. obliquus internus abdominus 2.0%

Neck thorax/ lumbar abdominal 14.9%

Contribution of approximately 70 smallest muscles in carcass (22.9% of total muscle in carcass)

Thoracic limb 7.0%

Lumbar abdominal 4.3%

M. serratus ventralis 4.0%
M. pectorales profundus 3.7%
Mm. intercostales externi et interni 2.5%
M. latissimum dorsi 2.1%
M. spinalis et semispinalis 2.0%
M. semispinalis capitis 1.5%
Mm. pectorales superficialies 1.5%
M. longus colli 1.4%
M. brachio ephalicos 1.4%
M. trapezius 1.3%

neck and thorax 22.4%

Pelvic limb 28.5%

M. glutieobiceps 6.8%
M. semimembranosis 4.7%
M. gluteous medius 3.7%
M. vastus lateralis 2.6%
M. semitendinosus 2.3%
M. rectus femoris 2.0%
M. gastrocnemius 1.9%
M. adductor 1.6%
M. vastus medialis et intermedius 1.5%
M. gracilis 1.4%

Fig. 3.10. Contribution of the 30 largest muscles of the beef carcass to the total muscle in the carcass and their location (based on Brown *et al.*, 1978).

connective tissue which surrounds the muscle fibres themselves (the endomysium). The major blood vessels of the muscle are enmeshed within and supported by the perimysium (Fig. 3.11).

Connective tissue is dealt with elsewhere (see section 3.3.4) but a ground substance containing several cell varieties, collagen, reticular and elastic fibres, support and regulate the environment in which the muscle cells must function.

Adipocytes found in muscle are located in the perimysial spaces and are

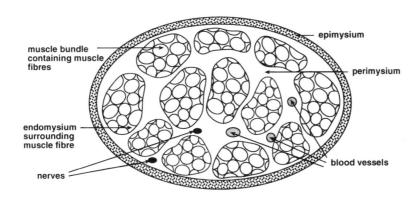

muscle bundle containing muscle fibres

epimysium

perimysium

endomysium surrounding muscle fibre

nerves

blood vessels

Fig. 3.11. Diagrammatic representation of cross-section of a skeletal muscle.

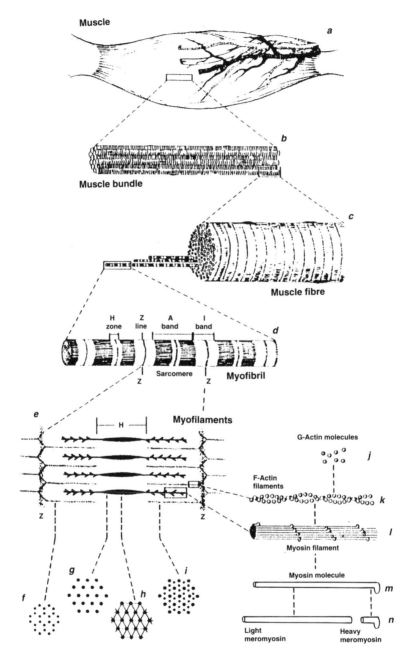

Fig. 3.12. Diagram of the organization of skeletal muscle from the gross structure to
the molecular level. **a**, Skeletal muscle, **b**, A bundle of muscle fibres. **c**, A muscle fibre,
showing the myofibrils. **d**, A myofibril, showing the sarcomere and its various bands and
lines. **e**, A sarcomere, showing the position of the myofilaments in the myofibril. **f-i**,
Cross-sections showing the arrangement of the myofilaments at various locations in the

extrafasicular. They are found scattered singly or in groups in loose connective tissue, particularly near blood vessels. Some lipid may be found within muscle bundles associated with membrane structure or as free lipid droplets within muscle fibre. Fatty tissue found within muscles is referred to as intramuscular or marbling adipose tissue and the adipocyte size tends to increase as the number of cells within a conglomerate of adipocytes increases.

The epimysium is the channel through which the blood vessels and nerves enter the muscle. Subsequently they branch and follow the strands of perimysial connective tissue. The muscle spindle is the receptor situated in skeletal muscle which furnishes information to the CNS with respect to movement and position. Thus individual muscles and groups of muscles contract in a coordinated manner to produce smooth and effective movement as a result of guidance received at the nerve centres by the peripheral spindle.

A schematic representation of the organization of skeletal muscle from the gross structure to the molecular level is given in Fig. 3.12 and the detail outlined will be described in the following sections.

The muscle fibre

Muscle fibres are multinucleated, unbranched tubular-like cells which taper slightly at both ends and which appear earthworm-like in external appearance (Fig. 3.13). The fibre diameter varies from 10 to 100 µm but is dependent on such factors as health, species, breed, sex, age and plane of nutrition (Lawrie, 1979). Within a muscle, fibres vary in diameter with smaller fibres tending to be more peripheral than larger fibres. Fibre length varies greatly but in most cases does not equal the length of the muscle. Depending on the individual muscle the fibres may run either in a direction longitudinal to the long axis of the muscle or at an angle to it. The presence in the fibre of many thin cross-striated myofibrils gives the characteristic cross-striated appearance (see below). The peripheral positioning of the nuclei is a characteristic of striated skeletal muscle.

The sarcolemma, lying immediately beneath the endomysium, encases the muscle fibre (Fig. 3.13). It has a complex structure of connective tissue filaments, basement membrane polysaccharides and a lipoprotein plasma membrane. It is to some extent elastic as it has to stretch and contract with the muscle fibres. Inside the sarcolemma the myofibrils are surrounded by the semi-fluid cytoplasm of the fibre known as the sarcoplasm. The muscle cell nuclei lie

---

Fig. 3.12. *contd* sarcomere. **j**, G-actin molecules. **k**, An actin filament, composed of two F-actin chains coiled about each other. **l**, A myosin filament, showing the relationship of the heads to the filament. **m**, myosin filament showing the head and tail regions. **n**, The light meromyosin (LMM) and heavy meromyosin (HMM) portions of the myosin molecule. (Modified after Bloom and Fawcett, *A Textbook of Histology,* 9th edn, W.B. Saunders Company, Philadelphia, p. 273, 1968.) From: *Principles of Meat Science* by Forrest, *et al.* Copyright © 1975 by W.H. Freeman and Company. Reprinted with permission.

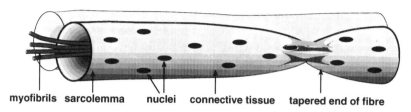

myofibrils sarcolemma   nuclei   connective tissue   tapered end of fibre

Fig. 3.13. Structural features of skeletal muscle fibres.

just below the sarcolemma. The variation in the length of muscle fibres means that the numbers of nuclei vary from fibre to fibre. A fibre several centimetres long may have several hundred nuclei with a regular distribution except near tendinous attachments where there is an increase in number and a more irregular distribution. They are also more numerous at the myo-neural junction, where the nerve fibre endings terminate on the sarcolemma.

The myofibrils are the elongated contractile elements of the fibres responsible for imparting the characteristic banded or striated appearance. They are rod shaped, between 1 and 2 µm in diameter and in mammals have axes that run parallel to the axes of the fibres themselves. In meat animals a muscle fibre with a diameter of 50 µm will have a minimum of 1000, and a maximum of 2000, myofibrils bathed in its sarcoplasm.

Except in smooth muscle (see next section) the myofibrils are in turn composed of two types of myofilaments: thick and thin. The thick filaments, which are about 14–16 nm in diameter and which contain myosin, are aligned parallel to each other but across the myofibrils. The thin filaments, which are about 6–8 nm in diameter and which contain actin, are also arranged across the myofibrils and parallel to each other and to the thick filaments. The arrangement of these filaments forming regular regions of overlap is responsible for the repeating pattern of cross-striations (Fig. 3.14) and forms the molecular basis for muscular contraction (Figs 3.15 and 3.16). The region which contains only thin filaments is called the I band while the region called the H zone contains only myosin filaments. The A band on either side of the H zone is an interdigitated mixture of both thick and thin filaments and is anisotropic and dark in colour. The myofibril is composed of alternating I and A bands (Fig. 3.15). The I band is isotropic, is light in colour and is bisected by a dense line, the Z line, which links together the mid-points of the thin filaments. The unit from one Z line to the next is called a sarcomere and consists of two half I bands with a complete A band between them. Sarcomeres are repeating structural units of myofibrils and vary in length but in mammalian muscle may commonly be about 2.5 µm. The relationship of the two filaments can be seen to be six thin filaments to one thick filament arranged in a hexagonal pattern as shown in transverse section (Fig. 3.17). Throughout the muscle there is a network of tubular systems which act in the coupling of excitation to contraction to relaxation. These are the sarcoplasmic reticulum and transverse tubular systems. A further bisecting line

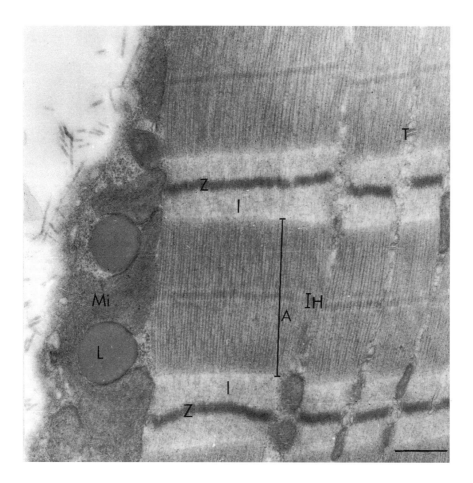

Fig. 3.14. Longitudinal section through rat soleus muscle showing repeating pattern of cross-striations. Scale bar = 500 nm; Z = Z line; A = A band; I = I band; H = H zone; T = triad (T system + sarcoplasmic reticulum) Mi = mitochondrion; L = lipid (by courtesy of Dr Charlotte A. Maltin, Rowett Research Institute, Aberdeen).

is the M line which has both a pseudo and a complete H zone around it (Fig. 3.15).

The contractile proteins actin and myosin account proportionately for between 0.75 and 0.80 of the total protein in myofibrils with actin accounting proportionately for about 0.25 of the total. The remainder are known as regulatory proteins because they regulate functions of the adenosine triphosphate–myosin complex. Myosin is a highly charged molecule; actin has a relatively low charge, is more fibrous in nature and has the higher proline content of the two. Trypsin splits myosin into light and heavy meromysin. The protein structure is, however, extremely complex and further details are to be found in Price and Schweigert (1971), Forrest *et al.* (1975) and Lawrie (1979).

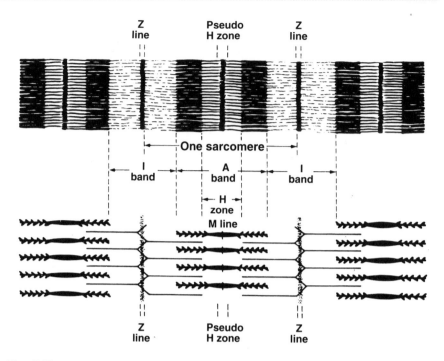

Fig. 3.15. A drawing adapted from an electron photomicrograph, showing portions of two myofibrils and a sarcomere (× 15, 333) and a diagram corresponding to the sarcomere, identifying its various bands, zones and lines. (Modified from H.E. Huxley, 'The Mechanism of Muscular Contraction'. Copyright © 1965 by Scientific American, Inc. All rights reserved.) From: *Principles of Meat Science* by Forrest *et al.* Copyright © 1975 by W.H. Freeman and Company. Reprinted with permission.

Historically, the fact that different muscles were predominantly either cherry red in colour or pale pink or white was used as a basis for classification and muscles were typed as being red or white. This classification was later extended to include the types of individual fibres which were characteristic of each muscle type. Histochemical techniques have shown that this classification is rather narrow and that most are composed of a mixture of fibre types. Standard staining techniques can be used to classify muscle fibres into at least four types with distinct physiological characteristics which are metabolic (oxidative or glycolytic) and contractile (fast or slow twitch) (Table 3.16). Slow twitch fibres have a slower, sustained mode of contraction and are important in maintaining posture. These fibres usually have an oxidative type metabolism, are small in size and contain large numbers of mitochondria. Muscles which are composed of large numbers of slow twitch oxidative fibres are highly vascularized and have high myoglobin contents, hence the original name 'red fibre'. Fast twitch fibres with a glycolytic metabolism contract rapidly in short bursts under anaerobic conditions but become easily fatigued and tend to be white in colour. Fast twitch fibres with capacity for both oxidative and

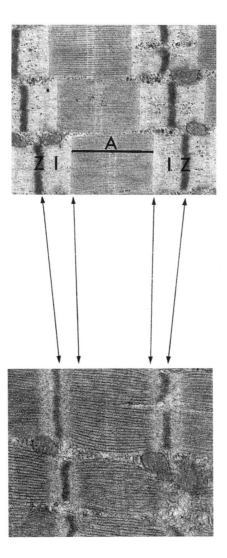

Fig. 3.16. Longitudinal section through parts of rat soleus muscles to illustrate the difference between relaxed (top) and contracted (bottom) states. A = A band; I = I band (by courtesy of Dr Charlotte A. Maltin, Rowett Research Institute, Aberdeen).

glycolytic metabolism are intermediate in nature and intermediate in colour between the extremes of red and white. Serial transverse sections through a rat gastrocnemius muscle are shown in Fig. 3.18 and the four different straining reactions of each fibre, as outlined in Table 3.16, are evident.

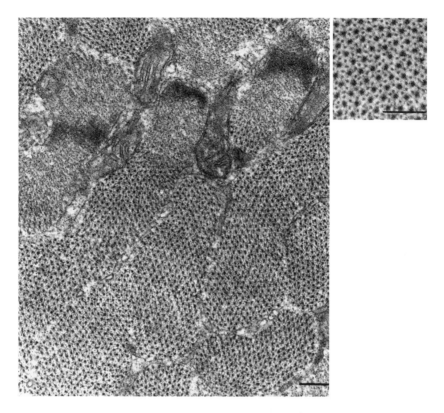

Fig. 3.17. Transverse section through a rat soleus muscle (scale bar = 200 nm). Insert: myofilaments at higher magnification reveal hexagonal pattern (scale bar = 100 nm) (by courtesy of Dr Charlotte A. Maltin, Rowett Research Institute, Aberdeen).

The muscle-tendon junction

The force of the contracting muscle myofibrils has to be transmitted to the muscle tendon. This is probably effected by the ends of the myofibrils dove-tailing with the tendon fibrils to form a cohesive joint. There is a decrease in the striated appearance of the myofibrils as they approach the tapered ends in this area. The tendons are attached in turn to bones but some muscles are attached by connective tissue fasciae to structures such as skin, ligaments and other muscles. In some cases muscles are attached to bones by fasciae. The fasciae are composed mostly of white collagen fibres but contain also small proportions of elastin fibres.

Table 3.16. Histochemical description of fibre in rat muscle.

Enzyme stained	Slow twitch oxidative (SO)	Fast twitch oxidative glycolytic (FOG)	Fast twitch glycolytic (FG)	Fast twitch oxidative (FO)
Ca^{2+} activated myofibrillar ATPase	+	+++	++	++(+)
NADH-Diaphorase	++	++(+)	+	++
α-Glycerophosphate dehydrogenase	+	+(+)	+++	+
L-Glucan phosphorylase	+	++	+++	+

+, Low activity; ++, moderate activity; +++, high activity.

3.4.3. Types

General

Three types of muscle are found in the animal body. Of these types skeletal muscles are by far the most important quantitatively in terms of meat production although, of course, in the live animal they act to control posture and movement. Two other types of muscle, cardiac and smooth, are essential to the basic physiology of the animal. Skeletal and cardiac muscle are often referred to as striated muscle because of the transverse banding pattern that can be observed microscopically.

Smooth muscle

In cross-sectional shape smooth muscles vary greatly, from being extremely flattened elipsoids to shapes which are markedly triangular or polyhedral. They are most numerous in the walls of arteries, in lymph vessels and in the walls of the gastrointestinal and reproductive tracts and compared with skeletal muscle have a poor blood supply. The cells are small and spindle shaped, 2–5 µm in diameter, with a single nucleus situated mid-way along their length. Each cell is surrounded by its plasma membrane, the sarcolemma, and has no protoplasmic continuity with its neighbours. However cell to cell contact is achieved at certain points along the cell surface where the plasma membranes of adjacent cells come into close approximation (gap junctions) or fuse (tight junctions). The intercellular space is small and contains blood vessels, nerve fibres, extracellular matrix and reticular fibres.

The contractile proteins or myofilaments are not divided into myofibrils and the light and dark banding pattern characteristic of striated muscle is absent, hence the name 'smooth muscle'. Electron microscope studies reveal that the myofibrils are 5–8 nm wide and about 1 µm long. They may occur in tracts or

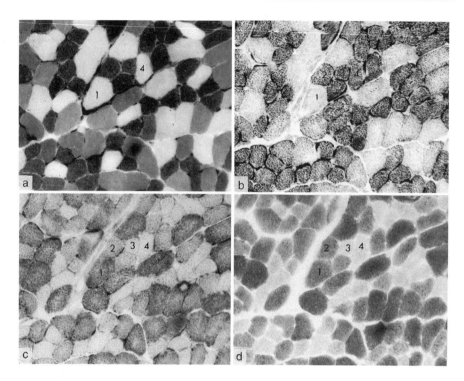

Fig. 3.18. Serial transverse sections through a rat gastrocnemius muscle. a, Section reacted for Ca^{2+} activated myofibrillar ATPase. b, Section reacted for NADH-tetrozolium reductase. c, Section reacted for α-glycerophosphate dehydrogenase. d, Section reacted for L-glucan phosphorylase.

Fibre typing based on the characteristic staining reactions of each fibre as outlined in Table 3.16.

Fibre no.
on plate   Fibre type

Fibre no. on plate	Fibre type
1	Fast twitch glycolytic (FG)
2	Fast twitch oxidative glycolytic (FOG)
3	Fast twitch oxidative (FO)
4	Slow twitch oxidative (SO)

All plates × 270 (by courtesy of Dr Charlotte A. Maltin, Rowett Research Institute, Aberdeen).

bundles within the cell but are not arranged in any obvious pattern that suggests how they function during contraction. In contrast to skeletal muscles, smooth muscles contract and relax slowly.

Cardiac muscle

Cardiac muscle possesses the unique property of rhythmic contractility and the myocardium, which is the contractile layer of the heart, contains the bulk of this tissue. Cardiac muscle consists of three types of tissue: nodal tissues found in the pacemaker region where autorhythmicity originates, Purkinje tissue where the impulse is rapidly conducted and myocardial muscle which has the highest degree of contractile power. Cardiac muscle fibres have an average diameter of 15 µm and have the same basic organization of myofilament as skeletal muscle, giving a cross-striated appearance (Fig. 3.14). However there are three main differences:

1. At irregular intervals along the length of the fibre, thick transverse bands called intercalated discs replace the Z lines. These discs are thought to be involved in cell to cell contact and the propagation of the rhythmic impulse.
2. The cells do not form a multinucleate syncytium but exist in cellular units with a single nucleus, usually in the centre of the cell.
3. The fibres are not cylindrical units but often divide and connect with adjacent fibres, thereby forming a complex three-dimensional structure.

Skeletal muscle

There are slightly more than 600 individual skeletal muscles in the animal body and there are big differences in shape, size and function. Most muscles are connected directly to bones but others have primary anchorage points in skin, cartilage and fascia, whilst others are attached to ligaments. In these cases the connections to bone are indirect. Details of the structure of skeletal muscle have already been given in the previous sections. Nevertheless, it is important to remember that all muscles are covered with a thin connective tissue sheath which is an integral part of the connective tissue found within the muscle and that this network of connective tissue is the base for the vascular vessels and nerve fibres. In skeletal muscle the dominance of the muscle fibres is evidenced by the fact that they account proportionately for between 0.75 and 0.92 of the total muscle volume, the remainder being composed of connective tissue, blood vessels, nerve fibres and extracellular fluid.

3.4.4. Chemical composition of muscles

The average composition of skeletal muscle is shown in Fig. 3.19. The quantitative importance of water is striking but not surprising if its importance as the medium for the transport of substances between the vascular network and muscle fibres, and as the principle constituent of extracellular fluid, are reflected upon carefully. The principle amino acids are α-alanine, glycine, glutamic acid and histidine.

As has been pointed out already the essential unit of skeletal muscle is the

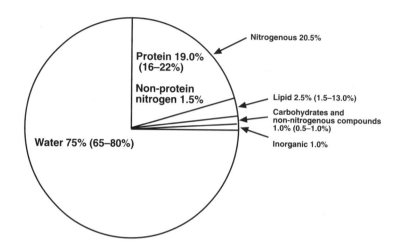

Fig. 3.19. Average (and range) percentage composition of mammalian skeletal muscle on a fresh weight basis.

muscle fibre. The composition of the protein and of the other components of the muscle are given in Fig. 3.20. Basically, the proteins divide on the basis of their solubility in different solvents. The sarcoplasmic proteins are composed mostly (proportionately about 0.92) of soluble proteins and the proteins of mitochondrial enzymes. The remaining proportion of 0.08 is mostly myoglobin together with smaller amounts of haemoglobin and cytochrome proteins. All are very complex and may contain up to 50 components, many of which are enzymes of the glycolytic cycle (Lawrie, 1979). Actin is richer in proline than the other major myofibrillar protein myosin. The other proteins are tropomyosin, troponin, M-and C-proteins, and β-actinin, and Lawrie (1979) gives further details of these. Of the insoluble stroma proteins proportionately about 0.50 is collagen and 0.03 is elastin, the remaining 0.47 being a mixture of various proteins such as reticulin.

Lipid is the most variable of the major chemical components of skeletal muscle and contains considerably more phospholipid and unsaponifiable constituents, such as cholesterol, than does the lipid in most other adipose tissues. The variation in fatty acid composition has been discussed previously (see section 3.3.4).

The carbohydrate content of muscle is relatively small with glycogen predominating (proportionately 0.80 on average but with a range from 0.50 to 0.87). The remainder is glucose (0.01) and various other substances such as lactic and citric acids.

Potassium is the most important inorganic element with phosphorus and sulphur occupying the second position. Other minerals (Fig. 3.20), magnesium, calcium, iron, cobalt, copper, zinc and manganese plus very minute traces of minerals such as nickel, account for the remainder (proportionately 0.10).

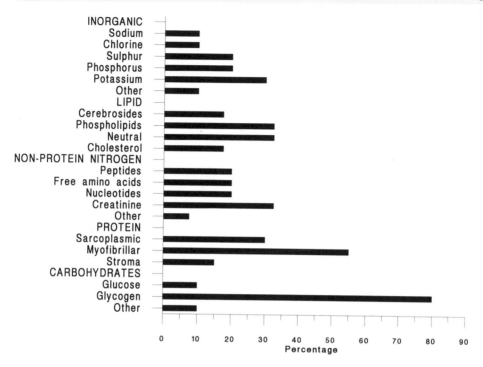

Fig. 3.20. Proportionate composition of major chemical components of skeletal muscle.

3.4.5. Muscle growth

Normal

In the embryo, skeletal muscle arises by mitotic division from the third germinal layer, that is from the mesodermic somites. The process is known as myogenesis and the somitic cells are discernible along each side of the embryonic axis between 2 and 3 weeks after conception. Muscle cells arise from these in about 40 groups known as myotomes (Lawrie, 1979). Two types of cell develop from these undifferentiated cells, one acquires the morphology of primitive connective tissue cells and the other that of primitive muscle cells or myoblasts. Myoblasts fuse with each other to form myotubes and these in turn form myofibrils by synthesizing myosin and actin. The myotubes are elongated multinucleated structures with central cores filled with cytoplasm, the myofibrils forming a complete cylinder on the outside. The number of myofibrils within a muscle fibre increase during embryonic development from a single original fibril. The first myofibrils are unstriated but as myogenesis progresses those immediately beneath the sarcolemma are the first to become striated. The numbers increase by longitudinal fission within each myotube. The muscle nuclei migrate from their central position to the periphery and increase in number by a further fusion of myoblasts. They also become flattened against the sarcolemma as the number of myofibrils increases. The typical cross-striations of skeletal and cardiac muscle develop as the myofilaments become aligned within the myofibrils.

It is generally agreed that the muscle mass of an animal is determined mainly by the number and size of its constituent myofibres (e.g. Hooper, 1982). During the embryonic and fetal stages, growth in muscle is characterized by increases in muscle fibre numbers and their grouping into bundles, that is by hyperplasia, but there are differences between species. For example, in the rat, hyperplasia does continue after birth and in some ways may be regarded as an extension of the hyperplastic activities of the embryonic and fetal stages. This does not occur either in the mouse or in the pig and may be interpreted as reflecting differences in the state of maturity of the muscle at birth. In the pig the total fibre number, that is both primary and secondary, appears to be fixed by birth (Wigmore and Stickland, 1983). Primary fibres are those formed during the initial stages of myoblast fusion and act as a framework on which the secondary fibres, which are smaller, develop. Differentiation between the two populations is effected by measuring ATPase activity and this is possible in the pig from 90 days onwards postconception. Because primary fibres are resistant to environmental influences they appear to be the main cause of inter-litter and inter-strain variations in total fibre numbers (Stickland and Handel, 1986; Dwyer and Stickland, 1991). Secondary fibres appear to be less resistant to environmental influence, for example possibly to nutrition *in utero* (Wigmore and Stickland, 1983), and must therefore be responsible for some of the within-litter variation in total fibre numbers (Dwyer and Stickland, 1991). Handel and Stickland (1987) found, in Large White pigs, that the average numbers of total and primary fibres for the m. semitendinosus muscle were 414,760 (SD 90,750) and 17,460 (SD 3870) respectively and also evidence to support the idea that there is no hyperplasia in pig muscle after birth from studies of the cellularity of the muscle of light and heavy birth weight littermates. In this comparison the total fibre number was proportionately between 0.15 and 0.25 less in light, compared with heavy, birth weight littermates and this difference did not change throughout the postnatal period even though the light weight pigs were given every chance to exhibit any propensities which they had for hyperplasia by being fed liberally. However, a reduced muscle fibre number was not in all individual cases associated with a low birth weight, but when it was it reflected a reduction in the secondary to primary fibre number ratio. Primary fibre number tended to be unaffected by birth weight except when, in extreme individual cases, birth weights were more than 2.5 SD below the mean weight of their litter and were proportionately less than 0.50 of their large littermates.

On the whole, in fast twitch muscles small fibres are rich in mitochondria and contain few nuclei. They have a comparatively small cross-sectional area relative to each nucleus. It is possible that fibres such as these remain small because of a strong propensity to oxidize certain amino acids which would otherwise be used for protein synthesis (Burleigh, 1980).

In postnatal life growth is characterized principally by increases in fibre cross-sectional area and by increases in length effected by the addition of complete sarcomere units to the ends of existing myofibrils. It follows that compared with older muscle fibres, younger fibres will be smaller in cross-sectional area and will contain less nuclei with each nucleus associated with a smaller volume. The proliferation of the myofibrils within the muscle fibres is largely responsible for increases in their diameter. The number of myofibrils in a single muscle fibre may increase by a factor of 10–15 during the life of the animal, but the age at which maximum diameter is achieved varies according to a number of factors including age at maturity, species, breed, sex, nutrition and activity. For example, pigs and cattle usually have larger diameter fibres than sheep, intact males usually have larger diameter fibres than females and castrated males, whilst those of older and well fed animals are larger than those of younger and poorly fed animals. Thus hypertrophy, rather than hyperplasia, is largely responsible for increases in muscle mass in postnatal life with the process characterized by an increase in cross-sectional area and length rather than in muscle fibre number (Goldspink, 1974).

The growth of a muscle reflects in its mass the number of cells present and the amount of protein accumulated in each cell. The quantities of nucleic acid (DNA and RNA) present give some indication of, respectively, the numbers of cells and ribosomes. Thus as muscle grows there is an increase in its content of nucleic acid and the amount of protein accumulated per unit of nucleic acid (Table 3.17), the former increasing more sharply than the latter when expressed on a quantitative basis for the entire muscle. The increasing ratio of protein to nucleic acid with increasing age is explained, because in essence the protein growth curve is the integral of the growth curve of nucleic acid. But again there are differences between species. Burleigh (1980; see Table 3.17) provides evidence that postnatal increases in muscle DNA are likely to be higher in chickens and pigs compared with cattle and sheep. Generally, more muscular animals have relatively low ratios of muscle protein to DNA and in some, but not in all, cases they have narrower fibres.

However, as muscle tissue is important in animal movement, growth cannot be considered purely in passive terms. Dynamic factors may play an important part. In antenatal life it may be argued that the primary stimuli to muscle growth are the tensions which result from skeletal elongation, whilst in the immediate postnatal, prepubertal and adolescent phases, functional demand is the most important stimulus to growth (Berg and Butterfield, 1976). There is evidence that the growth of muscle in length is positively and highly correlated with the bone to which it is attached but that once the bone is near the zenith of its growth curve then the continued growth of muscle assumes an independence within the confines of the genotype and the nutrition which the animal is receiving (Shahin and Berg, 1987).

Table 3.17. Increase in nucleic acid (DNA and RNA) and ratios of protein to nucleic acid in different animals (adapted from Burleigh, 1980).

Species	Period of growth	Proportional increase in:		Proportional increase in protein or muscle weight per unit of:	
		DNA	RNA	DNA	RNA
Sheep	70-140*	23.3-35.4	27.9-56.8	3.9-5.2	3.3
Pigs	20-140†	1.8-2.9	2.9-3.8	2.3-2.7	1.7-1.8
	0-120‡	>10	24.0-36.0	3.2-3.9	1.5-2.1
	1-235‡	21.1-25.7	42.0-45.4	12.6-15.7	7.1-7.9
	1-84‡	15.4	25.3	4.9	3.0
	1-120‡	32.7-36.0	83.5	4.1-4.5	1.7-1.8
	10-365‡	6.3-7.7	—	4.4-4.5	—
	0-50‡	8.8	7.6	2.9	3.4
Cattle	0-450‡	—	—	3.7-4.2	—
	145-665‡	2.4-2.5	2.5-3.0	1.5-1.6	1.2-1.5
Fowl	0-70‡	13.6-69.7	14.0-126.2	2.0-5.2	2.0-5.2
	0-266‡	18.0-96.2	—	3.2-7.9	—
	10.5-196	25.5	—	2.7	—
Human	(1-3.5) to (16-17)‖	3.7-5.0	—	1.4-1.6	—
	>1-16‡	20	—	—	—

*Days post-conception.
†Weight in kg.
‡Days from birth.
‖Years.

General

Both pathological and non-pathological factors may cause abnormal growth in muscle. Only the non-pathological factors will be considered here. Within this group of factors those of genetic and nutritional origin are probably most important. In some cases muscular hypertrophy is induced, in other cases muscular dystrophy. The classic cases of genetic influence on hypertrophy are in cattle where the 'double-muscle' and 'dwarfism' conditions have been long recorded. From the point of view of nutrition, white muscle disease, which is found in young sheep, cattle and pigs, is the most frequently cited condition of abnormal muscle growth representing muscular dystrophy.

Dwarfism

It is debatable if the several types of dwarfism found in cattle should, strictly speaking, be included under this heading, as whilst there is no doubt that muscle growth must be stunted because of the small, stocky form that is achieved ultimately, the effect primarily is on longitudinal bone growth and vertebral development, particularly in the lumbar region and in male animals. As in the case of double muscling the condition is of genetic origin.

Double muscling

Cattle showing signs of double muscling are referred to by different names in different countries: 'double muscled' in the UK and the USA, 'dopplender types' in Germany, 'culards' in France and 'a groppa doppia' in Italy. The condition is found in a number of breeds but assumes a considerable prominence in the Charolais and Belgian Blue breeds and is controlled by a recessive gene with incomplete penetrance. The animals exhibiting this condition generally give the impression of being compact but show a marked muscular dystrophy, particularly in the hindquarters where the muscles are enlarged and where creases appear, particularly between the m. semitendinosus and the m. biceps femoris muscles and to a lesser extent between the m. biceps femoris and m. vastus lateralis muscles. Muscle on either side of the sacrum is distinctly humped in appearance due to considerable hypertrophy in the triceps muscles (Fig. 3.21).

There are considerable difficulties in comparing the degree of hypertrophy of individual muscles with those in normal animals and Boccard (1981) suggests that as the weight of any one muscle is strongly linked to the weight of the carcass and to the whole musculature, relative growth rates and sizes of individual muscles must be compared at the same musculature weight. Using this approach Boccard and Dumont (1974) found that in double-muscled animals some individual muscles were lighter, and some were heavier, compared with the same muscles in normal animals (Table 3.18). Hypertrophy is therefore not generalized and it would appear that outside muscles with large surface areas are most affected.

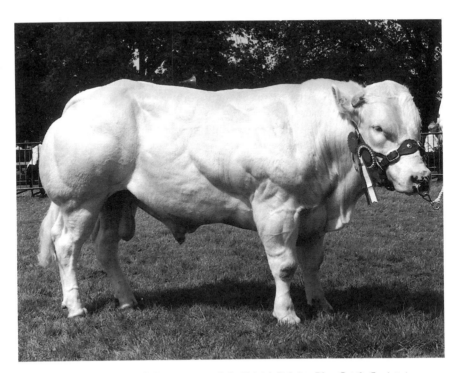

Fig. 3.21. Belgian Blue bull (by courtesy of the British Belgian Blue Cattle Society).

The external appearance of excessive musculature is reflected in the total muscle found in the carcass. The total muscle is greater than in the carcasses from normal animals (proportionately about 0.80 compared with 0.65 or just above) and the muscle to bone ratio considerably higher (about 7 : 1 in most cases but up to 9 : 1 in extreme cases compared with 5 : 1). These ratios are highest in the proximal parts, and lowest in the distal parts, of both the fore and hind limbs, and indicate that hypertrophy tends to be most pronounced in muscles with low proportions of connective tissue (Shahin and Berg, 1987).

The killing-out proportions of double-muscled cattle usually are proportionately 0.05 greater than normal cattle of the same breed, sex and live weight and this is largely a consequence of considerable hypotrophy in the organs which comprise the offal (the heart, the lungs, the gastrointestinal tract, the spleen and the liver) and because of lighter skin weights.

The term 'double muscled' is misleading because compared with normal animals the numbers of muscles are the same, but at the same carcass weight those muscles which are hypertrophied have more fibres than the same muscles in normal animals. There are other differences in the structures of the affected muscles compared with those that are normal. For example, they tend to have higher proportions of white, or fast twitch, fibres which are adapted for glycolytic metabolism. The greater number of these fibres and their tendency to

Table 3.18. Weights of main muscles and their relative development in double-muscled animals compared with normal animals at the same musculature weight of 140 kg (adapted from Boccard and Dumont, 1974).

	Absolute weights (g)		
	Hypertrophied (H)	Normal (N)	Ratio (H-N)/N
Heavier muscles than in normal animal			
m. cutaneus trunci	3346	2493	+0.342
m. tensor fasciae latae	2198	1827	+0.203
m. latissimus dorsi	3553	3028	+0.173
m. triceps brachii caput laterale	931	798	+0.167
m. pectoralis profundus	6264	5419	+0.156
m. semimembranosus	7151	6239	+0.146
m. vastus lateralis	3564	3127	+0.139
m. biceps femoris	11200	10029	+0.117
m. semitendinosus	3610	3282	+0.099
m. trapezius	2165	2005	+0.079
m. rectus femoris	3059	2847	+0.074
m. triceps brachii caput longum	4945	4652	+0.063
m. pectineus	814	772	+0.054
m. gracilis	1932	1836	+0.052
m. psoas major	2421	2305	+0.050
m. teres major	616	594	+0.037
m. obliquus externus abdominus	3136	3068	+0.022
m. deltoideus	754	741	+0.017
Lighter muscles than in normal animal			
m. gastrocnemius	2649	2669	-0.007
m. longissimus dorsi	9350	9447	-0.010
m. adductor femoris	2220	2245	-0.011
m. brachialis	552	564	-0.021
m. rhomboideus	2237	2288	-0.022
m. supra spinatus	1966	2013	-0.023
m. gluteus medius	5196	5332	-0.025
m. gluteus profundus	320	339	-0.056
m. biceps brachii	736	796	-0.075
m. infra spinatus	2598	2811	-0.076
m. sub scapularis	1438	1591	-0.096
m. transversus abdominis	1883	2087	-0.098
m. rectus abdominis	2928	3259	-0.102
m. iliacus	954	1125	-0.152
m. diaphragma	726	885	-0.180
m. obliquus internus abdominis	1350	1670	-0.192
m. teres minor	176	220	-0.200
m. semi spinalis capitis	2097	2836	-0.261
m. splenius	1369	1970	-0.305
m. vastus medialis	1073	1729	-0.379

be larger are two factors which play an important part in the increased size of the muscle. The intramuscular connective tissue framework of the hypertrophied muscle is finer and wider meshed than in normal animals and overall the muscle appears coarser in cross-section. This is a result of a reduced content of collagen and Boccard and Dumont (1974) propose that this reduced collagen presence is one of the prime reasons for the hypertrophy which takes place. They suggest that the reduced content of collagen decreases the tensions to which the muscles are usually subjected during development, tensions which consist of external constraints, represented by the skin, the fascia and the muscle sheath, and internal constraints represented by the perimysium. They argue that this could be responsible for the differences in hypertrophy found between the deep-seated and the more superficially located muscles referred to above.

Nutritional muscular dystrophy

This condition is characterized by some muscles exhibiting white or grey areas of degeneration. The areas of degeneration are localized in extent and involve a large group of fibres. The dystrophic muscles may also be exudative and the condition, which is often linked to a vitamin E deficiency, is associated with a greater capacity for proteolytic breakdown in the muscle. Also, affected muscle has a lower capacity for respiration, a greater content of connective tissue, protein, fat and water, a lower content of total nitrogen and exhibits structural changes in its myosin.

3.4.6. Muscle quality

As with adipose tissue, the various factors which contribute to concepts of quality vary according to the individual who is making the assessment. Basically the factors to be considered are those of the colour and appearance of the muscle and its texture relative to, in particular, how tough the muscle is for the teeth and for the muscles of the jaw of the human to contend with in the mastication process. Toughness, colour and general appearance are related to a number of factors of animal origin and to factors in the immediate pre- and post-slaughter environments.

Connective tissue is present in all muscles in the form of collagen in the epimysial, perimysial and endomysial components which surround the muscle fibres. The perimysial fibres at the periphery are quantitatively dissimilar but qualitatively similar to the endomysial fibres, being composed of large bundles of individual fibres between 600 and 800 nm in diameter. Nearer to the actual muscle fibres themselves, although the diameter is smaller (about 120 nm) and although this is very similar to that in the endomysial fibres, the two differ in terms of their morphology. In fact the various types of collagen (I, II, III, IV and V; see section 3.3) have been found to be differentially distributed between the epi-, peri- and endomysial components of individual muscles as well as between different muscles, and it is thought possible that a higher proportion of type III collagen may be associated with an increased toughness in any one muscle, and between different muscles, although the extent and type of cross-linking between the different types of collagen are generally thought to be of greater importance than is either the proportion of any one type of collagen, or of total collagen, *per se*. The increasing cross-linking and insolubility of the collagen which occurs with increasing age therefore imparts to muscle, all other things being equal, an increased toughness but the toughness is linked also, in muscle that has been cooked, to the thermal contraction which occurs at temperatures above 65°C and which is in turn accentuated by increased cross-linking. Therefore, from the point of view of collagen content, type and cross-linking, muscle from younger animals will be less tough than will be muscle from older animals.

Important though collagen is in determining toughness and texture, the contractile systems of muscles, composed of interdigitating filaments of actin and myosin, forming the actomysin complexes, also have an important, though quantitatively indefinable, part to play. In the live animal, when muscle is contracting, the actin filaments attached to Z discs (see section 3.4.2) are pulled over the myosin filaments with the result that Z discs are pulled closer together. After the animal is slaughtered, but before rigor mortis sets in, the filaments are still capable of sliding over each other in this way. However when rigor mortis sets in, the two types of filament become fixed in position. If for any reason the muscle goes into rigor mortis in the contracted state then it is extremely tough. The degree of contraction and toughness is to a certain extent temperature dependent in that minimum contraction may take place at about 15°C or above and maximum contraction at temperatures below 10°C. This phenomenon is often termed 'cold-shortening' and is principally a function of the myofibrillar proteins, although a change in the crimp length of the collagen fibres of the perimysial and epimysial structures may contribute to it slightly.

As well as age and postslaughter treatment, toughness can depend on animal type within a species, the double-muscled types of cattle referred to in section 3.4.5 having muscles that are generally paler and less tasty than the muscles from normal cattle, while the concentration of water and of protein is higher and that of fat and of collagen lower. The lower proportion of collagen compared with that in the normal animal is very marked and the collagen itself is more soluble because of a smaller degree of polymerization of the constituent chains. These latter two points probably explain why the muscle is considerably more tender than in the normal animal.

The intensity of the red coloration of muscle is an important facet of quality. Most, but not all consumers, appear to favour a light red rather than a dark red colour and in this particular context genotype within a species and preslaughter treatment all play a part, even though the ultimate common pathway is the rate of change in the pH of the muscle. In this context Lister *et al.* (1981) state that the two most important meat (muscle) quality defects are pale soft exudative (PSE) muscle, mostly in pigs, and dark and dry or dark cutting muscle (DFD), both in pigs and in cattle, and that in both cases stress in the preslaughter period is a common causative agent.

Before the effects of stress may be understood the sequence of events which take place in the normal situation must be detailed. In this respect anaerobic glycolysis in muscle continues post-mortem until all the glycogen reserves have been used or until the accumulation of lactic acid is so great that enzyme function is limited. The times taken under normal circumstances for muscles to reach an ultimate pH of 5.5 after slaughter vary considerably between individual muscles but in pigs, sheep and cattle the average time intervals are 4–8, 12–24 and 24–28 hours respectively. However the concentration of glycogen in muscle when the animal dies is inversely related to the ultimate pH which is achieved in the muscle. Concentrations of less than 9–10 mg of glycogen in each gram of muscle will only allow glycolysis to proceed as far as a pH of 6 or slightly less. This is a critical pH around which distinct changes take place, and muscle which attains a pH of less than 6 whilst its temperature is above 30°C is likely to be pale, wet and tough.

In cattle, sheep and pigs when muscles in the carcass do not develop the levels of acidity in the times shown above, they become dark and dry (the DFD condition) and result in 'dark cutting meat' in cattle and sheep. Some of the components characteristic of high ultimate pH such as trimethylamine, ammonia and collidine are of significance in the context of flavour and therefore a high ultimate pH causes changes in flavour as well as in colour. There is a difference between individual muscles in the extent to which the intensity of the red coloration deepens. In pigs the shoulder muscles are likely to be most affected and in cattle the m. longissimus dorsi, the m. semitendinosus and the m. gluteus medius muscles are most prone to the effects of lower acidity. The condition may be induced by depleting the glycogen reserves before slaughter as may happen, for example, in stressful situations. In cattle this is much more difficult to achieve than in pigs because of higher natural reserves of glycogen. Thus fasting, exercise and/or long-distance haulage and long periods in lairages may have little or no effect but the mixing together of stress-prone animals such as strange young bulls may have a significant effect. In pigs, long-distance haulage and/or lengthy periods in lairage and/or the mixing of strange pigs in large groups may deplete glycogen reserves to give high muscle pH levels, and there is evidence from pigs that have been transported over long distances that the resultant high pH may affect not only the colour of the muscle but also the eating quality (Dransfield *et al.*, 1985). The feeding of sugar solutions to pigs 16–20 hours before slaughter has been shown to build up glycogen reserves sufficiently to arrest these types of deleterious decline with their consequential effects on muscle colour and quality, the quality defects being related to a greater susceptibility to spoilage and to undesirable processing characteristics linked to the greater growth of microorganisms during storage at a high pH compared with a slow growth on PSE muscle.

The PSE condition is associated more with pigs than with cattle or with sheep because of the greater susceptibility of certain types and breeds, particularly the Pietrain and Poland China breeds and certain strains of Landrace (see also chapter 4). It is one symptom of a very much wider syndrome, the porcine stress syndrome (PSS), and it is likely that the greater responsiveness to stress is mediated through greater circulating concentrations of catecholamines conditioning muscle to be very highly sensitive to stimulation. In pig muscle that completes rigor mortis in a very short period of time the muscle becomes pale in colour (P), soft to touch (S) and often has copious quantities of fluid exuding from its cut surface (E) – hence the terminology PSE. These changes are brought about by the combined effects of both temperature and pH acting on the physical and chemical components of the muscle, particularly in the context of the muscle proteins which in acidifying at temperatures greater than 30°C lose their water-binding properties and thus allow the free release of fluid. The greater reflectance of light from the surface of the cut muscle causes its paleness and this too is related to changes in protein structure which interfere with the optical properties of the surface layers and cause a reduced translucency. Different muscles are more or less susceptible according to their anatomical location; for example, the muscles of the more bulky parts of the leg and of the loin cool at a slower rate than the muscles of the flank and of the shoulder. In terms of carcass yield and eating quality, there is a direct loss of weight in the former case, and a slightly greater degree of tenderness in the latter case. In cattle muscle the incidence of PSE is likely to be higher if rigor mortis is completed while the carcass temperature is still high and the PSE muscle is likely to be tougher than normal muscle. Some work suggests that the condition reflects a defect in the structure of the muscle cell membranes, including the sarcolemma, which results in a leakage of calcium ions (Heffron, 1987).

In pigs one further aspect of the PSE syndrome is the effect of the stunning process. Electrical and carbon dioxide anaesthesia are the two most commonly used techniques and, of the two, carbon dioxide induces more rapid post-mortem changes and poorer mucle quality. Pig genotype has a very profound effect on the susceptibility of individual animals to this condition and the reader is referred to chapter 4, section 4.3.4, where aspects of gene action are presented.

3.5. Epithelial Tissue

3.5.1. Types and structure

The external appearance of animals as perceived by the eye is markedly influenced by epithelial tissue which may be divided conveniently into two types: that which covers and lines membranes and that which forms glands. In the first category is the epithelial tissue that covers the skin and intestine (the mesothelium) and the epithelial tissue that lines the body cavity and the intestines. In the second category there are two broad types of gland: the exocrine glands in which the ducts convey products away from the point of secretion to the epithelial surface and the endocrine glands which are ductless and in which substances are secreted directly into the body substance via, usually, capillaries (see also chapter 4). Some glands, for example the liver and the pancreas, combine features of both types. In fact the epithelial tissues surpass all other tissues in their propensity to differentiate into a great variety of forms.

Epithelial membranes consist entirely of cells, are avascular and separated from connective tissue on which they lie by a non-living basement membrane. If the epithelial membrane consists of a single layer of cells it is known as a stratified epithelium. This latter type of epithelium can withstand generally more wear and tear than a simple membrane.

Epithelial tissues are renewing tissues and therefore are constantly growing. Some of the most intriguing renewing tissues are those that form the taste buds of the tongue. The average life span of a taste cell is about 10 days and new cells are added every 10 hours.

Various histological textbooks (e.g. Ham, 1974) give detailed accounts of all the epithelial tissues. Here only the integument and the epithelial appendages hair and wool, which arise from it, will be considered in some detail.

3.5.2. Integument

The epithelial tissue lining the peritoneal cavity is formed from the mesoderm, that lining the gastrointestinal tract is formed from the entoderm but the epithelial tissue of the integument or skin arises from the ectoderm. During embryonic development ectoderm cells grow down into the dermis and form gland-like structures, including sweat glands, as well as hair follicles (giving in turn hair and sebaceous glands), hoof and horn. The tissue of horn is an insensitive cornified layer of epidermis covering the distal ends of digits. Horn is formed over the horn process, which is a bony core that projects from the frontal bone of the skull.

The skin consists of two layers of completely different types of tissue attached to each other over their entire extent but varying in thickness over different parts of the body. The outer layer or epidermis is stratified, squamous, keratinizing epithelium containing no blood vessels and is nourished from the vascular inner, dense, irregular connective tissue layer known as the dermis or corium. The epidermis can scarcely exist without the dermal substrate and linkage between the dermis and the epidermis is nowhere more important than in the development of epidermal appendages such as hair, wool and feathers.

The skin assists in regulating body temperature (sweating), it acts as an excretory organ, as a manufacturing centre for cholicalciferol (vitamin D), and as a receptor for stimuli that evoke sensations such as touch and pressure. When the animal is slaughtered, shorn or plucked it yields products which are valuable and highly sought after by man.

The pigmentation of the skin is due to the presence in the epidermis of the pigment known as melanin. In the animal kingdom this pigment is responsible for skin colours ranging from yellow, through various shades of brown, to black. The dispersion of the pigment is conditioned by a melanocyte-stimulating hormone released by the pituitary gland. Melanin not only imparts colour to the skin, and therefore in some cases camouflages the animal, but also protects the deepest layers of the epidermis from ultraviolet light.

With the exception of the skin all epithelial tissue surfaces must be kept wet on the outside. In the case of the skin the membranes on the dry surface become stratified and their outermost cells become converted into a non-living material called keratin. Keratin is a tough, fibrous, nitrogen-rich material which will not allow the passage of some chemicals. It is produced by specialized epithelial cells known as keratinocytes, occurs in hair follicles as well as in skin and is often of two types, hard and soft. The soft type covers the skin as a whole; the hard type contains more sulphur, does not desquamate and is the chief component of nails, claws, feathers, hoof and horn. The soft keratin of the skin is continually worn away and shed from the surface and therefore must be continually replaced. There is thus a continual growth process which is facilitated by the architecture of the epithelium wherein columns of keratinized cells appear directly above columns of their progenitor epithelial cells. As a result the cells furthest from the dermis are transformed into keratinized tissue and this then desquamates from the surface.

3.5.3. Hair and wool

The previous discussion should have left the reader in no doubt that the skin is very much a living, growing tissue. Hair, wool and feathers have a common pattern of development from the epidermis. In the early stages of embryonic life and throughout the life of the animal they grow vigorously, though spasmodically. For example, hair growth is intermittent with cycles of growth and cycles of rest dictated by many factors including photoperiod. Whilst growth is less in the older animal, regenerating feathers are amongst the most rapidly growing of all epidermal appendages with new pin feathers appearing in less than a week and with full dimensions being realized in only a few weeks.

A different type of hair growth is found in the so-called velvet on deer antler. Velvet consists of hairs distinct from those found elsewhere in the body in that although the follicles have accompanying sebaceous glands they are not accompanied by arrector pili muscles. In this case the shedding of an old antler is followed by the raw surface of the pedicle healing and then, subsequently, giving rise to new velvet on the regenerating antler.

Hair, wool and feathers develop from structures in the skin known as follicles. The follicles are derived from invasions of epidermal tissue into the dermis (Fig. 3.22). The layer of dermis that develops the follicle is known as the papillary layer and is richly supplied with blood vessels and nerves. For part of its length the hair follicle is surrounded by a cellular tubular sheath, known as the internal root sheath, which is formed of soft keratin. The hair itself consists of a central region known as a medulla and is composed of soft keratin, which is in turn surrounded by a cortex and, on the outside, by a cuticle, both being composed of hard keratin. A connective tissue sheath lines the follicle and from this the sebaceous gland of the hair is formed. A small bundle of smooth muscle fibres, the arrector pili, is attached to the connective tissue sheath and passes upwards slantingly from beneath the sebaceous gland. Contraction of these fibres squeezes out the oily secretions of the sebaceous gland and also erects the hair in the skin (Fig. 3.23).

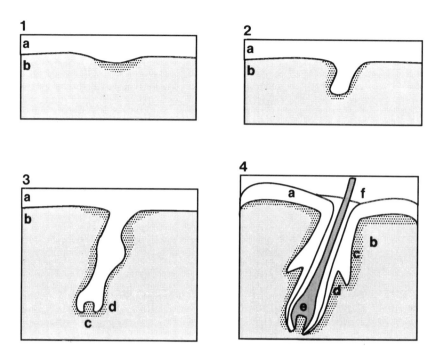

Fig. 3.22. Diagrammatic representation of development of a hair follicle and a sebaceous gland. 1, Epidermis (a) thickening over dermis (b) and in 2 invading the dermis. 3, Connective tissue papilla (c) covered by cap formed from invading epidermis (d). 4, Invading epidermis bulges to form a sebaceous gland (c) and an external root sheath (d); the cap of the epidermis differentiates into a hair (f) and an internal root sheath (e) (adapted from Ham, 1974).

True wool fibres lack a medulla and have small amounts of connective tissue in the follicle. They are found in the fleece of sheep together with two other types of fibre: kemp and hair. Kemp fibres are the coarsest sheep fibres and are fairly short. A unit of wool production consists of a follicle group composed of primary and secondary wool follicles. Primary wool follicles are the largest and are often arranged in the skin as trios. Secondary follicles are more numerous, often lie to one side and always produce finer fibres than primary follicles. Both types of follicle have sebaceous (or wax or grease) glands but only the primary follicles have a sweat gland and an erector muscle. All primary follicles are formed and growing by the time the lamb is born. Nearly all secondary follicles are formed at birth but do not grow until postnatal life. Detailed descriptions of wool growth are to be found in the book written by Ryder and Stephenson (1968), whilst if the reader wishes to study further the variation on the basic theme for hair which is exemplified in feather growth, any of the basic textbooks on the anatomy of domesticated animals will furnish the necessary information (e.g. Sissons and Grossman, 1975).

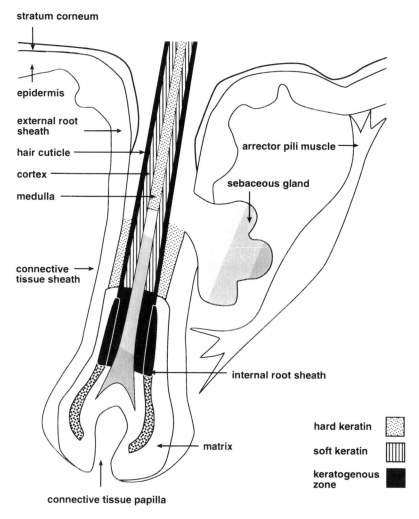

stratum corneum

epidermis

external root sheath

hair cuticle

cortex

medulla

connective tissue sheath

arrector pili muscle

sebaceous gland

internal root sheath

matrix

connective tissue papilla

hard keratin

soft keratin

keratogenous zone

Fig. **3.23**. Diagram of a hair follicle, showing the distribution of hard keratin.

References

Berg, R.T. and Butterfield, R.M. (1976) *New Concepts of Cattle Growth*. University of Sydney Press, Sydney.

Boccard, R. (1981) Facts and reflections on muscular hypertrophy in cattle: double muscling or culard. In: Lawrie, R. (ed.) *Developments in Meat Science – 2*. Applied Science, London, pp. 1–28.

Boccard, R. and Dumont, B.L. (1974) *Annales de Genetique Selection Animales* 6, 177.

Broad, T.E., Davies, A.S. and Tan, G.Y. (1980) *Animal Production*, 31, 73–79.

Brown, A.J., Coates, H.E. and Speight, B.S. (1978) *A Photographic Guide to the Muscular and Skeletal Anatomy of the Beef Carcass*. Meat Research Institute, Bristol.

Burleigh, I.G. (1980) Growth curves in muscle nucleic acid and protein: problems of interpretation at the level of the muscle cell. In: Lawrence, T.L.J. (ed.) *Growth in Animals*. Butterworths, London, pp. 101–136.

Dransfield, E. and Kempster, A.J. (1988) *Animal Production* 46, 50 (abstract).

Dransfield, E., Nute, G.R., Mottram, D.S., Rowan, T.G. and Lawrence, T.L.J. (1985) *Journal of the Science of Food and Agriculture* 36, 546–556.

Dutson, T. (1976) *Proceedings 29th Annual Recip Meat Conference, Provo, Utah* p. 336.

Dwyer, C.M. and Stickland, N.C. (1991) *Animal Production* 52, 527–533.

Enser, M. (1984) The chemistry, biochemistry and nutritional importance of animal fats. In: Wiseman, J. (ed.) *Fats in Animal Nutrition*. Butterworths, London, pp. 23–52.

Fletcher, T.V. and Short, R.V. (1974) *Nature, London* 248, 616–618.

Forrest, J.C., Aberle, E.D., Hedrick, H.B., Judge, M.D. and Merkel, R.A. (1975) *Principles of Meat Science*. W.H. Freeman, San Francisco.

Frandson, R.D. (1981) *Anatomy and Physiology of Farm Animals*. Lea and Febiger, Philadelphia.

Fuller, M.F., Duncan, W.R.H. and Boyne, A.W. (1974) *Journal of the Science of Food and Agriculture* 25, 205–210.

Garton, G.A. (1976) Physiological significance of lipids. In: Lister, D., Rhodes, D.N., Fowler, V.R. and Fuller, M.J. (eds) *Meat Animals: growth and productivity*. Plenum Press, New York, pp. 159–176.

Gemmell, R.T., Bell, A.W. and Alexander, G. (1972) *American Journal of Anatomy* 133, 143–164.

Girardier, L. (1983) Brown fat: an energy dissipating tissue. In: Girardier, L. and Stock, M.J. (eds) *Mammalian Thermogenesis*. Chapman and Hall, London, pp. 50–98.

Goldspink, G. (ed.) (1974) In: *Differentiation and Growth of Cells in Vertebrate Tissues*. Chapman and Hall, London, pp. 69–99.

Goss, R.J. (1978) *The Physiology of Growth*. Academic, New York.

Gross, J. (1961) *Scientific American* May, 120.

Ham, A.W. (1974) *Histology*, 7th edn. J.B. Lippincott, Philadelphia.

Handel, S.E. and Stickland, N.C. (1987) *Animal Production* 4, 311–318.

Heffron, J.J.A. (1987) Calcium releasing systems in mitochondria and sarcoplasmic reticulum with respect to the aetiology of malignant hypothermia: a review. In: Tarrant, P.V., Eikelenboom, G. and Monin, G. (eds) *Evaluation and Control of Meat Quality in Pigs*. Martinus Nijoff, Dordrecht, pp. 17–26.

Hood, R.L. and Allen, C.E. (1977) *Journal of Lipid Research* 18, 275–283.

Hooper, A.C.B. (1982) *Journal of Muscle Research and Cell Motility* 3, 113 (abstract).

Kauffman, R.G. and St Clair, L.E. (1965) Porcine myology. *Bulletin Illinois Agricultural Experimental Station* (715).

Kauffman, R.G., St Clair, L.E. and Reber, R.J. (1963) Ovine myology. *Bulletin Illinois Agricultural Experimental Station* (698).

Kempster, A.J., Cuthbertson, A. and Harrington, G. (1982) *Carcass Evaluation in Livestock Breeding Production and Marketing*. Granada, London.

Lacroix, P. (1971) The internal remodelling of bones. In: Bourne, G.H. (ed.) *The Biochemistry and Physiology of Bone*, vol. III, 2nd edn. Academic, New York, pp. 119–144.

Lawrie, R.A. (1979) *Meat Science*, 3rd edn. Pergamon, Oxford.

Leat, W.M.F. and Cox, R.W. (1980) Fundamental aspects of adipose tissue growth. In: Lawrence, T.L.J. (ed.) *Growth in Animals*. Butterworths, London, pp. 137–174.

Lister, D., Gregory, N.G. and Warriss, P.D. (1981) Stress in meat animals. In: Lawrie, R. (ed.) *Developments in Meat Science – 2*. Applied Science Publishers, London, pp. 61–92.

Pond, C.M. (1984) *Symposium of the Zoological Society of London* 51, 1–32.

Prescott, N.J. and Wood, J.D. (1988) *Animal Production* 46, 502 (abstract).

Price, J.F. and Schweigert, B.S. (1971) *The Science of Meat and Meat Products*. W.H. Freeman, San Francisco.

Priest, R.E. and Davies, L.M. (1969) *Laboratory Investigations* 21, 138–142.

Pyle, C.A., Bass, J.J., Duganzich, D.M. and Payne, E. (1977) *Journal of Agricultural Science* 89, 571–574.

Robelin, J. (1981) *Journal of Lipid Research* 22, 452–457.

Robelin, J. (1985) *Reproduction Nutrition Development* 25, 211–214.

Robelin, J. (1986) *Livestock Production Science* 14, 349–363.

Ryder, M.L. and Stephenson, S.K. (1968) *Wool Growth*. Butterworths, London.

Shahin, K.A. and Berg, R.T. (1987) *Animal Production* 44, 219–226.

Short, R.V. (1980) The hormonal control of growth at puberty. In: Lawrence, T.L.J. (ed.) *Growth in Animals*. Butterworths, London, pp. 25–46.

Simms, T.J. and Bailey, A.J. (1981) Connective tissue. In: Lawrie, R. (ed.) *Developments in Meat Science*. Applied Science Publishers, London, pp. 29–60.

Sinnett-Smith, P.A. and Woolliams, J.A. (1988) *Animal Production* 47, 263–270.

Sissons, S.S. and Grossman, J.D. (1975) *Sissons and Grossmans Anatomy of Domestic Animals*, 5th edn. W.B. Saunders, Philadelphia.

Slee, J., Simpson, S.P. and Wilson, S.B. (1987) *Animal Production* 45, 61–68.

Stickland, N.C. and Handel, S.E. (1986) *Journal of Animal Science* 147, 181–189.

Stott, A.W. and Slee, J. (1985) *Animal Production* 41, 341–347.

Taylor, St.C.S. (1980) *Animal Production* 30, 167–175.

Thompson, J.M. and Butterfield, R.M. (1988) *Animal Production* 46, 387–394.

Trayhurn, P. (1989) *Proceedings of the Nutrition Society* 48, 209–219.

Trayhurn, P., Temple, N.J. and Van Aerde, J. (1990) *Proceedings of the Nutrition Society* 49, 132A.

Trustcott, T.G. (1980) PhD Thesis, University of Bristol.

Trustcott, T.G., Wood, J.D. and MacFie, F.J.H. (1983a) *Journal of Agricultural Science* 100, 257–270.

Trustcott, T.G., Wood, J.D. and Denny, H.R. (1983b) *Journal of Agricultural Science* 100, 271–276.

Vaughan, J. (1975) *The Physiology of Bone*. Clarendon Press, Oxford.

Vaughan, J. (1980) Bone growth and modelling. In: Lawrence, T.L.J. (ed.) *Growth in Animals*. Butterworths, London, pp. 83–100.

Vernon, R.G. (1977) *Biology of the Neonate* 32, 15–23.

Vernon, R.G. (1986) The growth and metabolism of adipocytes. In: Buttery, P.J., Haynes, N.B. and Lindsay, D.B. (eds) *Control and Manipulation of Growth*. Butterworths, London, pp. 67–84.

Widdowson, E.M. (1980) Definitions of Growth. In: Lawrence, T.L.J. (ed.) *Growth in Animals*. Butterworths, London, pp. 1–10.

Wigmore, P.M.C. and Stickland, N.C. (1983) *Journal of Anatomy* 137, 235–245.

Wood, J.D. (1984) Fat deposition and the quality of fat tissue in meat animals. In: Wiseman, J. (ed.) *Fats in Animal Nutrition*. Butterworths, London, pp. 407–436.

Wood, J.D. and Enser, M. (1982) *Animal Production* 35, 65–74.

Wood, J.D., Whelehan, O.P., Ellis, M., Smith, W.C. and Laird, R.C. (1983) *Animal Production* 36, 389–397.

4 Hormonal, Genetic and Immunological Influences on Growth

4.1. Introduction

The cellular and tissue components of animals were described in the two previous chapters. A great number of endogenous and exogenous factors influence the way in which both cells, and therefore tissues, grow and in so doing change the weight and the shape of the animal. Environmental influence mediated via nutrition and housing is of major importance in affecting growth and development and the influence of the former of these two exogenous factors will receive special attention in chapter 6. Such exogenous factors must not, however, be regarded as isolated in their effects as they can influence many of the endogenous factors. The purpose of this chapter is to consider the ways in which hormones, genes and the immune system may influence the growth and development of animals. All of these endogenous factors may interact both with exogenous factors and between themselves. The complexity of the growth process, involving the deposition and removal of substances from cells and increases in cell size and in number at different rates and at different points in time, in essence reflects an intricate balance between the effects of all of these factors.

4.2. Hormones

4.2.1. Hormones and metabolism: modes of action

There are three chemical transmission systems in the animal body: endocrine, neural and paracrine. The endocrine system involves hormones of varied nature including glycoproteins and peptides, and steroids. Tissue parts are stimulated by the hormones travelling in the blood stream from their sites of manufacture to the target tissue. This type of transmission is often closely linked to neural communication. Perhaps the best example here is the stimulation of gastrin release in the stomach, the vagus nerve stimulating the acid-secreting parietal cells of the stomach directly. Brain activity is influenced by hormones and the control of food intake, which is integrated within the central nervous system, is probably mediated via the effects of neurotransmitters, such as noradrenaline and acetylcholine, and several peptide hormones (see also section 4.2.3). The latter are interesting in the sense that a number of the peptide hormones found in the gastrointestinal tract endocrine cells have also been identified in both central and peripheral neurons. Thus not only is the understanding of the control of digestion by gastrin, secretin and cholecystokinin widened to include these regulatory peptides, but also possible avenues of understanding of causal effects underlying the relationships between hormones and metabolism, relative to food intake and utilization and growth, become apparent. Oddy and Lindsay (1986) discussed metabolism × hormonal interactions and Dockray *et al.* (1984) listed some of those polypeptides that have been found in mucosal endocrine cells and point out that not all of these have a hormonal function because they exert their influences locally within their neighbouring cells and therefore are paracrine in their mode of action. An example is the polypeptide somatostatin, the importance of which in influencing growth will become clear later. Paracrine chemical communication therefore involves transmission within spaces between cells located at various distances apart within an organ or tissue. Probably these 'local' hormones, or autocoids, operate within a system in which the prostaglandins play a significant part in modifying nervous and hormonal activity (Blair, 1983). It is likely, however, that any one chemical messenger is not necessarily specific to one communication system. In this particular context catecholamines may function either as neurotransmitters at nerve terminals or as hormones released into the circulation from the adrenal medulla.

Growth control by hormones is not only dependent on interactions between the hormones themselves but is totally dependent on the presence of receptors. Effective control of growth is only possible if there are receptors present through which the hormones can exert their influence. Therefore, the ability of a hormone to influence tissue metabolism and growth depends on the circulating levels of that hormone, its rate of delivery to the target tissue, the number and affinity of hormone receptors present and the responsiveness of postreceptor events to hormone action.

Receptors occur on cell membranes and in the cytoplasm and nuclei of cells. The way in which chemical transmitters and receptors work is still imperfectly understood and it is unlikely that the sensitivity of a receptor to a circulating hormone is constant. It may change in response to the presence of other chemical transmitters produced by the endocrine, paracrine or nervous systems. The receptor is likely to be protein in nature and it is thought that the binding of the hormone to it may induce some changes in conformation, which in turn induces cellular messengers to modify cellular activity. The result of this sequence of events is the response characteristic of the target cell. It appears possible that the three types of receptor function in the following ways. In the case of cell membrane receptors an intracellular messenger may be stimulated. For example, the receptor complex may activate a membrane-bound enzyme to produce from ATP the nucleotide adenosine-3,5'-monophosphate, with the involvement of ionic calcium and magnesium and other regulatory proteins. Nuclear receptors are involved in the binding of thyroid hormones, in the actual nucleus itself. As the production of mRNA for growth hormone appears to be under the control of thyroid hormones, the importance of nuclear receptors to the growth process as a whole becomes very clear. The production of mRNA, and therefore protein synthesis by the cell and the growth of soft tissues, are extremely dependent on the presence and functioning of cytoplasmic receptors. The cytoplasmic receptors bind steroid hormones to give transmitter–receptor complexes that bind to nuclear chromatin and therefore modulate the production of RNA.

The changing priorities of different tissues for available nutrients within the overall growth patterns of animals may be met by a higher order of endocrine regulation than that provided by homeostatic mechanisms, which basically control complex compensatory actions to preserve constancy in function relative to challenges from the external environment. This hypothesis has been proposed by Bauman *et al.* (1982) and appears to be based on the work and ideas of Kennedy (1967), who postulated that the long-term regulation of the growth process operated in union with the acute regulation of thermoneutral stability. Kennedy introduced the term homeorhesis to differentiate the control of growth from the control of, for example, body temperature by homeostatic mechanisms. Bauman and his colleagues proposed that homeorhesis is the coordination of the metabolism in body tissues in support of a dominant physiological process with the direct partitioning of nutrients to support developmental processes. Therefore a chronic or long-term control, compared with a short-term control under homeostic mechanisms to redress potentially dangerous imbalances, is envisaged and it is proposed that prolactin and growth hormone, both of which will receive attention later, act as chronic coordinators of nutrient partitioning. A schematic representation of homeostatic and homeorhetic regulation is given in Fig. 4.1.

The homeorhetic regulation of growth involving a 'high order system of control' assumes greater feasibility when consideration is given to the enormous difficulties experienced in progressing knowledge, and applying that knowledge in practice, in the area of any one hormone. The incestuous-like relationships between hormones clearly indicate that a most complex and sophisticated system is operating to control growth processes. This should become abundantly clear in the next sections where the importance of the hypothalamus as both a direct and an indirect centre of control of growth will become evident.

4.2.2. Individual hormones and growth

Growth hormone

It is tempting to regard growth hormone, a protein consisting of 191 amino acids, as the most important hormone regulating growth processes if for no other reason than its importance has prompted man to use the nomenclature

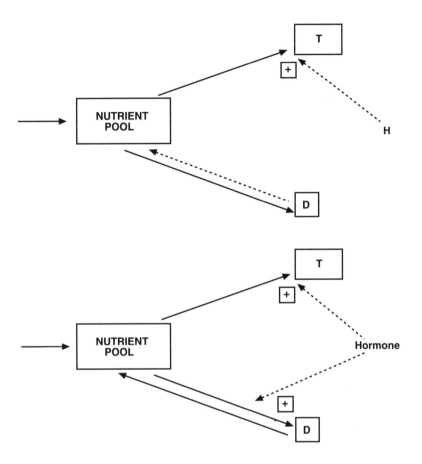

Fig. 4.1. Schematic representation of a homeorhetic regulatory mechanism. Top, Nutrients are removed for anabolic or secretory processes by tissue (T) under the influence of a stimulatory hormone (H). Depot tissue (D) buffers the central nutrient pool fluxes regulated homeostatically in response to sensors in the central pool. Botom, A homeorhetic hormone forcibly mobilizes reserves from D in support of the processes being promoted simultaneously in T (based on Bauman *et al.*, 1982).

that he has to describe it! However, although there can be no doubt that it is of great importance in controlling growth, the temptation to regard it as of singular importance should be resisted because of its influence on, and the way in which it is influenced by, other hormones and factors. It cannot be considered alone as being of primary importance because its growth-promoting role is not solely of direct action at tissue level but also as a mediator of other factors that act at that level. In turn other factors mediate in its primary growth-promoting role.

It is sometimes referred to as the pituitary growth hormone and this description is indicative of its site of origin, that is the anterior pituitary gland. As will be discussed below, the release of growth hormone is intricately influenced by the effects which it exerts on other hormones and by the effects which yet other hormones in turn exert on it. The overall result is an episodic release pattern which may have a physiological significance. This non-static pattern of release in humans, in cattle, in sheep and in rats is characterized by peaks or spikes of secretion apparently occurring at random. Only in ruminant animals, but not in humans and rats, is there evidence of diurnal fluctuations (Davis *et al.*, 1984). Baseline concentrations in the blood are therefore difficult to assess and can only be realized by recording the frequency of the secretory spikes (number per unit of time) and their amplitude (mean of maximum values) and then taking the mean of all observations less those which are not part of a spike. The other hormones and the various exogenous factors which control the episodic release and the modes of action of growth hormone are represented diagrammatically in Fig. 4.2.

The immensely complex web that is woven indicates that growth hormone may have both anabolic and catabolic functions in the animal body. Possibly different parts of the same molecule play different roles in metabolism (Spencer, 1985). As an anabolic agent it obviously has a vital role to play in controlling nutrient partitioning in the processes of growth and lactation. Contrarily, it can be lipolytic and diabetogenic and therefore clearly catabolic in its function.

The release of growth hormone from the anterior pituitary is controlled by neural peptide releasing factors, sometimes referred to as the hypothalamo-hypophyseotropic factors, secreted by the hypothalamus. Also, in some circumstances, thyrotrophin releasing hormone may affect the release of growth hormone. The amount of growth hormone released is conditioned by the relative concentrations of these three releasing factors in the hypophyseal portal blood, together with the sensitivity of the somatotrophs to each of the individual releasing factors. The adrenergic nerves of the hypothalamus control the releasing factors through adrenergic receptors. Two hypothalamic hormones are important in controlling the release of growth hormone, a growth hormone releasing factor known as somatocrinin and a growth hormone inhibiting factor known as somatostatin. The latter probably influences the release of not only growth hormone but also insulin, thyrotrophin and a wide range of other pituitary, pancreatic and gut hormones as well. It is postulated that because somatostatin controls the release of growth hormone and other hormones which are responsible for potentiating the release and actions of somatomedins, it may have a very central role to play in controlling growth processes. The essential and very intricate mediator links between somatomedins and growth hormone in controlling growth are discussed in the next section but at this point it is important to appreciate that insulin regulates the ability of growth hormone to control somatomedin production in the first instance. Insulin itself is considered later.

In its anabolic and catabolic roles growth hormone has a widespread effect on many aspects of metabolism through its effect on the metabolic fuels carbohydrates, lipids and proteins. Nevertheless, the most important function is its influence on the formation of proteins and nucleic acids that are not to leave the body in products such as hair, wool, eggs and milk (Buttery, 1983). The

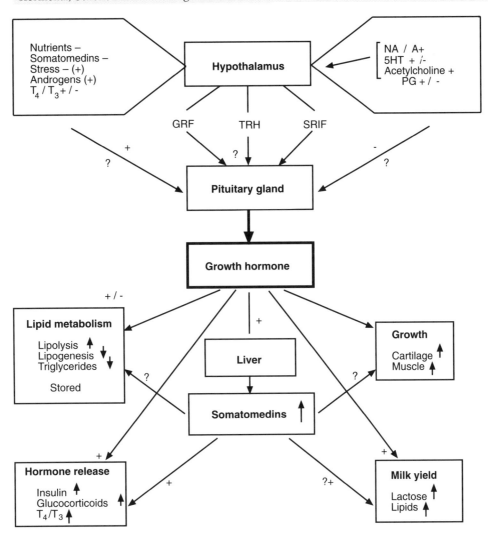

Fig. 4.2. Diagrammatic representation of the control of growth hormone (GH) secretion and its actions in domestic animals. TRH = thyrotrophin releasing hormone; GRF = growth hormone releasing factor; SRIF = somatostatin inhibitory release factor; T_3 = triiodothyronine; T_4 = thyroxine; NA = noradrenaline; A = adrenaline; 5HT = 5-hydroxytryptamine; ↑ = increase; ↓ = decrease (based on Scanes, 1984).

stimulus given to the tissues is probably a result of an increase in the activity of ribosomes engaged in the translation process (Kostyo and Isaksson, 1977). This effect is not immediate. There is a time lag and this may indicate that the activation of the ribosomes represents the culmination of a series of molecular processes triggered by the hormone. After the initial stimulation of ribosomal activity the synthesis of several types of RNA in cells may be stimulated. The acceleration of ribosomal RNA synthesis increases the number of protein synthesizing units in cells and therefore the effect on protein anabolism depends on the stimulating effect on RNA synthesis in the first instance. Relative to these protein anabolic effects, growth hormone has a stimulatory effect on membrane transport of amino acids and sugars into various target tissues and therein stimulates the uptake and the incorporation of amino acids into growing skeletal muscle. In addition to its direct effects on skeletal muscle growth, it also exerts a direct influence on the liver, the heart, fibroblasts, lymphoid organs and the placenta, and receptors are found in these organs and tissues. Receptors are also found in cartilage but the effect here is indirect via somatomedins. Adipose tissue is also directly affected by growth hormone and its receptors cast the hormone in its catabolic role.

The interplay between insulin and growth hormone has already been mentioned briefly above. The receptors in the liver for growth hormone are also regulated by insulin and it is the interplay between the two which is fundamental to the catabolic role of growth hormone. If the animal is undernourished, growth hormone concentrations in the plasma increase and it is used for lipolysis to allow an increased use of lipid as an energy source. However, such a reduction of adipose tissue by the breakdown of triglycerides is not the only way in which the amount of lipid in adipose tissue may be reduced, because if energy is in deficit in the diet to a lesser extent, then fatty acid synthesis, or lipogenesis, will also be reduced because the low insulin levels associated with this situation will deflect energy to the immediate needs for survival rather than into tissue for longer term storage (Spencer, 1985). The low insulin levels may then decrease the receptors for growth hormone in the liver and therefore reduce somatomedin production and growth. Therefore the effects of growth hormone on adipose tissue may be indirect.

The positive role of growth hormone in lactation is at first sight perhaps a little surprising until the control of the mobilization of body reserves in early lactation is considered in the context of hormonal status. The review of Johnsson and Hart (1986), and the work of others subsequently, leaves no doubt that whilst yields of milk, milk energy, lactose, protein and fat can be increased by the exogenous administration of growth hormone to lactating cows, the reasons underlying this are still very imperfectly understood. There are also higher endogenous plasma concentrations of growth hormone in high, compared with low, yielding cows (Scanes, 1984) and in calves at young ages with high predicted breeding values compared with low (Woolliams *et al.*, 1993). Weekes (1983) points out that the increased rate of lipid mobilization in early lactation is associated with an increase in the number of β-adrenergic receptors per adipocyte. Relative to this the lipolytic actions of growth hormone, and also possibly of prolactin (see later), as homeorhetic signals controlling the partitioning of nutrients in lactating animals therefore become apparent and some understanding of the role of growth hormone, in concert with insulin and prolactin in lactation, begins to emerge.

In vertebrates, growth hormone in its anabolic role exerts little or no influence on the growth of the fetus but the tissues become increasingly sensitive in postnatal life. In cattle the results of Trenkle (1977) imply that the basal secretion of growth hormone by the anterior pituitary is related to the total growth hormone in the body and that as animals grow there is gradually less hormone available per unit of body weight. In that work the calculated proportion of the total growth hormone in the anterior pituitary at all times was greater than 0.99 and the calculated hourly pituitary release, independent of body weight, was 0.0058 of its total content. Further work (Trenkle and Topel, 1978) added to this hypothesis and it appears that more growth hormone per unit of body weight may be found in younger, smaller cattle than in those of a heavier weight. Also, in the work of Trenkle, the growth hormone status of the animal was positively related to the amount of carcass muscle and the proportion of RNA in the muscle and negatively correlated to adipose tissue levels. There is good evidence in sheep, in cattle and in pigs to indicate that there is a positive correlation between the daily secretion of growth hormone and lean tissue, and a negative correlation with carcass adipose tissue.

A number of other exogenous factors may affect the growth hormone status of the animal. The special effects of photoperiod will be considered later. This apart, the influence of plane of nutrition appears to be particularly striking. It has already been pointed out that plasma concentrations of growth hormone increase in the malnourished animal. The concentrations are decreased when the animal is given more food. As Davis *et al.* (1984) point out, this may be a very much oversimplified picture because the changes recorded may not reflect true changes in the clearance of growth hormone from the blood stream. Notwithstanding this complication, true short-term changes in growth hormone concentrations in the blood stream do occur in ruminant animals in response to changes in food intake. The ingestion of food is associated with an immediate decline in plasma concentration followed within minutes by an increase. In sheep the response may be related to age as the initial decrease lasts longer and the ultimate rise is less in mature, compared with young, animals.

The exogenous administration of growth hormone to lactating animals has been shown to exert positive effects on milk yield as described above. Extrapolation might therefore suggest that similar exogenous administration to growing animals would bring growth responses largely through increases in skeletal and muscle tissue growth. The data available from early experiments using this approach did not support this hypothesis consistently. Certainly there were data which showed improvements in nitrogen retention, and in carcass quality (more muscle and less adipose tissue) in cattle, sheep and pigs, and the reader is referred to the paper by Hart and Johnsson (1986) for individual references of work conducted up to that time. But in many cases exogenous administration had been without effect, or with only a small effect, on growth rate although food conversion efficiency, presumably because of increased muscle growth, had been improved. Spencer (1985) postulated that in the majority of cases where there had been lack of improvement in growth in normal animals, this was because the exogenous growth hormone had upset the equilibrium that the body was trying to maintain and that in consequence the hormonal balance was also upset with the result that there had been changes in the levels of other hormones, such as insulin, which had decreased. He postulated further that by giving additional growth hormone additional receptors would be required if its effect were to be realized in a significant growth response. Bauman *et al.* (1982) suggested a further possibility to explain response failures, namely that the highly purified growth hormone used in the exogenous treatment of animals in early studies may have lacked sulphation factor activity and may have been devoid of hyperinsulinaemic activities.

Another factor which may explain, in some cases, the lack of growth response to exogenously administered growth hormone is the pattern of administration. Whether the hormone is injected in single or multiple doses, the dosage level and over what period of time may be of fundamental importance. Greater attention to these aspects and the availability of vastly improved preparations of growth hormone, derived from recombinant processes, have given more consistently positive results in later work. Indeed, in many cases the results in terms of muscle growth have been spectacular and some implications of the successes achieved from the exogenous administration of modern preparations are discussed in chapter 11. However, the future use of recombinantly derived growth hormone may not be limited solely to changing the growth rates of tissues of economic importance in the carcasses of meat animals. A more thorough understanding of its mode of action suggests potential use in animals (and in humans) in the following areas: reversing some aspects of the ageing process, improving wound healing, controlling the level of fat in adults and improving the performance of animals used for draught and racing purposes.

The possibilities of differences in growth rate between breeds within species, and between genotypes within breeds, being related to endogenous growth hormone production, receives attention in section 4.3.6. Also the fascinating role which growth hormone may play in immunoregulation is given some consideration in section 4.4.3.

Somatomedins

The fact that the major effect of growth hormone is mediated via somatomedins has been pointed out already. However, the somatomedins form only one of a group of polypeptides which are sometimes referred to as insulin-like growth factors. Overall they differ from 'classical' hormones although they are controlled by them, mostly from the liver but also from other tissues such as those of the kidney. Somatomedins are regarded as the most important of this group, and were originally discovered as the 'sulphate factors' because they regulated chondroitin sulphate formation in cartilage. From that single discovery of function, knowledge has accumulated which suggests that they have a direct influence on the growth process.

Somatomedins are present in relatively high concentrations in the blood stream and in a wide range of tissues exhibit hypertrophic (insulin-like) and hyperplastic properties. Their greatest influence may be in stimulating epiphyseal cartilage and bone growth, and therefore in stimulating the longitudinal growth of the skeleton, but their receptors are present in skeletal muscle, in the placenta and in the liver and in other visceral organs and their influence must accordingly be assumed to be strong in a variety of areas. Although several insulin-like growth factors have been identified, at the time of writing it would appear that IGF1 and IGF2, possibly working in concert with others, are the two most important somatomedins which provide the ultimate endocrine link in the chain of hormones regulating cell growth. Large quantities of IGF1 are synthesized by the liver but there is also a significant local release at many other centres in the body and it seems highly likely that both the systemic and local releases have a role to play in mediating the action of growth hormone. Also, the most recent findings on mode of action point to IGF1 as having a role *per se* to play in promoting growth processes as well as in mediating the anabolic role of growth hormone, and the reader is referred to the review of Pell and Bates (1990) for further information in this area.

Although, compared with the responses of different species to growth hormone, growth responses to somatomedins are less restricted, the evidence for ruminant species is less conclusive than for simple-stomached species. In ruminant species there is currently a lack of critical evidence which can allow confidence to be placed in somatomedin activity relative to growth rate and plasma growth hormone concentrations, even though in sheep relative body weight growth has been found to be positively correlated with serum somato-medin activity. Similarly, higher somatomedin activity has been recorded in faster-growing breeds of pigs, cattle and poultry (Falconer, 1981).

A possible axis of growth hormone–somatomedin action is presented in Fig. 4.3.

Insulin

Insulin and glucagon (see section 4.2.3) are the two primary pancreatic hormones but it is a moot point as to whether or not insulin should, strictly speaking, be classified as a true growth hormone because it plays a supportive, rather than a direct, role in influencing growth. This concept casts insulin, together with glucocorticoids, parathyroid hormone, calcitonin and prolactin, in the role of a supportive agent without direct influence on growth but in the presence of which growth may proceed at a normal rate. Muscle growth is therefore dependent primarily on the growth hormone–somatomedin link, with insulin playing a secondary supportive role in regulating the ability of growth hormone to control somatomedin production in the first instance. Irrespective of this point, insulin is clearly anabolic in its mode of action and acts in a permissive manner by influencing the levels of other hormones and by being of major importance, through its strong lipogenic properties, and therefore through its order of decreasing effect from adipose, to muscle, to bone tissue, in affecting body composition. In many ways if insulin is to be regarded as a hormone it should be regarded as an antilipolytic hormone, notwithstanding the fact that it can also influence lipolytic events. The central position of insulin in regulating carbohydrate metabolism and tissue growth in growing animals has been discussed fully by Weekes (1986).

The number and affinity of insulin receptors are regulated by a number of

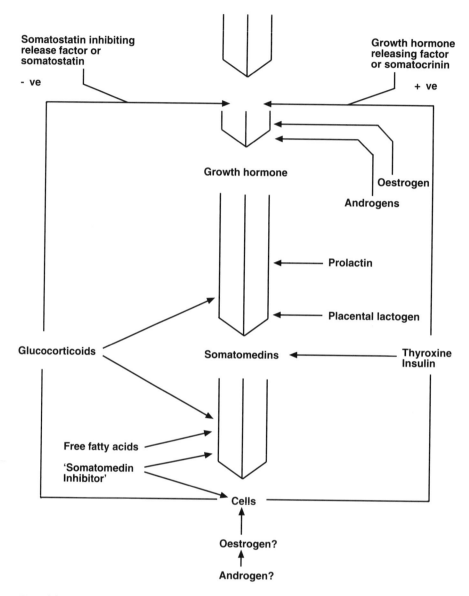

Fig. 4.3. A possible growth hormone–somatomedin axis (based on Spencer, 1985).

factors including nutritional status, the quantity of adipose tissue in the body, disease, growth hormone status and glucocorticoid levels. In itself insulin is a very important receptor in that receptor concentrations are inversely related to circulating plasma insulin levels, but it is possible that a gastric inhibitory peptide helps to ensure that potentially harmful quantities in the gut are not released in the presence of insufficient substrate concentrations. Also insulin regulates the binding of somatomedin to adipocytes. The links between different receptors in effecting the partitioning of nutrients may be mediated via the effects which insulin can have on food intake. Severe hypoglycaemia induced by insulin injections is associated with an increase in food intake and Baile *et al.* (1983) propose that this may reflect insulin acting as a 'body adiposity signal'. Plasma insulin concentrations tend to be positively correlated with adiposity and it is therefore possible that insulin may play a role in the maintenance of body weight by its action on the insulin receptors present in the hypothalamus of the brain. The insulin of the cerebrospinal fluid could form an integral link between the metabolic state of the adipose tissue and centres in the brain, particularly the hypothalamus, which are concerned with the reception of peripheral signals in the control of appetite. In this case insulin could be regarded as a primary hormone involved in the maintenance of energy balance or body weight.

Whilst insulin can have very pronounced effects on carbohydrate and protein metabolism, the effects differ between different species of animals. Overall the effects of insulin are less important in ruminant animals in regulating glucose and carbohydrate metabolism because, in contrast to simple-stomached animals, they absorb little glucose directly from the gastrointestinal tract because of the low activities of ATP-citrate lyase and, probably more importantly, of malate dehydrogenase. In consequence the effects on lipogenesis and lipolysis are fairly small with acetate acting as the major source of carbon. However in the ruminant animal insulin may stimulate lipid deposition by increasing the permeability of the membranes of adipocytes to glucose with the subsequent metabolism to α-glycerolphosphoric acid and the consequent stimulus to fatty acid esterification (Prior and Smith, 1982). The supply of fatty acids for esterification in adipose tissue may be further enhanced by insulin stimulating lipoprotein lipase in that tissue. Glucose production itself in the liver may also be altered indirectly through a reduction in the release of gluconeogenic precursors from the peripheral tissues. Also insulin either alone, or with glucose, exerts a marked effect in regulating plasma levels of branched-chained amino acids, possibly by enhancing intake or by decreasing their metabolism by muscle tissue. The adipocyte of the adult ruminant exhibits a responsiveness which is affected by both reproductive state and by nutritional status. For example, ruminant animals given diets based on cereals exhibit a greater rate of lipogenesis than do those given diets with no cereals, whilst lipogenic activity is increased when a liberal supply of food is given after periods of food restriction.

Taking the pig as an example of a simple-stomached animal to compare with the ruminant animal, the differences in the potency of insulin in conditioning nutrient partitioning and, in particular, in regulating lipid metabolism become very clear. The fact that there are big differences in the body compositions of breeds of pig lends this species particularly well to studies on the endocrine control of fat metabolism (Weekes, 1983). The numerically important and well known Large White breed has a lower lipogenic capacity and more muscle nuclei compared with the relatively unknown Ossabaw pig, which is slower growing and becomes obese with almost frightening ease. The Ossabaw also has a lower growth hormone secretion rate and a greater insulin response to arginine injection (Wangsness *et al.*, 1977). The Pietrain breed provides an interesting and striking contrast. This is a stress-sensitive breed which, compared with the Large White, is relatively lean but has a greater rate of fat mobilization during fasting. Also it differs in having a lower tissue sensitivity to insulin and lower plasma insulin responses to a number of stimuli, but this latter characteristic contrasts with the inverse relationship between insulin secretion and tissue sensitivity to insulin which occurs in forms of obesity such as that found in the Ossabaw breed.

The differences betwen pigs on the one hand and cattle and sheep on the other, and the differences in turn between breeds of pigs, in body composition relative to insulin levels, may be illustrated by the fat partition index proposed by Lister (1976). This index (Fig. 4.4) separates adipose tissue into internal and subcutaneous components and the hypothesis that insulin levels are higher in fatter animals with a high fat partition index, that is those breeds with a higher proportion of subcutaneous to internal fat, is supported by some work with pigs (Gregory, 1977) but not in other work with cattle (Gregory *et al.*, 1980).

Thyroid hormones

The states of hypothyroidism and hyperthyroidism in the postnatal life of the animal leave no doubt that thyroid hormones are extremely important in influencing growth at this stage but there is less certainty of the importance of the thyroid gland in fetal life. Two principal metabolically active compounds are produced: thyroxine (T_4) and 3,5,3′–triiodothyronine (T_3). Although T_4 is the predominant form secreted, T_3 is probably the more active of the two having both a wider distribution in tissues and a higher affinity for nuclear binding sites. In fact in peripheral tissues T_4 is deiodinated to T_3.

The binding of thyroid hormones to several plasma and nuclear proteins is central to their transport, to their distribution within the body and to their metabolic clearance. The receptors responsible for the binding are located in the actual nuclei of cells themselves. Their activity is translated through these receptors and they are transported via the plasma proteins. They stimulate oxidative metabolism and anabolic functions of cells in virtually all tissues by regulating oxygen consumption, mineral balance and the synthesis and metabolism of proteins, carbohydrates and lipids. Their effect on muscle growth is very potent and of both an anabolic and a catabolic nature. If they are deficient in postnatal life there can be a severe retardation of the growth of many systems, including the central nervous system, and of body growth in length and weight. In the case of muscle growth, the hypothyroidic state leads to a reduction in the fractional synthesizing and degradation rates. Administration of exogenous hormone can restore nearly normal protein turnover and growth but there are limits to the extent to which the massive catabolic effects characterizing the hyperthyroidic state may be reduced, with the responses to a certain extent being dose dependent.

The effect of thyroid hormones may be permissive in affecting growth and

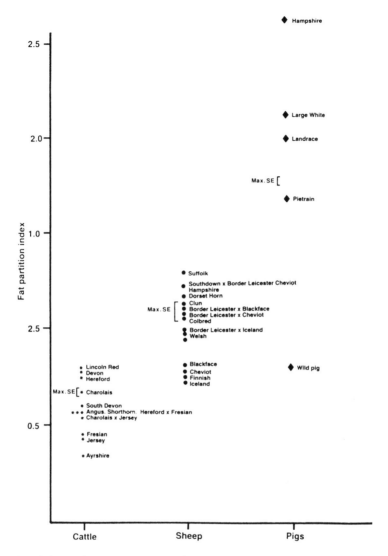

Fig. 4.4. The fat partition index in farm animals. The index is the quotient obtained by dividing the weight of dissectible subcutaneous fat by the sum of the weights of intermuscular, perinepheric and inguinal fat in a carcass (reproduced from Lister (1976) by kind permission of the author and the copyright holder, Cambridge University Press).

appetite in that they appear to influence growth hormone synthesis and possibly somatomedin levels. The inability of thyroid hormone administered exogenously to restore fully normal growth in hypothyroidic animals may rest in the refractoriness of tissues to stimulation by somatomedins. Growth stimulation by thyroid hormones is therefore probably via the regulation of somatomedin receptors.

Glucocorticoids

Cortisol and corticosterone inhibit the synthesis of DNA in some skeletal muscles and in the liver, the kidney and the heart. Different muscle types respond differently to exogenous treatment of the animal with synthetic glucocorticoids such as dexamethasone. Smooth muscle and white, fast twitch skeletal muscle exhibit reduced protein content whereas cardiac muscle exhibits an increased protein storage and red, slow twitch skeletal muscle shows little change (Kelly and Goldspink, 1982). An interesting hypothesis for the reasons underlying this difference in susceptibility is proposed by Sharpe *et al.* (1986): that it is a survival adaptation which protects the more physiologically active muscles against the effects of glucocorticoids. Overall, however, glucocorticoids are catabolic in their action (Buttery, 1983) and it is likely that insulin acts to counteract this effect (Odedra and Millward, 1982).

Inhibition of growth hormone secretion can be induced in adult, but not in young growing, animals by giving large doses of cortisol but it is unlikely that endogenous glucocorticoids inhibit the actions of somatomedins. At the moment it is anticipated that glucocorticoids inhibit growth more by a direct action on target tissues than by affecting other hormone levels. Nevertheless, high levels of glucocorticoids in animals are not always associated with slow growth rates but they can have a marked effect on body composition because as well as having a catabolic orientation for proteins they can be strongly lipogenic in some species, notably in cattle.

If, overall at a physiological level, glucocorticoids inhibit growth, then it should be clear that if this inhibition could be removed, if only partially, then growth rate would be enhanced. It is possible that some exogenously administered anabolic agents given to enhance growth rate, and to increase muscle mass, may exert their effects by removing some of the inhibitory effects of glucocorticoids. It is thought that this may be true in the case of trenbolone acetate. However, the growth-promoting and -repartitioning effects of β-agonists may be mediated not via similar effects but via the glucocorticoids having a permissive effect on their growth-promoting actions (Sharpe *et al.*, 1986).

Sex steroids

The androgen testosterone and oestrogens act as potent anabolic agents in the body. As such they have been used widely as exogenous stimulants, in particular for young cattle and sheep. The purpose of this section is not primarily to consider such a usage but to seek to explain their endogenous mode of action and their overall effects on animals.

Short (1980), in discussing the hormonal control of growth at puberty, suggests that the theories of Charles Darwin should be borne in mind when considering the influence of hormones on growth processes. Darwin (1871) was the first to point out that in polygamous species, the male tends to achieve a larger adult size than the female as a result of the intense competition between males for females. In monogamous species the competition for females is balanced and therefore there is no size dimorphism between the two sexes. On the basis of this fundamental difference, Short proposes a very important hypothesis: 'any somatic characteristics by which the male differs from the female will ultimately be determined by the sex hormones, not by the sex chromosomes'.

In mammals the male is the heterogametic sex with the result that the majority of dimorphisms result from testicular androgen secretion whilst the minority result from ovarian oestrogen secretion. In birds the female is the heterogametic sex so that most dimorphisms result from ovarian oestrogen secretion and the minority from testicular androgen secretion. Therefore male/female dimorphisms are most conspicuous in polygamous species and any characteristics that are more pronounced in the male will be androgen dependent.

There can be no doubt that testosterone is an extremely potent growth stimulant contributing to the superior growth rates of entire males, compared with castrated males, in cattle, sheep, pigs and poultry. It is difficult to give an average which has any meaning without the qualification of a standard error, but the averages given by Seidman *et al.* (1982) give some idea of the differences often found between the sexes. They cite average proportionate improvements in growth rate of 0.17 for bulls, compared with steers, and 0.15 for rams, compared with wethers. In the case of the pig the superiority of the boar over the castrated male in the weight range 20–100kg is well established, and is generally in the order of 0.10–0.12 but may depend on whether *ad libitum* or restricted feeding is practised, because the castrated male has a greater appetite than the boar (Rhodes, 1969; Fowler *et al.*, 1981). This can reduce, or even reverse, the superiority exhibited by the boar over the castrated male on some feeding regimes, but in so doing it will mask one of the important roles of testosterone, that is of stimulating muscle growth, for the castrated male will in part have narrowed or reversed the difference compared with the boar by depositing more adipose tissue. In all species the female exhibits a ranking which is inferior to the entire male and ultimately the entire male attains a greater absolute size.

Relative to the above practical realization, it is now appropriate to consider the existing evidence on the modes of action of both androgens and oestrogens. The gonadal steroid hormones are particularly important in stimulating the increased growth which is apparent in all animals at puberty and as anabolic agents they increase the efficiency of utilization of nitrogen from the diet. Heitzman (1981) proposes two possible modes of action in muscle cells. The first is through a direct effect on protein synthesis and/or degradation, mediated by a direct entry into the muscle cells. The second is an indirect effect, in which entry is into other endrocrine organs, the hypothalamus, the gonads, the pancreas or the thyroid, with the result that the synthesis, metabolism or secretion of other hormones, which in turn exert an anabolic effect in muscle and also affect intermediary metabolism in other tissues including the liver and adipose tissues, are altered. These possible modes of action are represented in

Fig. 4.5. Heitzman further postulates that as androgen and oestrogen receptors are present in the hypothalamus–pituitary complex, in the reproductive organs and, in smaller numbers in muscle and adipose tissue cells, it is more difficult to accept the reality of the first possibility than it is the second. There may, however, be species differences and in the first possibility one of two control routes is possible for androgens. The rate of nucleic acid synthesis may be altered to favour an increased rate of protein synthesis by the androgen–receptor complex. Alternatively corticosteroids, which may be acting as catabolic agents by controlling protein degradation rates, may be replaced by androgens from the receptor sites in muscle cells. The action in stimulating muscle protein synthesis may in part follow aromatization to oestradiol (Buttery, 1983). It is interesting in this context that adipose tissue contains the enzyme to aromatize testosterone and that testosterone reduces fat deposition.

Proposed actions of androgens and oestrogens are presented in Fig. 4.6. The indications are that protein synthesis in muscles is stimulated by somatomedins and growth hormone and regulated by insulin and thyroid hormones. As muscle cells possess receptors for these hormones, anabolic androgens may act directly on the cell receptors to alter their concentrations. The links with growth hormone may be very strong in ruminant animals, although it is likely that testosterone is a more potent stimulant of growth hormone than are the ovarian steroids (Davis *et al.*, 1984).

The effects of androgens are not solely those of stimulating muscle growth, they have a marked effect on skeletal growth too. This can be disadvantageous if they are used exogenously to stimulate prepubertal growth as they induce a premature closure of the epiphyseal plates of long bones and therefore they may

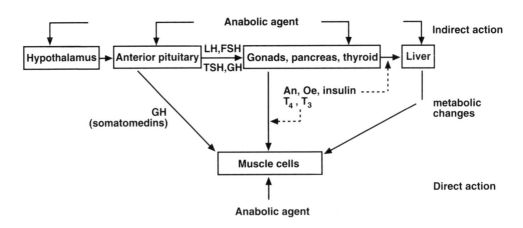

Fig. 4.5. Possible direct and indirect actions of anabolic agents in muscle cells proposed by Heitzman (1981). An = androgen; Oe = oestrogen; T_4 = thyroxine; T_3 = triiodothyronine; LH = luteinizing hormone; FSH = follicle stimulating hormone; TSH = thyroid stimulating hormone; GH = growth hormone.

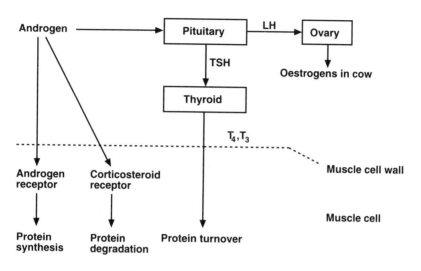

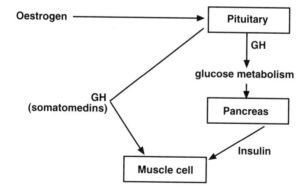

Fig. 4.6. Proposed actions of androgens and oestrogens (proposed by Heitzman, 1981). LH = luteinizing hormone; TSH = thyroid stimulating hormone; T_4 = thyroxine; T_3 = triiodothyronine; GH = growth hormone.

reduce adult height. In animals intended for meat production this will not be disadvantageous as the main aim of using androgens exogenously, or allowing their effect to be manifest naturally by leaving the male animal uncastrated, is to reap the reward of the anabolic effects at a young age where mature skeletal size is unimportant. The overall effect on skeletal growth may be dose dependent and Short (1980) points out that similar overall growth rate responses may result from relatively high or relatively low exogenous doses of testosterone, but that epiphyseal plate fusion will occur only from the high dose treatment. The castrated animal will have bones which are smaller in diameter than those of the intact animal but the length may actually be increased if the animal is retained until mature.

The net effect of oestrogens overall is probably one of inhibiting total body growth. In young animals, initially growth rate may be stimulated but eventually oestrogens accelerate ephiphyseal plate closure and the overall effect on muscle growth is less pronounced than is that of androgens. These effects may be most pronounced in young cattle and sheep and the overall poorer response from oestrogens may relate to the absence of specific oestrogen receptors in muscle cells.

Prolactin

The considerations of growth hormone earlier suggested a possible role for prolactin in partitioning nutrients for growth even though its major target organ is the mammary gland. It is possible that in a similar mode of action to growth hormone, prolactin acts as a homeorhetic control and may alter the capacity for net protein accretion in muscle as well as altering the metabolism of other tissues, in particular adipose tissue. If prolactin does act in this way then clearly there is the distinct possibility that it influences growth processes by a partitioning of nutrients. Relative to this possibility Bauman and his colleagues (1982) postulate that prolactin may either directly or indirectly decrease lipoprotein lipase activity in adipose tissue, decrease *de novo* fatty acid synthesis in the liver and in adipose tissue and increase lipolysis in the latter. They suggest that all are consistent with a direct partitioning of nutrients through reciprocal tissue changes. Further suggestion of the effects of growth hormone being mediated via prolactin is indicated by the increases recorded in both growth hormone and prolactin levels in the blood resulting from the exogenous use, with growing lambs and heifers, of thyrotrophin releasing hormones and synthetic oestrogens such as diethylstilboestrol. In contrast to concentrations of plasma growth hormone, insulin and thyroxin, plasma prolactin concentrations vary directly with temperature and day length and this may be yet another reason to suspect a role in mediating growth process.

4.2.3. Hormones and the control of food intake

As food intake is a vital determinant of growth it is appropriate to consider the roles which hormones may play in affecting appetite and satiety.

Cholecystokinin and glucagon are probably the two most important hormones affecting satiety. Cholecystokinin is a hormone of both intestinal and brain origin and with it two other brain peptides, bombesin and calcitonin, and a group of peptides known as the opiate peptides, may be involved in the control of food intake and in regulating energy balance. Glucagon is a pancreatic hormone and probably acts as a satiety factor in several different metabolic effects. It is probably the strongest peripheral satiety factor.

The receptors involved in the satiety response to cholecystokinin are located in the stomach wall. The direct control of feeding behaviour by the brain-derived cholecystokinin is reasonably well established in sheep and in pigs but not in other species. The complexity of the control process becomes only too apparent when it is considered that there are at least five forms of cholecystokinin present in the brain and that of these cholecystokinin-8 appears to be most prominently involved in satiety control in sheep but that this particular variant, and possibly the other four, may have their effect mediated through the release of the other brain peptides, such as calcitonin, and/or neurotransmitters such as noradrenaline.

The degree of complexity deepens when the possible roles of the opiate hormones and calcitonin are considered. Calcitonin is present in the hypothalamus, where it has specific sites to which it binds, and in cerebrospinal fluid, and is also secreted by the thyroid. Its primary involvement is in skeletal calcium metabolism and whilst it is as yet not completely certain that it plays a physiological role in the control of feeding behaviour, food intake and satiety, the evidence available is regarded by many as strongly supportive (e.g. Baile *et al.*, 1983). Opiate involvement presents an even greater problem to unravel. The site of receptors remains to be found but an insight into the possible significance in controlling feeding behaviour, and possibly hibernation and body weight regulation as well, is provided by the stimulation that has been found to feeding behaviour when endorphine is administered to satiated sheep.

Lastly somatostatin may also have a role to play in this very complex and still relatively poorly understood area. Again this is found both in the brain and in the gastrointestinal tract, as well as in other organs, and may play a role by virtue of its inhibitory influence on many peptide hormones, including those of the endocrine pancreas and the gastrointestinal tract. Support for this hypothesis comes from studies where cholecystokinin injections have been found to increase plasma somatostatin concentrations.

4.2.4. Hormones and the photoperiodic control of growth

Photoperiodic control of growth is most evident in animals which breed seasonally and in which growth and development are seasonally interrelated. In sheep there is a marked seasonality in food intake and growth: both are highest in the summer and lowest in the winter, even when the diet has been entirely of concentrates given for up to 4 years (Brinklow and Forbes, 1984). These same workers also found that the castrated male exhibited a lower overall level of intake and a smaller amplitude of the appetite cycle than did the entire male. In deer, hinds exhibit a lower overall intake of food with a smaller amplitude fluctuation than do stags.

The pineal gland is central to the control of events such as those described above. In sheep this gland is involved in many events, including the secretion of reproductive hormones and prolactin. Ultimately these events are mediated by the secretion of melatonin during periods of darkness. Where positive photoperiodic effects have been recorded on growth, pinealectomy has been shown to remove them. An involvement in controlling cortisol and testosterone secretion is also evident where skeletal photoperiods, that is those that alternate two unequal dark and two unequal light periods within an overall period of 24 hours, are imposed. In the case of the anabolic testosterone, increases are evident. Artificially long daylengths, compared with short daylengths, in sheep (Brinklow and Forbes, 1984) and in goats (Terqui *et al.*, 1984), increase serum prolactin and growth hormone levels and increase growth rate but in the present state of knowledge it would be wrong necessarily to assume causal, rather than casual, relationship between these events.

In terms of actual lengths of photoperiod, artificial long photoperiods (16 h light and 8 h dark) and skeletal photoperiods (7 h light, 10 h dark, 1 h light and 6 h dark), compared with 8 h light and 16 h dark, have been shown to elicit growth increases in lambs, particularly on *ad libitum* feeding (Brinklow and Forbes, 1984), whilst flashlighting can be as effective as the long photoperiod (Schanbacher, 1984). Less work has been done with aseasonal breeding animals but in cattle the long type of photoperiod described above has generally tended to improve growth rate more in peripubertal animals than in those in the pre- and postpubertal phases. However, the overall effects have been small usually, although the length of the previous photoperiod and the speed of transition of light intensity at dawn and at dusk are factors that may affect the magnitude of the growth responses obtained from the induced change in photoperiod (Zinn *et al.*, 1988). In this particular context the findings of Mossberg and Jonsson (1996) are of particular interest. Using nearly 500 bulls over a 3-year period they attempted to separate the effects of temperature, daylength *per se* and changing daylength on food intake and growth rate. The results showed that the rate of change in daylength had a profound effect on growth rate, such that peak growth rates were not found on the longest day of the year even though they and food intake increased as photoperiod increased. Herein lies a cautionary note in interpreting results from artificially imposed skeletal periods in animals such as cattle, which are basically photoresponsive in the natural state but which under domestication rarely are allowed the chance to exhibit this phenomenon fully (see also chapter 9). In such cases rate of change of photoperiod could be a more important variable than the photoperiod *per se* in determining the overall responses obtained. Thus conflicting results of effects of changing photoperiod on growth rate in cattle (negative effects: Zinn *et al.*, 1986b; no clear effects: Petitclerc *et al.*, 1984; positive effects: Zinn *et al.*, 1986a) may hide important responses to rate of change in daylengths.

4.3. Genes

4.3.1. Introduction

A gene is part of the DNA molecule as described in chapter 2. Genes are responsible for laying down the templates in the animal that dictate the limits of what is, or is not, possible in the various growth processes. Man has for many years attempted to select animals for improved growth performance and reproductive and carcass characteristics allied to that performance. In the quest for the superior genotype and an understanding of how various characteristics are passed from generation to generation, the investigation of the inheritance of the undesirable and desirable results of gene action has been necessary. The undesirable results are reflected in those conditions where there are growth and size depressions and abnormalities, often leading to death soon after birth, and where there are effects in postnatal life on specific attributes of, for example, carcass quality. Individual effects are usually controlled by a single gene or a small group of genes. The mode of inheritance has in most cases not been too difficult to elucidate on the basis of simple Mendelian principles. The desirable characteristics of fast growth rate, good food conversion efficiency and desirable carcass attributes are controlled by many genes and in consequence it has been an uphill task to achieve consistent improvements in growth and growth-related processes in breeding programmes. Progress has only been possible by understanding and applying the principles of population genetics, and in improving livestock for these characteristics the concepts of heritability, selection differential and generation interval have become all prevailing. However, in all cases, the mechanism(s) by which genes operate through biochemical and physiological parameters to control growth are at best very poorly understood and the purpose of this section is to consider the evidence that is available in this particular context and, also, to consider both the heritability of growth and growth-related parameters as well as those conditions in which gene action has impaired growth.

4.3.2. Heritability of growth and growth-related traits

Introduction

To understand the concept of heritability it is first necessary to appreciate the components of phenotypic variation in animals. The components of phenotypic variation are genetic variation, environmental variation and the interaction and association between them. Both the genetic and environmental components can be further divided. In the case of the former, the division is into additive genetic effects, dominance effects and epistatic effects; in the case of the latter, the division is between general environmental effects on the one hand and common environmental effects, that is those effects that are experienced by members of the same litter growing from conception to weaning in a common environment, on the other. Dominance effects reflect those conditions where one allele, that is any one of the alternative forms of a gene occupying the same locus on a chromosome, masks the effects of the other (recessive) allele. Epistatic effects occur when one gene appears to be dominant over another gene but they are not alleles. Additive genetic effects are stable, are regularly passed on from one generation to the next and are most important. Heritability is defined as the proportion which the additive genetic variance represents of the total pheno-typic variance. It is therefore a ratio and not an absolute value and for a given trait is the amount of the superiority of the parents above their contemporaries that is passed on to the offspring. Because heritability values cannot be absolute it is important to appreciate at the outset that estimates vary greatly according to many factors including the method of calculation, for example parent–offspring regression or correlation between paternal half sibs. They are always specific to the generation and population from which they were derived. The notation h^2 is given to heritability and is expressed on a scale from 0 to 1.0 or 0 to 100%.

With the above as a background it is obviously difficult to talk about heritabilities in other than relative and generalized terms. Acceptance of this leads Dalton (1980) to consider three broad groupings of h^2; low or weak, 0–0.10 (0–10%): medium or intermediate, 0.10–0.30 (10–30%) and high or strong, 0.30 and above (30% or above). The division between three such groupings must be very imprecise but the data presented in Tables 4.1 and 4.2 give some idea of the relative order of h^2 for growth rate and characteristics allied to it with particular reference to facets of carcass composition. These data suggest that early weights for age, such as birth weight and weaning weight, tend to have lower h^2 values than do later weights for age. Growth rates clearly fall into the medium/high categories defined by Dalton whilst carcass attributes are, on the whole, in the high category for h^2. The traits of higher h^2 are those that may be selected for with some degree of confidence by performance testing individual animals. In these cases concentrating selection on the sire rather than on the dam is desirable because sires have the propensity to carry their genetic effects to greater numbers of offspring than do dams. Progeny testing is more useful for those traits of lower h^2 and in some cases a combined performance and progeny test is operated. Progeny testing, as well as being useful for those traits of lower h^2, is also useful where traits are expressed in one sex (e.g. milk production) and for traits expressed after slaughter (e.g. carcass composition), save where acceptable estimates of body composition can be made on the live animal in performance testing schemes (see chapter 10). The main disadvantage compared with performance testing is that the time interval is long before results become available.

Selection differential, generation interval and genetic gain

If it is accepted that because of relatively high h^2 some progress may be achieved in selecting for some parameters of growth and growth-related attributes, it is next pertinent to consider the factors that control the rate or progress that may be possible in these directions. In this case the genetic progress will depend on the selection differential and the generation interval for the species in question.

Selection differential measures the superiority of the selected parents over the mean of the population from which they came. In other words it is a measure of how good the parents, selected for a given trait or traits, will be in producing the next generation. Selection intensity is the ratio of the selection differential to the phenotypic standard deviation and some examples of phenotypic standard deviations, that is the variation normally found in growth and growth-related traits in a particular population, in farm livestock, are given in Table 4.3. The example of Table 4.4 indicates how an actual selection differential may be calculated. To obtain a high selection differential necessitates having large numbers of animals with wide variability in the first instance.

Generation interval, $_pI$, is the time interval between generations and is defined as the mean age of the parents when their offspring first produce their own offspring. Examples are given in Table 4.5. Thus, h^2 and selection differential ($_sD$) are the determinants of the rate of progress, or genetic gain, that can be achieved in selecting for any one trait. The progress made per generation ($_pG$) will be the product of h_2 and $_sD$. Therefore the progress made per year (DG) will be described by the ratio $_pG$ to $_pI$ (i.e. D.G. = $_pG \div _pI$). Obviously a high h^2, a large $_sD$ and a short $_pI$ will give a maximum rate of progress. In selecting for growth rates in certain periods of the animal's life and for growth-related parameters, the contribution of the h^2 component to this two-factor equation has already been pointed out to be moderately high. However, clearly the generation interval varies between species, the decreasing order of $_pI$ being from cattle to sheep to pigs to poultry in the common farm animal species. The $_sD$ component will depend on a great number of factors, in particular and as pointed out above, the degree of homogeneity and size of the population. For growth and growth-related parameters a number of possible rates of progress may therefore be envisaged. If the $_sD$ is large and roughly the same for populations of cattle, sheep, pigs and poultry, then clearly genetic progress will be that of a decreasing order between the species as ranked above. Within a species, to obtain maximum progress the larger the size of $_sD$ the more rapid will be the rate of progress. This contrasts to the traits with lower h^2 values, such as birth weights and weights at relatively young ages, where even if $_sD$ is of the same size the rate of progress will be considerably less rapid.

Repeatability and breeding value

A good animal, genetically speaking, will perform consistently well in different periods of time if its life span is of sufficient length to allow merit to be shown. It will always be above average merit in spite of fluctuations in its environment. This reflects the concept of repeatability, that is the performance of the same animal is repeated from time period to time period. High repeatability is very valuable because with breeding animals predictions of performance can only be made early in life. Therefore if heretability gives information on how an animal will pass on a trait to the next generation, repeatability gives information on the extent to which an animal will repeat a trait during its lifetime. As in the case of heritability, a scale of 0–1.0 or 0–100% is used and some general estimates of repeatability for traits related to growth in farm animals are given in Table 4.6.

Although overall genetic progress will be retarded if lifetime records of breeding animals have to be compiled before decisions are taken on genetic merit, the h^2 of certain traits may apparently increase, because the temporary

Table 4.1. Examples of heritability estimates of growth and efficiency traits in farm livestock.

Species*	Reference[†]			
	L	W	D	B
Weights for age				
Birth weight				
C	30–40	25–40	20–59	40
S	30–35	10–30	—	10–35
P	—	—	—	15
Litter size at birth				
S	10–15	—	0–15	10
P	5–10	5–15	15	15
Weaning weight				
C	30–35	25–30	20–55	35
S	—	10–30	10–40	10–40
P	—	10–20	8	—
Litter size at weaning				
S	—	—	0–10	—
P	5–10	5–15	7	—
Litter weight at weaning				
S	—	—	30–40	—
P	15–20	10–20	—	—
Weight at 5/6 months				
P	20–25	—	—	—
Weight at 140/180 days				
P	—	20–30	—	—
Weight hogget at 10 months				
S	—	—	35	—
Yearling weight				
C	50–55	50–60	—	—
S	40–45	30–40	—	—
Weight at 18 months				
C	—	45–55	30–55	—
Final weight				
C	—	—	50–60	—
Mature weight				
C	—	50–80	50–70	70
S	—	40–60	—	20–40
P	—	—	—	25
Growth rates				
Birth to weaning				
C	—	25–30	—	—
Post weaning				
P	25–30	25–40	21–40	—

Table 4.1. Continued.

Species*	Reference[+]			
	L	W	D	B
Growth rates continued				
Feedlot				
C	50–55	45–50	45–60	–
S	40–45	30–40	–	–
Pasture				
C	–	25–30	30	–
Efficiency of live-weight gain				
C	35–40	40–50	40	–
S	20–25	–	–	–
P	25–30	–	20–48	–

*C = cattle; S = sheep; P = pigs.
[+]L = Lasley, 1978; W = Warwick and Legates, 1979; D = Dalton, 1980; B = Bowman, 1974.

environmental variation is reduced. The estimates of repeatability obtained may be used to build up a breeding value (BV). Breeding values are assessments of the future genetic potentials of animals. This basically is a way of building up confidence in records collected over a period of time relative to a particular trait, the h^2 and repeatability of which are known. Details of this concept are provided by many basic texts on genetics (e.g. Dalton, 1980), but the point to make here is that breeding values may be particularly useful in assisting the selection for weaning weights, and other weights at young ages, where the dam's influence can be recorded over a number of successive parities with, probably in the case of grazing animals, considerable variation in environmental factors from parity to parity.

4.3.3. Hybrid vigour

If animals of widely differing genetic constitutions are mated then the phenomenon of heterosis occurs. If the offspring of the mating are better than both parents the heterosis is positive but if the offspring are worse than both parents then the heterosis is negative. Positive heterosis is termed hybrid vigour and an animal is said to exhibit hybrid vigour if its performance is better than the mean of both of its parents. This is the compromise which is adopted because it is rare to find an offspring that is better than both of its individual parents. In plants the true definition – 'better than both individual parents' – is used to define hybrid vigour because progeny are often better than both individual parents.

Hybrid vigour is of some considerable interest in animal production and

Table 4.2. Examples of heritability estimates of carcass traits in cattle, sheep and pigs, of fleece characteristics in sheep, and of some growth-related traits in poultry.

Species*	Reference[†]			
	L	W	D	B
Carcass traits				
Killing out proportion				
C	35–40	—	—	—
P	—	25–35	26–40	—
Carcass length				
P	50–60	40–60	40–87	—
Backfat thickness				
C	30–35	—	45	—
S: constant age	20–25	15–30	—	—
S: constant weight	20–25	40–60	—	—
P	40–50	40–60	43–74	—
Eye muscle area				
C	55–60	—	70	—
S: constant age	40–45	20–30	—	—
P	45–55	40–60	35–49	—
S: constant weight	40–45	30–50	—	—
Lean proportion in carcass				
S: constant age	—	0–10	—	—
S: constant weight	—	25–40	—	—
P	—	—	45	—
Fleece characteristics of sheep				
Clean fleece weight	30–40	45–50	—	45
Greasy fleece weight	30–40	45–60	30–40	—
Staple length	40–60	40–45	30–60	—
Staple diameter	30–50	50–55	40–70	—
Poultry traits				
Egg weight	—	—	40–50	60
Live weight (fowl)	—	—	25–65	—
Live weight at 22 weeks (fowl)	—	—	—	25
Live weight at 24 weeks (turkey)	—	—	—	60
Body depth	—	—	20–53	—
Shank length	—	—	40–55	—
Keel length	—	—	30–57	—

*C = cattle; S = sheep; P = pigs.
[†]L = Lasley, 1978; W = Warwick and Legates, 1979; D = Dalton, 1980; B = Bowman, 1974.

Table 4.3. Examples of phenotypic deviations in farm livestock (Dalton, 1980).

	Beef cattle	Sheep	Pigs	Poultry
Birth weight (kg)	4–7	—	—	—
Weaning weight (kg)	20–26	3.6	—	—
Weight of lamb weaned (kg)	—	5.0	—	—
200-day weight (kg)	20–26	—	—	—
400-day weight (kg)	25–30	—	—	—
550-day weight (kg)	25–30	—	—	—
Hogget body weight (kg)	—	4.5	—	—
Preweaning growth (kg day^{-1})	0.10–0.15	—	—	—
Feedlot growth (kg day^{-1})	0.10	—	—	—
Pasture growth (kg day^{-1})	0.07–0.10	—	—	—
Daily growth (kg day^{-1})	—	—	0.06	—
Food conversion efficiency (kg food kg growth^{-1})	—	—	0.20	—
Carcass weight (kg)	—	—	1.28	—
Killing-out proportion	—	—	0.016	—
Backfat 'C' (mm)	—	—	2.5	—
Backfat 'K' (mm)	—	—	2.7	—
Ewe fleece weight (kg)	—	0.5	—	—
Age at sexual maturity (weeks)	—	—	—	3.7

some general estimates of growth and growth-related traits are given in Table 4.7. Compared with various reproductive traits the values for growth traits are clearly smaller. In any event hybrid vigour will be at a maximum in the F_1 generation and is halved in each subsequent backcross to either parent. Also it is very important that both parents and offspring are compared in the same environment, otherwise the true potential of the hybrid offspring may be confounded with environmental effects.

Hybrid vigour has proved to be an attractive concept to the poultry industry for a considerable period of time. Relatively more recently the pig industry in the UK has followed this lead to the point where the hybrid pig now

Table 4.4. Example of calculation of selection differentials using daily growth rates (kg day^{-1}) of a herd of beef cattle (Dalton, 1980).

	Males	Females
Mean growth rates of selected animals	2.00	0.75
Overall herd means	0.25	0.25
Selection differentials	1.75	0.50

Average of selection differentials = 2.25 ÷ 2 = 1.13.
NB: If no selection of females, i.e. selection differential = 0, average selection differential = 1.75 ÷ 2 = 0.88, therefore potential genetic gain has been reduced by 0.25.

Table 4.5. Average length of generation intervals (years) for different species of farm and other animals (from Lasley, 1978; Dalton, 1980).

	Lasley (1978)		Dalton (1980)
	Males	Females	
Beef cattle	3.0–4.0	4.5–6.0	4.5–5.0
Dairy cattle	3.0–4.0	4.5–6.0	4.0–5.0
Sheep	2.0–3.0	4.0–4.5	3.0–4.0
Pigs	1.5–2.0	1.5–2.0	2.0–2.5
Poultry	1.0–1.5	1.0–1.5	1.0–1.5
Horses	8.0–12.0	8.0–12.0	9.0–13.0
Dogs			3.0–4.0
Humans			25.0

Table 4.6. Estimates of repeatability for growth-related traits in farm animals (Dalton, 1980).

	Beef cattle	Sheep	Pigs	Poultry
Birth weight	0.20–0.30	0.30–0.37	0.18–0.40	–
Weaning weight	0.30–0.55	–	0.12–0.15	–
3-month weight	0.42	–	–	–
9-month weight	0.40	–	–	–
Yearling weight	0.25	–	–	–
Adult live weight	–	–	0.37	–
Farrowing weight	–	–	0.20	–
8-week body weight	–	–	–	0.55
18-week body weight	–	–	–	0.94
58-week body weight	–	–	–	0.88
Litter weight at birth	–	–	0.25–0.40	–
Litter weight at 3 weeks	–	–	0.15	–
Litter weight at 8 weeks	–	–	0.04–0.14	–
Weaning weight per piglet	–	–	0.12–0.15	–
Daily growth to weaning	0.18–0.20	–	–	–
Daily growth to yearling	0.07–0.10	–	–	–
Body measurements	0.70–0.90	–	–	–
Daily growth of lambs	–	0.38–0.48	–	–
Carcass traits (eye muscle area, lean:fat ratio)	–	–	0.95–0.98	–
Egg weight	–	–	–	0.80–0.95
Fleece weight	–	0.30–0.40	–	–

Table 4.7. Estimates of hybrid vigour for growth, growth-related and reproductive traits in farm animals (as percentages) (Dalton, 1980).

	Dairy cattle	Beef cattle	Sheep	Pigs
Growth traits				
Birth weight	3-6	2-10	6	—
Weaning weight	—	5-15	—	—
18-month weight	—	10-12	—	—
Growth to weaning	—	—	5-7	—
Postweaning growth	—	4-10	—	—
Growth	—	—	—	10
Carcass weight	—	—	10	—
Carcass traits	—	0-5	—	0-5
Fleece weight	—	—	10	—
Age at puberty (hybrid younger)	—	5-15	—	—
Reproductive traits				
Live offspring per parturition (no.)	2	—	—	2-5
Calving/lambing rate	—	7-16	—	19-20
Viability of offspring	—	3-10	10-15	—
Offspring weaned per dam	—	10-25	60	5-8
Total litter weight weaned	—	—	—	10-12

predominates, numerically speaking, in the national pig herd. The Meat and Livestock Commission in the UK, over a number of years in the late 1970s and early 1980s, tested centrally under carefully controlled conditions common to all animals, hybrids from different companies against each other and against pure-bred Large White control pigs derived from herds participating in its own Pig Improvement Scheme. As an example of the reproductive, growth and growth-related responses obtained from these varying genotypes, the data of Table 4.8 represent the results of the fourth test conducted. These data show in a clear light the superiority in reproductive potential of each hybrid over the sample of the Large White breed taken as the control. Equally clearly, however, the growth and growth-related responses in the postweaning period are superior for relatively few of the hybrids compared with the Large White controls. The differences between, on the one hand, reproductive traits and, on the other hand, growth and growth-related traits, probably reflects the relative differences which existed in the genotypes of the breeds used in creating the hybrids in the first place, the genetic diversity for growth traits not being sufficiently wide for hybrid vigour to be manifest in all cases and to the same degree as for reproductive traits. There is a greater potential for heterosis effects to be of a higher order for growth and growth-related traits in cattle and in sheep cross-breeding programmes because the differences in size and growth characteristics of many of the domesticated breeds are much greater than are those for the domesticated breeds of pig.

Table 4.8. Relative responses (where overall average = 100) for reproductive, growth and growth-related traits from the Fourth Commercial Pig Evaluation Test of the Meat and Livestock Commission (Meat and Livestock Commission, 1978).

	Large White	Company hybrids								
		A	B	C	D	E	F	G	H	I
Reproductive traits										
Average no. born per litter	90	100	102	100	99	105	101	104	95	104
Average no. alive at 5 weeks	90	97	105	102	92	103	108	106	89	108
Growth and related traits										
Adjusted piglet weight at 5 weeks	92	100	102	100	99	102	101	103	102	100
Postweaning daily growth:										
restricted feeding	103	97	102	97	99	98	104	102	99	100
ad libitum feeding	96	95	100	99	101	103	101	102	99	103
P_2 fat measurement	105	87	105	98	102	100	114	107	89	95
Proportion of lean in carcass	102	94	102	100	103	97	106	101	98	98

4.3.4. *Undesirable genetic effects on growth and related traits*

Undesirable traits in many instances surface in progeny where inbreeding has been too intense. Many of the defects cause death soon after birth and represent abnormal growth sequences under genetic control *in utero*. In cases where the effect is not lethal, postnatal growth may be severely retarded. An example in this category is the dwarfism conditions of cattle mentioned earlier in chapter 3. The cause of the abnormal growth is often, but not always, due to recessive gene action and therefore it is relatively easy to select against such conditions because the gene action is simple. In Table 4.9 a selection of genetic defects in animals reflected mostly in abnormal growth patterns *in utero* is presented. In the majority of cases the abnormalities can be seen to cause death soon after birth. In most cases where this does not occur, growth in postnatal life is severely retarded.

A fascinating gene action that has at one and the same time both desirable and undesirable effects is found in pigs. The so-called *hal* (halothane) gene is manifest in sensitivity in the animal to the anaesthetic gas halothane. Heritability is low and estimates from covariance analyses between half-sibs and from regression coefficients from offspring on sire and on dam have been determined as 0.07 (SE 0.06) for British Landrace and 0.16 (SE 0.12) for certain strains of Pietrain–Hampshire cross-breeds (Blasco and Webb, 1989). The sensitivity is controlled by a single autosomal locus and the mode of inheritance can be of complete recessivity in some breeds. The recessive zygotes give desirable increases in carcass lean content and better food conversion efficiencies compared with the normal homozygotes. Homozygosity for the single recessive gene can be detected as early as 8 weeks of age by sensitivity to halothane.

The undesirable effects are those of a greater susceptibility to stress giving sudden death, a higher incidence of pale, soft exudative (PSE) muscle (see also chapter 3) and shorter carcasses. The gene therefore appears to be additive for meat quantity but largely recessive for meat quality and stress susceptibility. Major differences appear to exist between breeds (Webb, 1980). From a survey of published work Webb found that the British Large White pig and its contemporaries in other countries have zero, or near zero incidences. Landrace pigs from all countries have a higher incidence. The British Landrace has an incidence of about 11%, the next lowest incidence is that of the Swedish Landrace (14%) and the highest incidences are found in the Belgian Landrace (88%) and in the Dutch Pietrain where they can reach 100%. Studies in the British Landrace (Simpson *et al.*, 1986; Webb and Simpson, 1986) suggest that the effects on carcass lean content are less pronounced than in certain lines of Pietrain–Hampshire cross-bred animals and that, in common with other studies, there is very little effect on actual live-weight gain.

Identification of the *hal* gene is important in breeding stock and some very basic detective work has given an important insight into how this may be effected and how the gene may function. In this work the *hal* gene was identified as an allele of the sarcoplasmic ryanodine receptor/calcium release channel gene (*ryr*1) (Fujii *et al.*, 1991). Mickelson *et al.* (1992) found that the biochemical and physiological responses to halothane in the homozygous halothane reactor animal were very highly correlated with the presence of this gene. The basic underlying biochemical changes were shown by Fujii *et al.* to be related to a single base pair mutation in the *ryr*1 gene, resulting in cytosine at base pair 1843 of the cDNA coding sequence being converted to thymidine with the consequential effect of arginine being substituted for cysteine at amino acid residue 615 of this particular protein.

4.3.5. Breeds

The common domesticated breeds within species represent gene pools deliberately established historically for the expression of specific characteristics, in some cases relative to particular environments. Differences in growth rate between breeds within species differ by varying margins. Generally speaking the larger the mature size of the breed the faster the growth rate of individuals within that breed, compared with individuals in breeds of smaller mature sizes. This general principle, and the way in which the various tissues develop, is given detailed consideration in other chapters but it should require little stretch of the imagination at this stage to appreciate that in species with widely differing sized breeds at maturity, greater differences in growth rate can be expected between breeds than in species which have smaller differences between breeds in mature size. On this basis greater differences in growth rate should be expected between breeds of cattle and sheep than between breeds of pigs. It is extremely difficult to present data in support of this expectation which does not oversimplify the complexities which such a generalization hides and the data of Fig. 4.7 and Table

Table 4.9. Some examples of genetic defects in farm animals.

Defect	Characteristics	Mode of inheritance	Species
Abracia	Fore limbs absent	Lethal recessive gene?	Horse
Achondroplasia 1	Shortened vertebral columns, very short legs, bulging foreheads, inguinal hernia	Partially dominant – two genes needed for lethal effect	Cattle
Anchondroplasia 3	Deformations of axial and appendicular skeletons in Jersey cattle	Recessive genes	Cattle
Amputed legs	Fore limbs and hind limbs have varying proportions of distal limbs missing	Recessive gene	Sheep
Atresia ani	No anus. In females colon opens into vagina	Two pairs of dominant genes (epistasis)?	Pigs
Atresia coli	Closure or part closure of ascending colon	Lethal recessive gene	Horse
Bulldog head (prognathism)	Skull broad, eye sockets large, nasal bones short and broad in Jersey cattle	Recessive gene	Cattle
Curved limbs	In Guernsey cattle hind legs are grossly deformed	Recessive gene?	Cattle
Ducklegged cattle	Body is of normal size but legs are greatly shortened in Hereford cattle	Recessive gene?	Cattle
Dwarfism	Parrot-mouth dwarfs in strains of Southdown sheep	Semilethal recessive gene	Sheep

Condition	Description	Genetic basis	Species
Hair whorls	Hair whorls appear on various parts of the body – undesirable but not lethal	Two pairs of dominant genes (epistasis)	Pigs
Hydrocephalus	Bulging forehead and enlargement of cranial vault. Limbs and other bones sometimes involved	Lethal recessive gene	Cattle, sheep and pigs
Mule foot	Hoof is solid as in mule	Non-lethal dominant gene	Pigs
Muscular hypertrophy	Thighs very thick and full. Fore and hind legs extended anteriorly and posteriorly	Recessive gene with variable expressivity	Cattle
Polydactylism	Extra toes on all feet	Dominant gene	Cattle
Short spine	Vertebral column is shortened by about one-half of normal length	Recessive gene	Cattle
Syndactylism	One rather than two toes on one or more of feet	Recessive gene?	Cattle
Wryneck	Contraction of cervical muscles to give a twisted neck	Lethal recessive (one form) gene	Horse

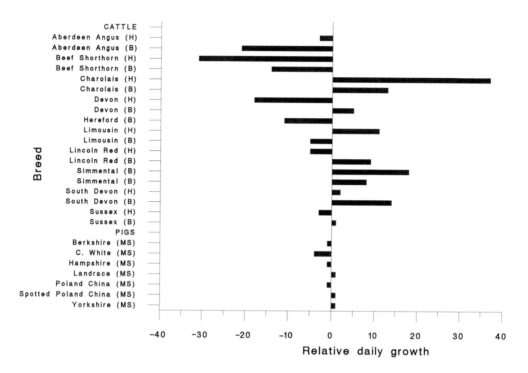

Fig. 4.7. Relative growth rates of eight breeds of pigs (based on data of Lasley, 1978) and ten breeds of beef cattle (based on data of The Meat and Livestock Commission cited by Allen and Kilkenny, 1980). For both sets of data mean = 0. For cattle data: B = bull and H = heifer. For pigs: MS = mixed sexes.

4.10 should be regarded with some degree of caution. The cattle data were obtained by interpolation between the 200- and the 400-day weights recorded under farm conditions by the Meat and Livestock Commission's pedigree recording scheme (cited by Allen and Kilkenny, 1980). Weight records were corrected for age of dam and whilst the 200-day weights may be taken as indicators of the mothering abilities of the dams, the 400-day weights, and therefore by interpolation the 200- to 400-day growth rates, may be taken as truer indicators of the genetic potential of the breeds for growth rate. The pig data relates to postweaning performance only. For sheep, because of the natural system of lamb rearing in the UK which leaves the lamb with its mother for a large proportion of its life to the point of slaughter at approximately 4–5 months of age, appropriate comparable data for a variety of the common breeds are difficult to find. The compromise adopted in Table 4.10 gives some indication of the general point being made but the variable influence of the dam in affecting growth rates of lambs in this period must be borne in mind.

To what extent in the future the present concepts of genotypes remain inviolate is a moot point, now that the possibilities of genetic manipulation to produce transgenic animals are being explored (see chapter 11).

Table 4.10. Mature breed weights (average of both sexes) of different breeds of sheep (Croston and Pollott, 1985) and 8-week live weights (average of both sexes) of single lambs (National Sheep Association, 1982).

	Mature breed weight		Lamb weight at 8 weeks	
Breed	kg	Relative value when mean = 100	kg	Relative value when mean = 100
Border Leicester	94	110	24.5	108
Dorset Down	77	90	20.9	92
Hampshire Down	78	91	22.7	100
North Country Cheviot	82	96	23.4	103
Oxford Down	100	117	25.4	112
Southdown	61	71	14.8	65
Suffolk	91	106	23.3	103
Texal	87	101	23.4	103
Wensleydale	101	118	25.8	114
Mean	85.7	100	22.7	100

4.3.6. Biochemical and physiological considerations of gene action in growth

The title of this section will perhaps lead the reader into thinking that the underlying causes of the outward manifestations of gene action described in the previous sections are clearly understood. Nothing could be further from the truth, for the molecular mechanisms and interactions involved in the bio-chemical and physiological genetics of growth are only just beginning to be understood and much work remains to be done before more than the most tentative of conclusions can be made in any one area. It is quite possible that a better understanding of the significance of the circulating levels of certain hormones in the blood stream may offer assistance to breeding programmes in the future, but the section which covered hormones in this chapter indicated all too clearly the complexity of the problems which have to be unravelled before such a possibility becomes a reality.

In reviewing the whole area of the biochemical and physiological aspects of gene action, Bulfield (1980) discussed the possibility of differences in circulating enzyme and hormone levels being manifest in growth processes through changes in nutrient partitioning. Much of the evidence which existed came from studies on the laboratory mouse where Bulfield suggested that the selection of lines on an age basis had done nothing more than change the relationship between developmental and chronological age, because most of the differences disappeared when selected and control lines are compared on the basis of similar live weight. However, in spite of this, in mice selected for high growth rate, definite differences have been found in lipid metabolism, the number and size of cells in organs have increased on an age, but not a weight, basis and the number and size of muscle fibres have increased although no differences, compared with control lines, have been found in circulating enzyme levels. In a rather different context pituitary hormone secretion and action appear to be under specific genetic control in that the condition of dwarfism in the mouse is caused by insufficient growth hormone for normal growth.

What is the state of knowledge with farm livestock? It would appear that some very weak correlations between circulating levels of growth hormone and insulin, and growth rate and tissue deposition, may exist. In cattle, the concentrations of circulating growth hormone have been shown to be positively correlated with carcass muscle content and RNA in the muscle, in Aberdeen Angus × Hereford cows mated to either Charolais or Aberdeen Angus bulls, although the breed of bull was without significant effect on any endocrine measurement (Trenkle and Topel, 1978). Other work has shown significant effects on growth hormone concentrations (Keller *et al.*, 1979). In this work, at all ages, the concentrations in the blood stream were greater in Aberdeen Angus, compared with Hereford, cattle and comparisons at the same age showed higher concentrations in Simmental than in Hereford bull calves. Also, in cross-bred steers of differing growth potentials, the average plasma growth hormone concentrations were found to be higher in larger breed types than in smaller breed types.

In Targhee ram lambs selected for fast growth rates, circulating levels of growth hormone have been found to be higher than in unselected control animals (Dodson *et al.*, 1983). The intricate relationships between somatomedins and growth hormones have been discussed already and it is interesting that in Suffolk sired cross-bred lambs, somatomedin activity has been found to be higher than in slower growing Finn sheep sired lambs (Wangsness *et al.*, 1981). Also Falconer *et al.* (1980) noted higher plasma somatomedin activity in Friesian, than in Angus × Friesian, bull calves.

It is quite possible that a greater ability to secrete growth hormone, in more rapidly growing breeds or lines within breeds, reflects a causal relationship which may be useful in predicting genetic growth potential in cattle. However, although the within-animal repeatability of growth hormone secretion is reasonably high at 0.79 (Davis *et al.*, 1979), the evidence is still too fragile to be used with any degree of certainty in a predictive role in the robust area of practical breeding.

The possible control of lipid metabolism by insulin has already been discussed. An interesting model in which investigations of this linkage have been made is that constituted by the performance test of selected and non-selected lines of Large White pigs developed at the University of Newcastle upon Tyne. The selected lines exhibited improvements over the non-selected lines in backfat thickness (reduced), in food conversion efficiency and in lean tissue growth rate (Whitemore *et al.*, 1982). From those animals, in comparisons with boars from other sources, although the insulin response to noradrenaline injections was significantly correlated with carcass fat measurements, the coefficients were small, ranging from 0.23 to 0.30, and the levels of basal insulin were not significantly correlated with these same measures of body adiposity.

From this and other work it appears that insulin alone is not, in the present state of knowledge, a factor that may be used in any equation to predict genetic control of tissue growth. However, in view of the fact that growth hormone can be both lipolytic and protein anabolic in pigs, using the concentrations of this in conjunction with insulin may provide a better prediction of tissue growth rates in the pig's body than if either is used alone.

4.4. The Immune System

4.4.1. Introduction

The immune system of the animal may be manipulated to give protection against disease. It may be used also to manipulate endocrine function. In this latter context the greatest initial progress in immunophysiological manipulation was in the control of ovulation rate and fecundity in female animals, particularly in sheep and to a lesser extent in pigs. Now immunophysiological manipulation offers the possibility of growth control as well as reproductive function control. Importantly, however, it is now realized that the immune system itself can be affected by parts of the endocrine system and their secretions, in particular but not exclusively by growth hormone, and this area of so-called 'cross-talk' between the two systems receives ever-increasing attention by research workers who continually unfold tantalizing glimpses of how in the future a better understanding of some growth processes may be realized.

The subject of immunology can be dealt with but briefly here with the aim of describing the fundamental principles of the subject in order that the relationships between growth and the immune system may be appreciated.

4.4.2. The immune system, disease and growth

Hyperfunction and hypofunction of the immune system can affect the animal in a number of different ways, but in both cases growth can be retarded. Hyperfunction is manifest in a number of allergic diseases and is probably less important in considerations here than is hypofunction, which may increase the incidence and severity of infections in animals and thereby affect growth. Before relationships between growth, disease and the functioning of the immune system can be understood, the main tissues constituting the system and the two basic ways in which it works must first be clarified.

The development and maintenance of the immune system is dependent on the thymus, the lymph nodes, the bone marrow and the spleen and may be regarded in two parts, although of course the two parts are interlinked. The two parts of the system rest on antibody production and cell-mediated immunity. Antibodies or immunoglobulins are of several types although built of similar units. They are large protein molecules specific for particular antigens. There are four main types, IgM, IgG, IgA and IgE, each with subclasses. Of the main types, IgM is regarded as the most important and is produced early during the primary response. IgG is the most important immunoglobulin produced in the secondary response, IgA protects seromucous surfaces and IgE may be important in resistance to certain parasites and is associated with allergic reactions. Cell-mediated immunity is based on the action of thymus-dependent (T) lymphocytes of several types: helper cells, killer cells and suppressor cells. Other antibacterial and antiviral cells, the macrophages and the polymorphs, also play a part in conditioning the response of the animal.

Passive immunity in the animal may result either from the administration of hyperimmune serum in the treatment of a specific disease or by transfer from the mother to her offspring. The transfer differs from species to species. In humans the transfer is entirely before birth. In carnivores and in some rodents transfer is both before birth and for some time afterwards via colostrum and milk, and in cattle, sheep, goats, pigs and horses the transfer is entirely from colostrum for the first 1 or 2 days after birth.

Diseases in animals can in many cases retard growth. The effects may be either direct, causing upsets in metabolism and/or absorption, which may lead to diarrhoea, or indirect, in which the animal will not or cannot eat sufficient food. The immune system will be important here in determining the extent of the growth retardation. If the immune system has been previously challenged sufficiently strongly, the infective vector may be unable to establish itself and to have any important effect on growth rate. Vaccination can be important therefore in not only reducing morbidity and mortality rates, but also in allowing optimal growth rates to be manifest. In the case of young pigs and young calves reared under intensive conditions, oral immunization with antigen from enteropathogenic bacteria has given these benefits (Porter *et al.*, 1975; Porter, 1976).

If the immune system can affect growth responses can the opposite work; that is can growth rates determine the response of the immune system to challenge by disease? This is a very difficult question to answer, in part because the effects of level of nutrition and growth rate are often confounded and inseparable. Furthermore, there may be a fundamental difference between the effects of a specific nutrient deficiency compared with a reduced nutrient intake overall from a completely balanced diet. In certain circumstances malnutrition may reduce antibody production in the human (Halliday, 1980). As antibodies are proteins it is therefore not beyond the bounds of possibility that their production is impaired. There is evidence for this from studies of the Kwashiorkor condition in the human child, which is thought to be caused by protein-deficient diets, but it is possible that it is a terminal effect after other body mechanisms have been affected. For example, the production of secretory IgA by the mucous membranes is likely to be depressed before other classes of antibody and it is thought that this may account for the greater frequency of diarrhoeal and respiratory diseases among malnourished children. The immunoglobulins appear to be resistant to the effects of malnutrition and in some cases actual increases in concentrations have been recorded. At the moment there is stronger evidence in support of cell-mediated response being much more susceptible to, and affected by, malnutrition. Again the Kwashiorkor condition has yielded much information in this area and has shown that the thymus and spleen, together with certain areas of the peripheral lymph tissue, become greatly atrophied. Halliday (1980), in reviewing the effects in animals other than humans, found evidence to suggest that malnutrition may generally produce a depressive effect on antibody production. Also he postulated that severe undernourishment is likely to give a marked atrophy of lymphoid tissue, to give decreases in the numbers and activity of lymphocytes and to give a decreased phagocytosis activity. Furthermore, it appears that specific vitamin or mineral deficiences may precipitate these changes as effectively as suboptimal intakes of well-balanced diets.

How does passive immunity fit into this picture, if at all? Can growth affect passive immunity or, alternatively, can passive immunity affect growth? Again the separation of nutrition and growth effects is difficult. There is some evidence to suggest that nutrition *per se* can influence antibody production in colostrum and therefore, in consequence, it might have some effect on passive immunity. Also there is evidence that if rapid growth and pregnancy requirements for nutrients overlap, then immunoglobulin production and levels in sucking lambs may be affected. Ewes mated in their first year of life when they are still growing rapidly, produce lower levels of immunoglobulins in colostrum than do ewes mated at later points on their growth curves when the nutrient needs for new tissue growth are waning (Halliday, 1976). Breeds of sheep may differ in respect to immunoglobulin concentrations in sucking lambs. Lambs from small breeds of sheep, such as the Welsh Mountain or the Soay, may have higher immuno-globulin concentrations than lambs from larger breeds, such as the Border Leicester or Oxford Down (Halliday, 1976). In terms of the effects of passive immunity on growth there is evidence of a significant positive relationship between the immunoglobulin concentrations of calves, lambs and piglets, and their growth rates during the suckling phase up to the point of weaning. Nevertheless, the effect is usually quite small and may be a reflection of keeping disease effects at very low levels.

Certain immunological disorders may also depress growth. Recurring bacterial and viral problems may be evident in immunodeficiency situations where lymphocyte population depletion allows these microorganisms foothold to cause diarrhoea and growth depression. Growth may also be impaired by autoimmune diseases. In these cases the individual's own antibodies act against its own tissues. The immunological relationships between the mother and the fetus can have an effect on the growth of the placenta and is a good example of this phenomenon. These relationships are manifest in the placenta weight exhibiting a proportional relationship with pregnancy number, the hypothesis here being one of a progressive increase in sensitization to antigens from the fetus.

Corticosteroids can also cause lymphoid tissue to atrophy and lymphocytes to become depleted, and somatotrophin is known to have a direct effect on the former. As pointed out previously in this chapter, corticosteroids stimulate protein catabolism and cause growth depressions and, possibly, muscle atrophy.

Thus the direct effect of corticosteroids in reducing growth rate may be accentuated if the animal is challenged with disease because the immune system is deleteriously affected as well. Because natural stresses in animals such as starvation, overcrowding and transportation can increase corticosteroid secretion, the animal may, through either direct or indirect effects, grow more slowly. The possible link between the endocrine and the immune systems is discussed next.

4.4.3. Endocrine and immune system interactions

From evidence that is fast accumulating from research carried out principally in the human field, it is becoming increasingly apparent that certain products of the endocrine system, in particular but not exclusively growth hormone, can have a major effect on the immune system. Therefore a better understanding of some growth processes might emerge in the future when the so-called 'crosstalk' between the two systems is interpreted further.

A positive effect of growth hormone in primary mononuclear phagocytes has been established by Edwards *et al.* (1988). *In vitro* work found that macrophages derived from alveolar tissue and from the blood could be activated by growth hormone to produce reactive oxygen intermediaries such as superoxide anion (O_2^-). Superoxide anion and other intermediaries are responsible for the killing of pathogenic microbes and it was found that *in vitro* treatment of hypophysectomized rats with physiological doses of either native or recombinantly derived porcine growth hormone primed peritoneal macrophages at the same time as inducing proportional increases in growth rate of between 0.10 and 0.40. This is probably the best demonstration so far of a physiological role for growth hormone in immunoregulation but the demonstration of a dual but simultaneous effect on growth and immunological events by no means exhausts the possibilities of the effects which growth hormone might have on the immune system. Kelly (1989) points to other possibilities which research work has shown glimpses of so far but which may gel as research work progresses. The full list of possibilities is given in Table 4.11.

Kelly (1989) suggests that of all the possibilities three lines of evidence have the strongest arguments for a physiological role for growth hormone in immunoregulation:

1. Because growth hormone has been shown to be synthesized by lymphoid cells there is a strong possibility that it may be synthesized locally in regional lymph nodes. The same may be true for prolactin. If this proves to be the case then the quantity of growth hormone in the serum will not be indicative of the immune response that might be expected in the animal.

2. Because a fraction of one of the thymic hormones stimulates the release of both growth hormone and prolactin from the pituitary, it is possible that growth hormone can itself stimulate not only the synthesis of some thymic hormones but also the size of the gland itself.

3. Because endotoxin is a potent stimulus in the release of growth hormone but an equally potent inhibitor to the release of prolactin, it is possible that a product derived from macrophages alters the release of pituitary hormones, in so doing alters the ratio of growth hormone to prolactin and thereby affects immunoregulation.

Further intriguing links between growth hormone and the immune system are indicated by Kelly in what may be described as 'abnormal growth

Table 4.11. Regulation by growth hormone of the activities of cells of the immune system (based on Kelly, 1989).

Growth hormone deficiencies and immunoregulation:
    Thymic atrophy and wasting in mice and dogs
    Reduced antibody synthesis in mice
    Delayed skin graft rejection in mice
    Normal lymphoid cell subsets and thymic histology with reduction in peripheral T and B
        cells
    Pituitary hypoplasia and thymic atrophy in humans
    X-linked growth hormone deficiency and complete inability to synthesize antibodies
    Reduction in activity of natural killer cells in humans
    Defective allogeneic mixed lymphocyte reaction
    Reduction in plasma thymulin in humans and mice
    Normal immunoglobulin concentrations and lymphoid cell subsets in humans
    Decreased insulin-induced growth hormone response in patients with telangiectasis and
        bowel disease
Growth hormone and the thymus gland:
    Increases thymic size and DNA synthesis in young rodents
    Improves thymic size and morphology in aged animals
    Increases plasma thymulin in humans and dogs
Growth hormone and lymphoid cells:
    Lymphocytes have receptors for growth hormone
    Augments antibody synthesis and reduces skin graft survival *in vivo*
    Increases lectin-induced T-cell proliferation and IL-2 synthesis *in vivo*
    Stimulates proliferation of human lymphoblastoid cells
    Augments basal lymphocyte proliferation *in vitro*
    Increases activity of cytotoxic T lymphocytes *in vitro*
    Augments activity of natural killer cells *in vivo*
    Synthesized by lymphoid cells
Growth hormone and phagocytic cells:
    Primes macrophages for superoxide anion release *in vitro* and *in vivo*
    Augments respiratory burst in neutrophils from growth hormone-deficient patients *in
        vivo*
    Increases basal respiratory burst of human neutrophils and inhibits activated burst *in
        vitro*
Growth hormone and haemopoiesis:
    Augments neutrophil differentiation *in vitro*
    Augments erythropoiesis

situations'. For example, immunoreactive growth hormone has been found in prostatic tumours and growth hormone levels have been found to be elevated in humans with cancer.

There can be no doubt from this short discourse that this area of research holds immense fascination for those interested in growth processes in animals. The future is awaited with great expectations but application of findings to practical situations may be very difficult to realize, if not totally impossible, in some circumstances.

4.4.4. The immune system and manipulation of endocrine function

The principle that underlines the manipulation of endocrine function is one of neutralizing or partially inhibiting the activity of a specific endogenous hormone. The possibility that many of the endogenous hormones act through homeorhetic mechanisms has been discussed earlier in this chapter and the physiological function of the hormone concerned will dictate the extent to which either immuneutralization or inhibition is effective in giving the desired response. Two areas have received most attention; first that of immunization against somatostatin and secondly that of immunization against gonadotrophin releasing hormone.

Having read the earlier parts of this chapter the reader should be aware of the great importance of pituitary growth hormone in influencing postnatal growth and of the inhibitory effect that somatostatin has on its secretion. Also the inhibitory effects of somatostatin do not stop with growth hormone, as it has inhibitory effects on the release of the other hormones which have a major effect on growth and metabolism, that is, insulin, glucagon and thyroid stimulating hormone. Autoimmunization against somatostin has been studied most in ruminant animals (Spencer *et al.*, 1983a, b, 1985). In these experiments immunization against somatostatin was effected early in the life of lambs so that the responses were measured over a period of high natural growth potential. Dutch Moor sheep and Suffolk × Scottish Half-bred sheep were used. The significant improvements in growth rate found were without apparent significant effects on the proportions of the major tissues of the carcass but were more pronounced in the Dutch Moor lambs than they were in the lambs from cross-bred ewes. An intriguing response was that of the increase in long bone length from the autoimmunization procedure (Fig. 4.8). These and other results hold fascination for the future but the autoimmunization procedures will need much refining and developing before they constitute a proposition worth considering in practice for farm animals.

The immensely powerful anabolic effects of testosterone in the male animal have received appropriate attention already in this chapter. Economically speaking these effects have extremely important implications for all species of farm animals, but in cattle the uncastrated animal creates problems because of the secondary sexual characteristics which develop and which cause management difficulties. Surgical castration is obviously not the answer to these

Fig. 4.8. Autoimmunized (left) and control sheep (right) (by courtesy of Dr G.S.G. Spencer).

problems as the removal of the testes will remove the source of the desirable anabolic steroid. Active immunization against releasing hormone can deprive the testes of the follicle stimulating and lutenizing hormones which are essential for folliculogenesis and spermatogenesis and is an attractive alternative for modifying the undesirable male characteristics without loss of growth potential. This allows the anabolic effects of the testes to be retained as long as possible, offers at the same time a humane alternative to surgical castration for manipulating growth and presents the facility of being able to rear bulls in systems incorporating a period of grazing. The results presented in Table 4.12 show the type of response that may be expected. The drawbacks at the moment revolve around the variation in responses between individuals and the uncertainty that a predictable response can be guaranteed for a period of 4–6 months from an initial one or two injections of the vaccine (Robertson *et al.*, 1984). Nevertheless, the potential for greater productivity from the autoimmunized animal, compared with the castrated animal, is apparent in the improved growth and food conversion efficiency responses and increased yields of lean meat.

Table 4.12. Effects of neutralizing lutenizing hormone releasing hormone on the performance of bulls, compared with steers. Mean values from 10 Friesian bull calves immunized at 28 weeks of age and compared with 10 Friesian steers (Robertson *et al.*, 1982).

	Immunized bulls	Steers
Weight at first injection (kg)	246.5	254.2
Weight at slaughter (kg)	630.6	629.2
Growth rate (kg day^{-1})	0.910	0.810
Dry matter conversion ratio (kg food kg gain^{-1})	6.88	7.85
Killing-out proportion	0.557	0.546
Cannon bone length (cm)	22.5	23.5
Tenth rib cut		
Lean proportion	0.556	0.451
Fat proportion	0.265	0.390
Bone proportion	0.154	0.143
Eye muscle area (cm^2)	76.0	50.6

References

Allen, D.M. and Kilkenny, B. (1980) *Planned Beef Production*. Granada, London.

Baile, C.A., Della-Fera, M.A. and McLaughlin, C.L. (1983) *Proceedings of the Nutrition Society* 42, 113–127.

Bauman, D.E., Eisemann, J.H. and Currie, W.B. (1982) *Federation Proceedings. Federation of American Societies for Experimental Biology* 41, 2538–2544.

Blair, E.L. (1983) *Proceedings of the Nutrition Society* 42, 103–111.

Blasco, A. and Webb, A.J. (1989) *Animal Production* 49, 117–122.

Bowman, J.C. (1974) *An Introduction to Animal Breeding*. Edward Arnold, London.

Brinklow, B.R. and Forbes, J.M. (1984) Effect of extended photoperiod on the growth of sheep. In: Roche, J.F. and O'Callaghan D. (eds) *Manipulation of Growth in Farm Animals*, Martinus Nijhoff, The Hague, pp. 260–273.

Bulfield, G.M. (1980) The biochemical determinants of selection for growth. In: Lawrence, T.L.J. (ed.) *Growth in Animals*. Butterworths, London, pp. 11–24.

Buttery, P.J. (1983) *Proceedings of the Nutrition Society* 42, 137–148.

Croston, D. and Pollott, G. (1985) *Planned Sheep Production*. Collins, London.

Dalton, D.C. (1980) *An Introduction to Practical Animal Breeding*. Granada, London.

Darwin, C. (1871) *The Descent of Man, and Selection in Relation to Sex*. John Murray, London.

Davis, S.L., Ohlson, D.L., Klindt, J. and Everson, D.O. (1979) *Journal of Animal Science* 49, 724–728.

Davis, S.L., Hossner, K.L. and Ohlson, D.L. (1984) Endocrine regulation of growth in ruminants. In: Roche, J.F. and O'Callaghan, D. (eds) *Manipulation of Growth in Farm Animals*. Martinus Nijhoff, The Hague, pp. 151–178.

Dockray, G.J., Sharkey, K.A. and Bu'lock, A.J. (1984) Peptides as intergrators of gastrointestinal function. In: Batt, R.M. and Lawrence, T.L.J. (eds) *Function and Dysfunction in the Small Intestine*, Liverpool University Press, Liverpool, pp. 39–54.

Dodson, M.V., Davis, S.L., Ohlson, D.L. and Ercanbrock, S.K. (1983) *Journal of Animal Science* 57, 338–342.

Edwards, C.K., Ghiasuddin, S.M., Schepper, J.M., Yunger, L.M. and Kelly, K.W. (1988) *Science* 229, 769–771.

Falconer, J.S. (1981) Somatomedin studies in animals. In: Spencer, G.S.G. (ed.) *Proceedings of a Colloquium on Somatomedins*. Somatomedins Club, Bristol, pp. 46–49.

Falconer, J., Forbes, J.M., Bines, J.H., Roy, J.H.B. and Hart, I.C. (1980) *Journal of Endocrinology* 72, 30P.

Fowler, V.R., McWilliam, R. and Aitken, R. (1981) *Animal Production* 32, 357 (abstract).

Fujii, J., Otsu, K., Zorato, F., DeLeon, S., Khanna, V.K., Weiler, J.E., O'Brien, P. and MacLennan, D.H. (1991) *Science, Washington* 253, 448–451.

Gregory, N.G. (1977) A physiological approach to some problems in meat production. PhD Thesis, University of Bristol.

Gregory, N.G., Truscott, T.G. and Wood, J.D. (1980) *Proceedings of the Nutrition Society* 39, 7A.

Halliday, R. (1976) *Research in Veterinary Science* 21, 331–334.

Halliday, R. (1980) Interrelationships between immunity and growth. In: Lawrence, T.L.J. (ed.) *Growth in Animals*. Butterworths, London, pp. 65–82.

Hart, I.C. and Johnsson, I.D. (1986) Growth hormone and growth in meat producing animals. In: Buttery, P.J., Haynes, N.B. and Lindsay, D.B. (eds) *Control and Manipulation of Animal Growth*. Butterworths, London, pp. 135–159.

Heitzman, R.J. (1981) Mode of action of anabolic agents. In: Forbes, J.M. and Lomax, M.A. (eds) *Hormones and Metabolism in Ruminants*. Agricultural Research Council, London, pp. 129–138.

Johnsson, I.D. and Hart, I.C. (1986) Manipulation of milk yield with growth hormone. In: Haresign, W. and Cole, D.J.A. (eds) *Recent Advances in Animal Nutrition. 1986*. Butterworths, London, pp. 105–123.

Keller, D.G., Smith, V.G., Coulter, G.H. and King, G.J. (1979) *Canadian Journal of Animal Science* 59, 367–373.

Kelly, F.J. and Goldspink, D.F. (1982) *Biochemical Journal* 208, 147–151.

Kelly, K.W. (1989) *Biochemical Pharmacology* 38, 705–713.

Kennedy, G.C. (1967). Ontogeny of mechanisms controlling food and water intake. In: *Handbook of Physiology. Section 6, Alimentary canal, Vol. 1 Control of food and water intake*. American Physiological Society, Bethesda, p. 337.

Kostyo, J.L and Isaksson, O. (1977) Growth hormone and the regulation of somatic growth. In: Greep, R.O. (ed.) *International Review of Physiology. Reproductive Physiology II*, vol. 13. University Park Press, Baltimore, pp. 255–274.

Lasley, J.F. (1978) *Genetics of Livestock Improvement*, 3rd edn. Prentice-Hall, New Jersey.

Lister, D. (1976) *Proceedings of the Nutrition Society* 35, 351–356.

Meat and Livestock Commission (1978) *Commercial Pig Evaluation. Fourth Test Report*. Meat and Livestock Commission, Bletchley, UK.

Mickelson, J.R., Knudson, C.M., Kennedy, C.F., Young, D.I., Litterer, L.A., Rempel, W.E., Campbell, K.P. and Louis, C.F. (1992) *FEBS Letters* 301, 49–52.

Mossberg, I. and Jonsson, H. (1996) *Animal Science* 62, 233–240.

National Sheep Association (1982) *British Sheep*, 6th edn. National Sheep Association, Horseheath, Cambridgeshire.

Oddy, V.H. and Lindsay, D.B. (1986) Metabolic and hormonal interactions and their potential effects on growth. In: Buttery, P.J., Lindsay, D.B. and Haynes, N.B. (eds)

Control and Manipulation of Growth. Butterworths, London, pp. 231–248.

Odedra, B.R. and Millward, D.J. (1982) *Biochemical Journal* 204, 663–672.

Pell, J.M. and Bates, P.C. (1990) *Nutrition Research Reviews* 3, 163–192.

Petitclerc, D., Chapin, L.T. and Tucker, H.A. (1984) *Journal of Animal Science* 58, 913–919.

Porter, P. (1976) *Proceedings of the Nutrition Society* 35, 273–282.

Porter, P., Kenworthy, R. and Thompson, I. (1975) *Veterinary Record* 97, 24–28.

Prior, R.L. and Smith, S.B. (1982) *Federation Proceedings. Federation of American Societies for Experimental Biology*, 41, 2545.

Rhodes, D.N. (1969) What do we want from the carcass. In: Lister, D. Rhodes, D.N., Fowler, V.R. and Fuller, M.F. (eds) *Meat Animals; growth and productivity*. Plenum, London, pp. 9–24.

Robertson, I.S., Fraser, H.M., Innes, G.M. and Jones, A.S. (1982) *Veterinary Record* 111, 529–531.

Robertson, I.S., Wilson, J.C., Fraser, H.M., Innes, G.M. and Jones, A.S. (1984) Immunological castration of young bulls for beef production. In: Roche, J.F. and O'Callaghan, D. (eds) *Manipulation of Growth in Farm Animals*. Martinus Nijoff, The Hague, pp. 137–145.

Scanes, C.G. (1984) Growth hormone in domestic animals. In: *Proceedings Monsanto Technical Symposium, Fresno, California, March 1984*, pp. 35–52.

Schanbacher, B.D. (1984) Hormonal and photoperiodic control of growth. In: Roche, J.F. and O'Callaghan, D. (eds) *Manipulation of Growth in Farm Animals*. Martinus Nijhoff, The Hague, pp. 275–286.

Seidman, S.C., Cross, H.R., Oltjen, R.R. and Schanbaher, B.D. (1982) *Journal of Animal Science* 55, 826–840.

Sharpe, P.M., Haynes, N.B. and Buttery, P.J. (1986) Glucocorticoid status and growth. In: Buttery, P.J., Haynes, N.B. and Lindsay, D.B. (eds) *Control and Manipulation of Animal Growth*. Butterworths, London, pp. 207–222.

Short, R.V. (1980) The hormonal control of growth at puberty. In: Lawrence, T.L.J. (ed.) *Growth in Animals*. Butterworths, London, pp. 25–46.

Simpson, S.P., Webb, A.J. and Wilmut, I. (1986) *Animal Production* 43, 485–492.

Spencer, G.S.G. (1985) *Livestock Production Science* 12, 31–46.

Spencer, G.S.G., Garssen, G.J. and Bergstrom, P.L. (1983a), *Livestock Production Science* 10, 469–478.

Spencer, G.S.G., Garssen G.J. and Hart, I.C. (1983b) *Livestock Production Science* 10, 25–38.

Spencer, G.S.G., Hallett, K.G. and Fadlalla, A.M. (1985) *Livestock Production Science* 13, 43–52.

Terqui, M., Delouis, C. and Ortovant, R. (1984) Photoperiodism and hormones in sheep and goats. In: Roche, J.F. and O'Callaghan, D. (eds) *Manipulation of Growth in Farm Animals*. Martinus Nijhoff, The Hague, pp. 246–257.

Trenkle, A. (1977) *Growth* 41, 241–247.

Trenkle, A. and Topel, D.G. (1978) *Journal of Animal Science* 46, 1604–1609.

Wangsness, P.J., Olsen, R.F. and Martin, R.J. (1981) *Journal of Animal Science* 52, 57–62.

Wangsness, P.J., Martin, R.J. and Gahagan, J.M. (1977) *American Journal of Physiology* 233, E104.

Warwick, J. and Legates, J.M. (1979) *Breeding and Improvement of Farm Animals*, TMH edn. McGraw-Hill, New York.

Webb, A.J. (1980) *Animal Production* 31, 101–106.

Webb, A.J. and Simpson, S.P. (1986) *Animal Production* 43, 493–504.

Weekes, T.E.C. (1983) *Proceedings of the Nutrition Society* 42, 129–136.

Weekes, T.E.C. (1986). Insulin and growth. In: Buttery, P.J., Haynes, N.B. and Lindsay, D.B. (eds) *Control and Manipulation of Animal Growth*, Butterworths, London, pp. 187–206.

Whitemore, C.T., Henderson, R., Ellis, M., Smith, W.C., Laird, R. and Wood, J.D. (1982) *Animal Production* 34, 380–381.

Woolliams, J.A., Angus, K.D. and Wilson, S.B. (1993) *Animal Production* 56, 1–8.

Zinn, S.A., Chapin, L.T. and Tucker, H.A. (1986a) *Journal of Animal Science* 62, 1273.

Zinn, S.A., Purches, R.W., Chapin, L.T., Petitclerc, D., Merkel, R.A., Bergen, W.G. and Tucker, H.A. (1986b) *Journal of Animal Science* 63, 1804.

Zinn, S.A., Chapin, L.T. and Tucker, H.A. (1988) *Animal Production* 46,300–303.

Gametes, Fertilization and Embryonic Growth

5.1. Introduction

The primary aim of this chapter is to consider the events immediately before, at and subsequent to fertilization, up to the point where the various organs and systems have differentiated and when the male and female can be identified first by the shape and position of their external genitalia. Such events include the stages of the terminal development of gametes and their activation, the act of fertilization and the resultant zygote formation and attachment and the early development of the embryo. The embryonic period is followed by a phase, often referred to as the fetal phase, in which the differentiated tissues, parts and organs exhibit increasingly rapid but differential growth rates up to the point of parturition (see chapter 6). Really there is no division between the two phases: one slips imperceptibly into the other. In considering the events of the embryonic stage the fertilized ovum is therefore considered as an embryo up to the point where differentiation of the organ systems and parts of the body can be identified and where growth is about to commence in these differentiated organs, parts and tissues. Therefore the multiplying stages of cell division and the morula, blastocyst, conceptus and early postconceptus stages are within the overall consideration. Broadly speaking the end of the embryonic period may be regarded as about 35 days postfertilization in both sheep and in pigs and about 45 days postfertilization in cattle.

5.2. Meiosis, Gametes and Fertilization

5.2.1. Introduction

In chapter 2 the division of single cells by the process of mitosis, to give exact replicates of themselves, was discussed. The continuity of animal life is not, however, dependent on this process as much as on the process known as meiosis or reduction division, wherein male and female gametes each contribute half of the chromosomes to a new being or zygote and therein maintain chromosomal constancy from one generation to the next. Compared with mitosis, in which all chromosomes in body cells duplicate themselves by a longitudinal division to produce daughter cells, each containing the same number of chromosomes as the parent cells, meiosis differs in several respects. First, it takes place in germ cells, that is in ova and spermatozoa, rather than in body cells. Secondly, it takes place during gametogenesis, that is in the formative stages of ova and spermatozoa development and reduces the somatic or diploid number of chromosomes to the haploid state. Therefore, in contrast to mitosis, each daughter germ cell receives only one member of each chromosome pair. This is essential for survival of the species in that if the mechanism of reduction division did not exist, chromosome numbers would increase geometrically, from generation to generation, with successive fertilization of ova, to the point where reproduction would cease because of the huge number of chromosomes produced. Thirdly, the process of meiosis not only reduces by half the somatic or diploid number of chromosomes but it also increases the genetic variability in the offspring by homologous pairs of chromosomes (homologous chromosomes are similar chromosomes, one contributed by the male and one by the female) 'crossing over' to give two new chromosomes, each different from its parent.

5.2.2. Meiosis and gametogenesis

An understanding of meiosis is not possible unless the process of gametogenesis is understood in the first place. In this process each gamete ultimately formed contains the diploid ($2n$, where n equals the numbers of pairs of chromosomes) complement of chromosomes, including the two sex chromosomes, XX in the ova of the female and XY in the spermatozoa of the male. The diploid and haploid complements of chromosomes for some of the most common species of domesticated animals, some other animals and of humans are given in Table 5.1, and the process of gametogenesis is represented diagrammatically in Fig. 5.1. It is clear that during the process of fertilization in mammals the genetic sex of the future embryo is fixed. The female of the species has an XX complement of chromosomes and her ova carry X chromosones only. In contrast, the male has an XY complement of chromosomes and his spermatozoa can carry either the X or the Y chromosome. Therefore the genetic sex of the embryo is determined by the chromosomes introduced by the fertilizing spermatozoon. Spermatozoa containing X and Y chromosomes are produced in equal numbers and both types are competent to fertilize ova. It follows that there is an equivalent chance of male and female embryos forming at fertilization and in the very young embryo there is a latent potential for sexual differentiation to proceed to give ultimately either a male or a female because of the existence of two sets of primitive genital tracts – the Wolfian and Mullerian ducts. If development of the former dominates, a male reproductive system emerges; if development of the latter dominates, a female reproductive system emerges. The

Table 5.1. Diploid and haploid chromosomal complements of some domesticated and other animals and of humans.

	Diploid (2*n*)	Haploid (*n*)
Domestic horse (*Equus caballus*)	64	32
Mongolian wild horse (*Equus przewalskii*)	66	33
Persian wild ass, onager (*Equus hemionus*)	56	28
Donkey (*Equus asinus*)	62	31
European domestic cattle (*Bos taurus*)	60	30
Zebu domestic cattle (*Bos indicus*)	60	30
American bison (*Bison bison*)	60	30
Domestic buffalo (*Bubalus bubalus*)	48	24
Musk ox (*Ovibus moschatus*)	48	24
Reindeer (*Rangifer tarandus*)	70	35
Domestic pig (*Sus scrofa*)	38	19
European wild pig (*Sus scrofa*)	36	18
Domestic sheep (*Ovis aries*)	54	27
Domestic goat (*Capra hircus*)	60	30
Domestic fowl (*Gallus domesticus*)*	78	39
Domestic rabbit (*Oryctolagus cuniculus*)	44	22
Mouse (*Mus musculus*)	40	20
Rat (*Rattus norvegicus*)	42	21
Dog (*Canis familiaris*)	78	39
Cat (*Felis catus*)	38	19
Man (*Homo sapiens*)	46	23

*The sex chromosomes are very small and it is therefore difficult to locate the centromere attachment in the metaphase stage of division.

influence of the sex chromosomes normally permits the development of one set of ducts and keeps the other in a vestigal form. However, the mechanism is not necessarily the simple result of the presence of two X chromosomes in the embryo. Rather it is a permissive state in the absence of a Y chromosome. The male condition is dominant in this situation and in the absence of both male sex hormones (androgens) and a Y chromosome, the Mullerian ducts develop predominantly at the expense of the rapidly regressing Wolfian ducts. However, irrespective of this, the fertilized egg's complement of chromosomes is restored to the diploid state as a result of the fusion of the ovum with the spermatozoon, each with its haploid chromosomal complement.

There is clearly one major difference between oogenesis and spermatogenesis in that in the former the reduction division produces polar bodies as well as ova, whereas in the latter each half produces a spermatozoon. This is because in oogenesis the longitudinal split of the chromosomes to give a pair of chromatids, leaving the egg with single chromosomal threads, is associated with equal nuclear divisions but unequal cytoplasmic division, the proportion of cytoplasm not required for the formation of the ovum being thrown out of the cell as a non-functional polar body. In spermatogenesis there is no comparable wastage.

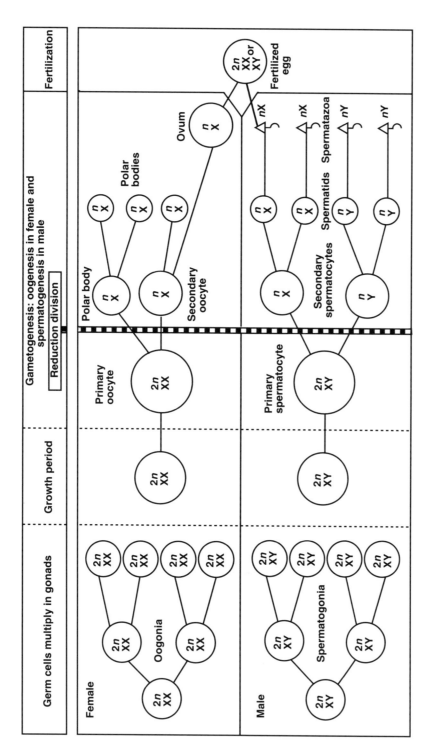

Fig. 5.1. Diagrammatic representation in mammals of oogenesis and spermatogenesis leading to the point of fertilization of ova by spermatozoa and to the establishment of the diploid complement of chromosomes in, and to the determination of the genetic sex of, the newly formed zygote. XX chromosomes present in the female and XY chromosomes present in the male, $2n$ represents the diploid complement of chromosomes and n represents the haploid complement.

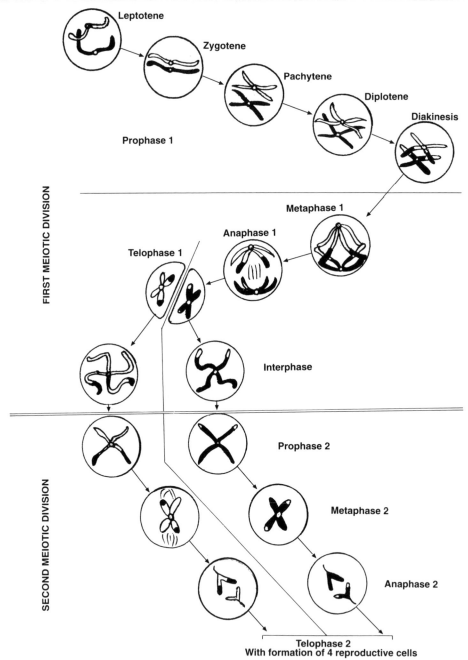

Fig. **5.2.** Stages of meiosis. *First meiotic division* - Prophase 1: leptotene, cell nuclei contain chromosomes in the form of very fine single threads; zygotene, pairing of matching or homologous chromosomes, one chromosome from each of the haploid sets derived from each parent; pachytene, two chromosomes shorten and thicken; diplotene, separation of homologous chromosomes begins revealing two daughter chromatids formed from each chromosome; diakinesis, chromosomes are shorter and thicker and

The various stages of meiosis or reduction division are shown in Fig. 5.2. It is apparent that the process consists of two cell divisions associated with the separate events of duplication and division. During both divisions the cell goes through the same stages as in mitosis: interphase (the interval between cycles where DNA replication occurs), prophase, metaphase, anaphase and telophase. The prophase stage of the first meiotic division is considerably longer than the comparable stage in mitosis but the main difference, compared with mitosis, is that the make-up of the chromosomes is changed beforehand by the chromatids crossing-over and exchanging similar areas (e.g. one end, middle portion or both ends) to give the two new chromosomes. Also, in anaphase of the first meiotic division, the chromatids do not separate before moving to each pole as happens in mitosis. As a result, when the cell divides in telophase, each daughter cell will have half as many chromosomes (n) as the parent cell ($2n$). In the second meiotic division which follows, the two daughter cells are simply duplicated to form four, much as in mitosis, but each chromatid separates into two chromosomes, one of which migrates to each pole to maintain the haploid number (n) in each of the new sex cells or gametes which contain also, relative to most other normal tissue cells of the body, reduced amounts of DNA. Fertilization then restores the original $2n$ number of chromosomes.

In the germ cells of female animals the first meiotic divisions occur at the primary oocyte stage with the result that the secondary oocytes obtain their haploid set of chromosomes just before they are realeased from the ovary. In most species meiotic prophase is completed by the diplotene stage shortly after birth and thereafter the cell enlarges greatly and the oocyte enters a long period of rest which finishes just before ovulation with preovulatory changes in the Graafian follicle. However, there are species differences: by the time of birth the ovaries of the ewe, the cow and the woman contain mainly oocytes which have reached the diplotene stage of meiosis, whilst in other species, for example the rabbit and the mink, the ovaries contain oogonia only and the prophase of meiosis is completed within the first few weeks following birth. In the pig the process of oogenesis occurs mostly in fetal life but extends into the postnatal period. In the cat the meiotic prophase may be found at any point up to the time of puberty.

Fig. 5.2. *contd* move further apart; metaphase 1, starts when the nucleoli and nuclear membranes have disappeared and the spindle formation is evident as in mitosis; anaphase 1, chromosomes move further apart; telophase 1, nuclear membrane forms around haploid set of chromosomes at each pole and cell divides into two daughter cells; interphase, a brief phase accentuating the events of telophase 1 and leading to the second meiotic division. *Second meiotic division* – this is similar to mitosis in so far as two chromatids join at a single centromere but different in that only half the normal number of chromosomes is present. Anaphase 2, centromeres divide and daughter chromosomes move to opposite poles; telophase 2, nuclear membranes form around four haploid nuclei which have originated from original diploid parent cell.

In spermatogenesis the meiotic prophase is the same as in oogenesis, but thereafter it differs in several respects. It does not begin until around the time of puberty but once it has started it then continues in an uninterrupted manner throughout adult life. There is no period of arrested development and diakinesis follows on immediately from the diplotene stage.

5.2.3. Gametes and fertilization

Ova

Shed ova retain their covering layers of follicle cells and the gelatinous material in which they were previously embedded. They are transported along the oviduct, in currents of secretory fluids which are kept moving by large numbers of cilia, to the point where they finally meet spermatozoa and are fertilized. The jelly-like substance of the ovum's outer layer is known as the cumulus oophorus. Immediately beneath this, the thin non-cellular layer known as the zona pellucida forms an outer covering to the cytoplasmic and nuclear components but is separated from them by the fluid-filled perivitelline space (Fig. 5.3).

Spermatozoa

Spermatozoa consist of two main regions, the head and the tail. The tail is composed of the middle, principal and end pieces (Fig. 5.3). The neck is the connecting piece between the head and the tail and the region between the neck and the annulus is known as the middle piece and is covered by a sheath of sausage-shaped mitochrondria arranged in a helical pattern around the longitudinal fibres of the tail. It is thought that these mitochrondria generate the energy needed for sperm motility. The major feature of the heads of spermatozoa of bulls, boars, rams, stallions and man are the oval, flattened nuclei containing condensed chromatin, mostly DNA complexed to specific nuclear proteins known as histones. As a result of the reduction divisions which occurred during spermatogenesis, the chromosome numbers and the DNA contents are half those of the somatic cells of the same species. The spermatozoa of all species do not have the characteristic oval and flattened shape referred to above. For example, the sperm head of the cockerel is needle shaped whilst that of the rat is crooked shaped. Irrespective of this, however, the chromatin of the head is protected by an essentially non-porous membrane and the anterior is covered by a double-layered membranous sac known as the acrosome. As discussed below, this is involved in the fertilization process.

Although spermatozoa are mature when they leave the testes they must undergo a final phase of maturation before they can penetrate the membranes of ova, effect fertilization and start new life. The changes of the final phase of maturation take place in the female tract and are known collectively as capacitation. The capacitation process, that is the time required by spermatozoa in the female tract to achieve penetrative competence, takes 4–5 hours in cattle,

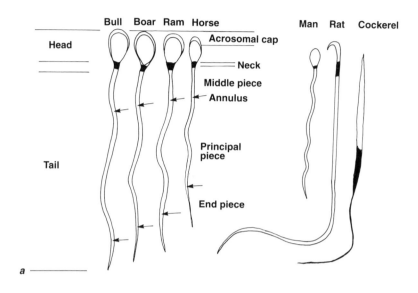

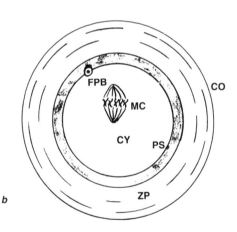

Fig. 5.3. a, Spermatozoa of farm and other animals; relative differences in size and shape are shown. b, Schematic representation of ovum of sow - not to scale relative to spermatozoa. FPB = first polar body; PS = perivitelline space; MC = mitochondria; CY = cytoplasm; ZP = zona pellucida; CO = corona radiata (a based on Garner and Hafez, 1980).

1–1½ hours in sheep and 2–3 hours in pigs. It is characterized by changes occurring in the membranes of the spermatozoa, particularly in the head regions. The term 'acrosome reaction' is used to describe these changes collectively and involves the formation of a number of small apertures in the plasma membrane covering the acrosome. These enable the contents of the acrosome, mostly hyaluronidase and a trypsin-like enzyme, to escape. In escaping from the spermatozoa heads they leave each spermatozoon with the capacity to dissolve the jelly-like substance of the ovum's outer layer, the cumulus oophorus, and thereby allow the spermatazoon to reach the surface of the zona pellucida of the ovum (Fig. 5.4).

Fertilization

New life starts at the moment of fertilization when the zygote is formed. In birds fertilization obviously must take place before laying; in mammals

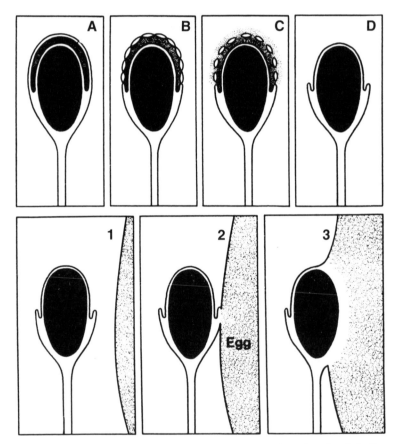

Fig. 5.4. Diagram showing the pattern of the acrosome reaction (above) and the first steps in spermatozoon–ovum fusion (below). A, intact acrosome; B and C, apertures forming in plasma membrane covering the acrosome and leading to the elimination of the outer membrane (D). Spermatozoon arriving at the surface of the ovum (1), penetrating the cumulus oophorus (2) and arriving at the surface of the zona pellucida (3) (reproduced from Austin (1972) by kind permission of the copyright holder, Cambridge University Press).

fertilization takes place within the female and the embryo develops within the uterus. In nearly all mammals the site of fertilization is a wide part of the oviduct or Fallopian tube known as the ampulla.

At coitus, semen is deposited by the penis of the male directly into the vagina of the female in the human, sheep, cattle and rabbits. In other species such as the horse and the pig, the penis deposits semen directly into the uterus by penetrating the cervix. By swimming and with the help of contractions of the female tract, which may be stimulated by the prostaglandins in the semen, spermatozoa finally reach the point of contact with ova or a single ovum depending on the species. Although 'fertile life' is a relative concept because fertility declines progressively over a period of hours, in most mammals spermatozoa and ova as individual entities have relatively short fertile life spans, generally only a few hours and not more than 24 hours in the case of ova (Table 5.2). The variation evident reflects the influence of a number of factors including the hormonal state of the female. The situation with the domestic hen differs in that spermatozoa are conserved in a fertile form in specialized crypts in the wall of the oviduct for about 3 weeks, or slightly more, and fertilize ova one by one as they progress down the oviduct.

As pointed out the capacitation process allows a spermatozoon to reach the surface of the thin non-cellular envelope, known as the zona pellucida, which surrounds the ovum. Once the individual spermatozoon has passed through this latter barrier, for which purpose enzymes appear to be unimportant, it enters the narrow fluid-filled perivitelline space that surrounds the cytoplasmic body and is then in a position to complete the act of germ cell fusion. The spermatozoon attaches itself to the cytoplasmic body of the egg and the contiguous plasma membranes then fuse. This is succeeded by fusion of the two gametes.

An ovum reacts quickly but in several different ways to penetration by a single spermatozoon. Overall the response is one of a vigorous activation,

Table 5.2. Estimates of the normal fertile life span of spermatozoa in the female reproductive tract and of ova from the point of ovulation.

	Normal fertile life span (h)	
	Spermatozoa	Ova
Cattle	30–48*†	10†/20*–12†/24*
Sheep	30–48*†	10†/16*–15†/24*
Pigs	24*†–42†/72*	8–10*†
Horse	72–120*	6–8*
Man	28–48*†	6–24*†

*McLaren, 1980.
†Hunter, 1982.

signifying the initiation of embryonic development. Within this overall activation there is a blocking to polyspermy, the resumption of the previously inhibited second meiotic division and the formation of the egg nucleus. The two nuclei appear very distinctive and the chromosomes appear as diffuse threads of chromatin. In this distinctive form the two nuclei are known as the male and female pronuclei. There follows a very rapid enlargement so that each pronucleus attains a very large size and contains very visible nucleoli varying in number between one and 30, or more. The pronuclei move towards each other and contact is established in the centre of the egg in about 12 hours in mammals. Contact induces further changes heralding the process of syngamy (the fusion of the two nuclei) but without at any time fusion actually taking place. It is first noticeable that the nucleoli diminish in number and decrease in size. Following this the nuclear membranes disappear to give two loose gatherings of chromosomes which then condense, entangle with each other and thus finally unite the genetic material from the male and the female. This signifies, simultaneously, the completion of syngamy, the enactment of the last scene of fertilization and the raising of the curtain on the next scene, the prophase of the first cleavage division (see section 5.3.1). Therefore, via the various stages of the fertilization process, development in the mammal passes from the stage of the zygote to the stage of the embryo. The general course of fertilization is represented in Fig. 5.5.

5.3. Embryonic Development

5.3.1. Cleavage

Mammalian ova are the largest cells in the body but are very small compared with ova from birds. Diameters (μm) of eggs, without their zona pellucida, for cows, ewes, sows and mares are given by Hafez (1980) as varying between 120 and 160, 140 and 185, 120 and 170 and 100 and 180 respectively. Their metabolic rate immediately after fertilization is very sluggish. Within 3 to 4 days of fertilization the metabolic rate has increased dramatically and whilst very young embryos may contain less than 100 cells, most will approximate in average size to those of the adult.

When the ovum is still at the one-cell stage it is characterized, because of its relatively large size, by a low ratio of nuclear to cytoplasmic material. The process of cleavage restores the ratio to that resembling the position in the adult animal. Cleavage is a process of several successive cell divisions which occur without any increase in the total mass of the very young embryo. In some ways there may be a type of negative growth during cleavage in that the total amounts of cellular material may decrease. McLaren (1972) suggests that these decreases may proportionately be about 0.20 in the cow and about 0.40 in the ewe.

The rate at which cleavage proceeds varies both between and within species and among the individual cells, known as blastomeres, of a single embryo. Therefore the initial synchrony of the cleaving embryo quickly disappears. In consequence, after the first cleavage division, in which the division of the cytoplasm gives a two-cell egg, the two- and four-cell stages of cleavage are more often encountered than are the three- and five-cell stages, with the eight-cell stage predominating on the following day. In addition, by this stage of cleavage there is a differential rate of cleavage between the inner and outer cells, with the latter dividing more slowly than the former and more slowly than those in the middle. The synthesis of DNA in the daughter cells succeeds each mitosis during the first cleavage divisions. The approximate times taken to reach various points are given in Table 5.3.

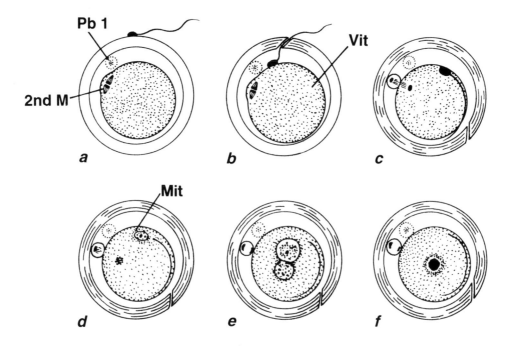

Fig. 5.5. The processes occurring during fertilization in the sow. **a**, Spermatozoon in contact with zona pellucida. The first polar body (Pb 1) has been extruded and the ovum is undergoing the second meiotic division (2nd M). **b**, The spermatozoon has penetrated the zona pellucida and is attached to the vitellus (Vit). This evokes the zona reaction which is illustrated by the shading of the zona pellucida. **c**, The spermatozoon has been taken almost entirely within the vitellus and the head has become markedly swollen but the vitellus has decreased in volume and the second polar body has been extruded. The zona pellucida has rotated relative to the vitellus. **d**, Male and female pronuclei develop. Mitochondria (Mit) gather around the pronuclei. **e**, The pronuclei are fully developed and contain numerous nucleoli. The male pronucleus is larger than the female. **f**, Fertilization is complete. The pronuclei have disappeared and have been replaced by chromosome groups which have united in the prophase of the first cleavage division (reproduced from McLaren (1980) by kind permission of the copyright holder).

Table 5.3. Stages of early embryonic development. Data for 8-cell stage from McLaren (1980); all other data from McLaren (1972) where times are measured from coitus for sheep and pig and from ovulation for cow and horse.

Species	2 cells (h)	4 cells (h)	8 cells (h)	16 cells (h)	Entry to uterus (days)	Blastocyst (days)	Gestation (fertilization to birth) (days)
Cow	27–42	50–83	70	96	3–4	8–9	275–290
Sheep	38–39	42	60	72	2–4	6–7	145–155
Pig	25–51	25–74	60	80–120	2–2.5	5–6	112–115
Horse	24	30–36	72	98–100	4–5	—	335–345

There is no uniformity about the way in which embryos migrate down the oviduct but survival is guarded by their passage being slowed to enable a previously hostile uterine environment to change to one which allows survival. Prior to this, before and around the time of fertilization, the environment was favourable for spermatozoa but not for ova survival, and approximately 24 hours or more have to pass before the environment changes to one of tolerance for the embryo. The embryo remains in the oviduct for 2–2.5 days in the case of the pig, up to 4 days in the case of both cattle and sheep, and up to 5 days in the case of the horse (Table 5.3). These time periods are conditioned by the influence of ovarian steroid hormones on the muscles of the oviduct wall. The corpus luteum produces increasing quantities of progesterone and this induces a progressive relaxation of the muscles, reduces oedema in the mucosa and increases the size of the lumen of the oviduct. However, this effect may be upset by changes in circulating hormones induced by extraneous events such as attempts by man to synchronize oestrus and by the animal ingesting oestrogenic plants.

Compared with mammals, in birds the food reserves of the yolk are of an enormous proportion. But of course there is no expectation of uterine attachment, and nourishment via this attachment, as in the mammal. Ova released from mammalian ovaries have considerable cytoplasmic reserves of yolk which are replaced quickly as a source of nourishment for the young developing embryo by the fluids of the reproductive tract. This situation changes only when the placenta is formed and there is access to the maternal blood supply. Nevertheless, not all elements of the oviduct fluids are available equally at all times and the ability of the embryo to use the oviduct substrate varies with its stage of development. In this particular context the intermediate compounds from Kreb's cycle change rapidly and whilst single-cell embryos appear to be able to incorporate pyruvate and lactate, eight-cell embryos can incorporate malate and glucose. The existence of a dynamic relationship between the embryo and its fluid environment is therefore apparent.

5.3.2. Blastocyst formation and hatching

By the processes of mitotic division embryos continue to develop after entering the uterus. When the embryo has reached the stage of containing 16 or more cells it is termed a morula. Subsequently, individual blastomeres secrete fluid into the intercellular spaces, after which they become arranged around a central fluid-filled space known as the blastocoele. This signifies the changing of the morula to a blastocyst which contains about 60 cells (Fig. 5.6). At the blastocyst stage the group of cells destined to form the embryo proper (the inner cell mass) becomes distinguishable from those that will form embryonic membranes (the trophoblast). The inner cell mass appears as a knob to one side of the central cavity and these cells are the progenitors of the ultimate adult oganism. The trophoblast layer is a single peripheral layer of large flattened cells which are the progenitors of the placenta and the embryonic membranes (Fig. 5.7). The

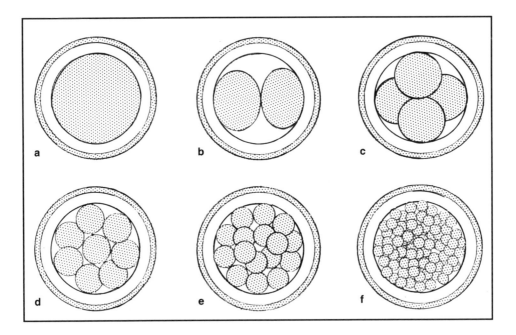

Fig. 5.6. The first stages of development of the embryo from the newly fertilized single-cell egg (a) through the 2- (b), 4- (c), 8- (d) and 16-cell (e) stages to that of a blastocyst with a fluid-filled cavity (f). Note that the zona pellucida still surrounds the embryo.

general view held is that the differentiation into trophoblast and inner cell mass is the most mysterious aspect of development in its entirety, from the single fertilized egg to the adult with its bulk of complex tissues and organs. The times taken to reach the blastocyst stage are given in Table 5.3.

Differentiation into trophoblast and inner cell mass is followed by shedding of the protective zona pellucida. The shedding process is known as hatching and postfertilization occurs between days 9 and 11 in cattle, days 7 and 8 in sheep and days 6 and 7 in pigs (Fig. 5.8). At hatching blastocysts contain about 175–180 cells and in the pig there is an increase to about 500 cells in the first posthatching day and to about 6000 cells over the next 2–3 days. Up to the hatching stage blastocysts of all mammals resemble each other closely. Thereafter, however, there are considerable differences in development. Overall a roughly spherical ball elongates to some 6–7 cm in the horse and achieves lengths of up to 20 cm in cattle and sheep before attachment within the uterus. In pigs there is a grossly accentuated elongation between the 8th and 12th day postfertilization with the result that the blastocyst attains the form of a very long, thread-like, zig-zag tube of up to 1 m in length (McLaren, 1980) before it becomes attached to the uterine wall (Fig. 5.9). The embryo in its newly assumed elongated form is often referred to as the conceptus.

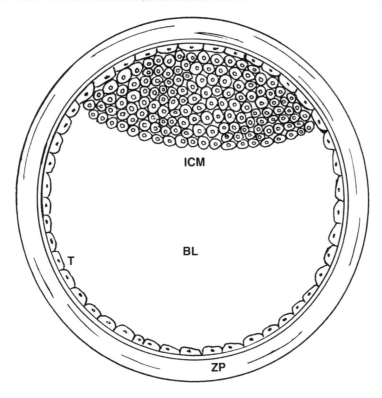

Fig. 5.7. Spheroidal blastocyst of pig immediately before hatching showing the inner cell mass (ICM) which will develop into the embryo and the single layer of flattened cells forming the trophoblast (T) (the future embryonic membranes) surrounding the fluid-filled blastocoele (BL). The zona pellucida (ZP) eventually disintegrates as the blastocyst expands. This is the stage at which splitting into two to give two identical embryos may be carried out.

5.3.3. Gastrulation and tubulation

Gastrulation is the stage of embryonic development which succeeds the formation of the blastocyst and which takes developmental processes, in a period of 4–6 days, a little further towards the point where some mature characteristics first become discernible. The process is essentially similar in both mammals and birds and consists of the movement of cells such that the embryo is converted from a two- to a three-layered structure and the future organ-forming regions are brought into their definitive positions. In mammals gastrulation involves cells of the embryonic disc only. The formation of the three germ layers of the embryo, the ectoderm, the mesoderm and the endoderm, from the original ball of cells, involves the processes of cell division, migration and induction whereby the juxtaposition of sheets of cells allows differentiation via cellular interactions.

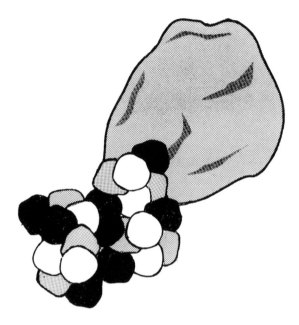

Fig. 5.8. Pig embryo at the blastocyst stage of hatching from the zona pellucida by expansion, due to active pumping of water into the blastocyst cavity - the blastocoele.

Embryos at the start of this period have the shape of an elliptical disc (Fig. 5.10). The disc has an elongated mark, known as the primitive streak, coincidental with its major axis but restricted to one end. This corresponds to the caudal or tail end, and the opposite end to the head or cranial end, of the future fetus and therefore of the future adult animal. In a period of 2–3 days three types of germ cell tissue differentiate, the ectoderm, the mesoderm and the endoderm, and from these all the embryonic membranes and fetal tissues develop (Fig. 5.11).

The notochord and mesoderm are formed by invagination of the cells in the region of the primitive streak (Fig. 5.12). The ectoderm above the notochord forms a groove which becomes the neural tube and then, eventually, the central nervous system. The endoderm or inner skin grows into the blastocoele to form the primitive gut whilst the mesoderm spreads and differentiates in a number of different ways. Future tissue and organ development relative to these three basic germ cell tissues is shown in Table 5.4. These changes, taking a further 2–3 days, transform the flattened disc embryo, with its three types of germ cell tissue, into a tubular embryo. The collective forces which effect this change are known as the tubulation forces and the tubulation process dramatically changes the flattened disc into an essentially cylindrical body, itself containing tubular

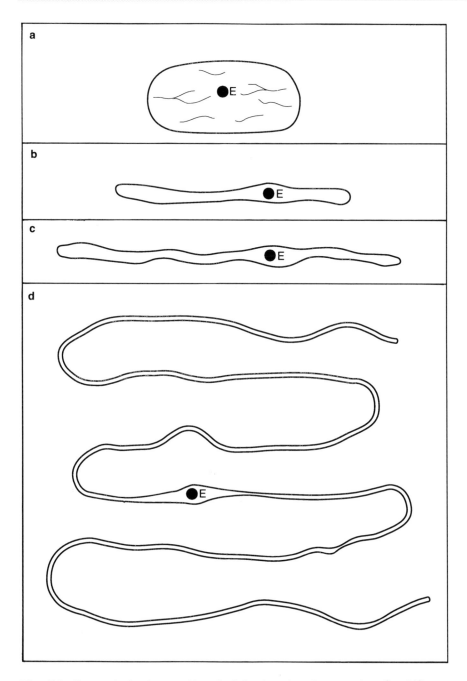

Fig. 5.9. Changes in the shape and length of the pig embryo between about 8 and 12 days postfertilization after hatching from the zona pellucida. a represents a roughly spherical embryo; b and c represent a progressively elongating trophoblast; d represents the thin, elongated zig-zag final form of embryo. E = area of embryonic disc.

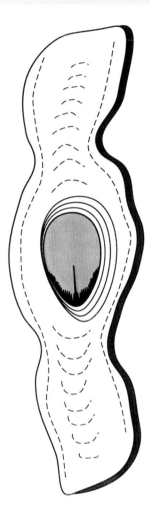

Fig. 5.10. Central portion of the filamentous blastocyst of the pig at about 10 days postfertilization. The embryonic disc is centrally positioned and the thin, straight, black line on it represents the primitive streak.

structures such as the primordial gut and the neural tube (Fig. 5.12). In the pig, in this form at 15 days postfertilization the tubular heart may show spasmodic contractions and start, in an erratic manner, to pump blood in a cranial direction. At this stage the coelomic cavity is not divided but at a later date will become partitioned to form the pericardial, pleural and peritoneal cavities to house the heart, the lungs and the intestinal tract, respectively.

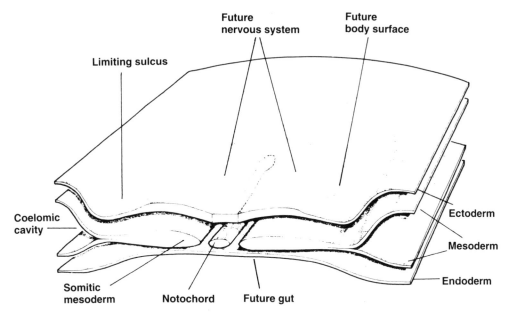

Fig. 5.11. Flat embryo of the pig at the beginning of tubulation at about 12 days postfertilization. Perspective view of the cranial part of the embryo showing the three germ layers at the cut surfaces. Schematic and not to scale (source: Marrable, 1971).

5.4. The Uterus, the Placenta and Embryonic Attachment

5.4.1. The uterus

The uterus is dominated by endocrine control from the ovaries. In particular ovarian progesterone has a major influence and this, with other hormonal participation, induces changes in the uterine environment for the successful reception of embryos at the times shown in Table 5.3 for the common farm animal species.

At the end of oestrus there is a major invasion of polymorphonuclear leucocytes, the function of which is to cleanse and sterilize the lumen of the uterus by ingesting dead spermatozoa, seminal waste and bacteria introduced at the time of coitus. At the same time the endometrium or glandular epithelium of the uterus, as a result of a coordinating interplay between the uterus and the ovary, has its receptivity heightened in preparation for the young embryo entering the uterus. Thereby the chances of successful attachment are optimized. The heightened receptivity is associated with an increased secretory activity which enables the embryo to find nourishment from this source after leaving the nutritionally important fluids of the oviduct.

The uterus itself exhibits extraordinary changes in size from one reproductive cycle to the next. It has the amazing capacity to grow in response to being stretched and the distension that occurs in pregnancy is responsible for

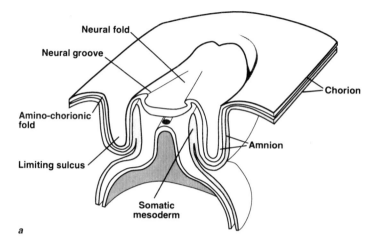

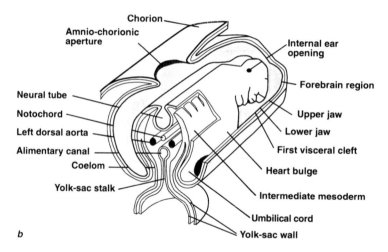

Fig. 5.12. a, Semitubular embryo of the pig at about 14 days postfertilization; perspective view of cranial part of embryo and surrounding membranes. b, Tubular embryo of the pig at about 15 days postfertilization; perspective view of cranial part of embryo and surrounding membranes. Both schematic and not to scale (after Marrable, 1971).

inducing hyperplasia of the smooth muscle fibres. The effect is most pronounced in the first third of pregnancy and most noticeable in the vicinity of the attachment site. This is followed in the last two-thirds of pregnancy by a massive hypertrophy of the smooth muscle fibres which may proportionately increase in diameter by as much as one-half of their original size. The distension of pregnancy is followed by an equally stunning regression after birth where the

Table 5.4. Origins of tissue from embryonic germ layers.

	System	Subsystem	Structure
Ectoderm	Nervous	Central nervous system	Brain, spinal cord
		Epithelial tissues of sense organs	Retina, internal ear, olfactory surface
	Alimentary	Digestive tract	Mouth, teeth, tongue, salivary glands*
	Respiratory		Nasal cavity, sinuses
	Urinary		Urethra*
	Genital	External	Scrotum, penis
	Integumentary		Cutaneous glands, hooves, hair, nails, lens, cornea, skin including mammal ear canal
Mesoderm	Alimentary	Digestive tract	Anal canal, stomach, intestines
		Ancillary organs	Salivary glands, liver[+], pancreas[+]
	Respiratory		Trachea, lungs
	Circulatory		Heart, arteries, capillaries, veins, blood, lymphatic vessels, lymph
	Urinary		Kidney, ureter, urethra
	Genital	Internal	Gonads, associated gonadal ducts, accessory glands
		External	Labia, clitoris
	Musculoskeletal		Muscles, bones, cartilage, tendons, connective tissue
Endoderm	Alimentary	Digestive tract	Pharynx, root of tongue, oesophagus
		Ancillary organs	Liver, pancreas
	Respiratory		Pharynx, larynx

*Jointly with mesoderm.
[+]Jointly with endoderm.

resilience of the uterus is illustrated by its involution. In this process the mass of the uterus is reduced abruptly during the first week following parturition. The regression is characterized by excess collagen disappearing and by the size of smooth muscle fibres decreasing.

The uterus not only exhibits these unique growth cycles but is also naturally endowed with an ability to repair injuries. In species that menstruate the decidual layers are lost but the endometrium is repaired quickly. In species that do not menstruate the attachment of the infantile embryo is at the expense of local traumatization of the endometrial epithelium and this elicits a growth response in the decidual tissues which eventually form the maternal components of the placenta. The ability to respond in this way is often referred to as the decidual reaction.

5.4.2. The placenta and embryonic attachment

The development of the placenta during early pregnancy is closely related to the extra-embryonic or fetal membranes that are differentiated into the amnion, allantois, chorion and yolk sac (Fig. 5.13). The amnion, which encloses the fetus in a fluid-filled cavity, is derived by a cavity forming in the inner cell mass. The chorion originates from the trophoblastic capsule of the blastocyst and encloses the embryo and other fetal membranes. In some animals, including all farm mammals, it fuses with the allantois, which is derived from the diverticulum of the hindgut and the blood vessels which connect the fetal and placental circulations, to form the chorioallantoic placenta. The yolk sac originates from the early endodermal layer, quickly becomes vestigial and in some species acts as a placenta. In yet other species the chorion remains separate and forms the placenta.

Chorioallantoic placentas are characterized by chorionic villi giving, through interdigitation with vascular foldings, a large surface area at the feto–maternal junction, and may be classified according to the characteristics given in Table 5.5. The shape of the placenta is determined by the distribution of villi over the chorionic surface. In the pig and in the horse, the epitheliochorial placenta is regarded as simple or diffuse with the epithelium and chorion lying in apposition. In the horse this simple apposition is replaced after 75–110 days with a more complex structure with the formation of microcotyledons. In sheep, cattle and goats there are specialized areas of attachment known as caruncles. These project from the uterine mucosa and the cotyledons of the allantochorion complex interdigitate with them to form the placenta, which is known as a cotyledonary placenta. The caruncles are therefore maternal structures belonging to the uterus, whilst the cotyledons are placental structures. The attachment complex is termed a placentome and is the site of functional exchange between the two sides of the placenta in the ruminant animal. There are between 90 and 100 placentomes in sheep and between 70 and 120 placentomes in cattle. There is considerable placentome growth in pregnancy and in cattle those placentomes above the gravid horn of the uterus develop to a larger size than those at the extremities. The several-fold increase in size is accompanied by a change in shape from flat, disc-like bodies to mushroom-like structures which, except for an area around the pedicle, are completely engulfed by the chorioallantois. The characteristics of the epitheliochorial placenta in the horse, in cattle, where the overall shape is convex, and in sheep, where the overall shape is concave, are given in Fig. 5.14.

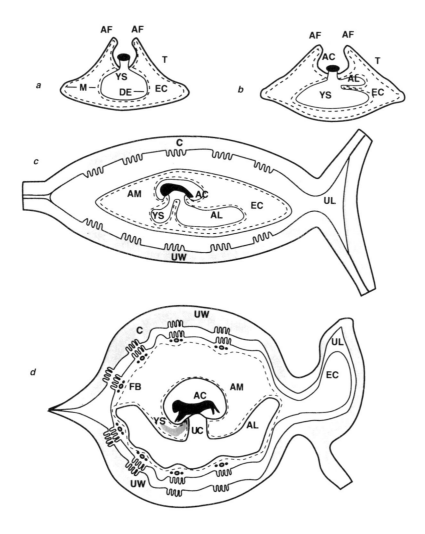

Fig. 5.13. Attachment sequence in the sheep. a, Elongated blastocyst with amniotic folds beginning to close over the embryo. b, Formation of yolk sac placenta with vascularized yolk sac wall (distal endoderm and mesoderm) closely apposed to chorion (trophoblast and mesoderm). c, Relationship of fetal membranes to uterine cavity. Amniotic folds have closed over the embryo, to enclose the amniotic cavity, which is expanding and pushing the enlarging allantois into line with the regressing yolk sac. d, Formation of chorioallantoic placenta. The expanding allantois presses the chorion against the uterine caruncles. AC = amniotic cavity; AF = amniotic folds; AL = allantois; AM = amniotic membrane; C = caruncle; DE = distal endoderm; EC = extra-embryonic coelom; FB = fetal blood vessels; M = mesoderm; T = trophoblast; UC = umbilical cord; UL = uterine lumen; UW = wall of uterus; YS = yolk sac (drawings based on those of McLaren, 1980).

Table 5.5. The classification of, and the tissues forming the placental barrier in, choriollantoic placentas (adapted from McLaren, 1974).

	Pig	Horse	Sheep, cattle and goat	Dog and cat	Man and monkey
Loss of maternal tissue at birth	None (non-deciduate)	None (non-deciduate)	None (non-deciduate)	Moderate (deciduate)	Extensive (deciduate)
Chorionic villous pattern	Diffuse	Diffuse – microcotyledonary	Cotyledonary	Zonary	Discoid
Maternal–fetal barrier	Epitheliochorial	Epitheliochorial	Epitheliochorial	Endotheliochorial	Haemochorial
Tissues forming placental barrier					
Maternal					
Endothelium	+	+	+	+	+
Connective tissue	+	+	+	–	–
Epithelium	+	+	+	–	–
Fetal					
Trophoblast	+	+	+	+	+
Connective tissue	+	+	+	+	+
Epithelium	+	+	+	+	+

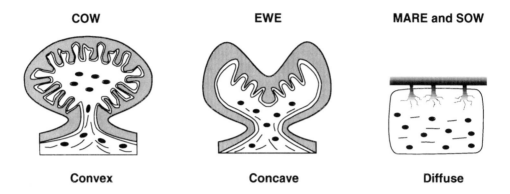

Fig. 5.14. Diagrammatic representation of epitheliochorial placentas of cow, ewe, mare and sow. In the cow and the ewe the apposition of fetal and maternal tissues is localized to give placentome structures with each placentome composed of fetal cotyledon and maternal caruncle. In the sow, and in the mare up to about 75 days postfertilization, the apposition is diffuse. After about 75 days this changes to a cotyledonary placenta as in the cow and in the ewe. In all cases villi from the chorioallantois (shaded) invade crypts in the maternal epithelium (based on drawings of Jainudeen and Hafez, 1974).

Because in farm animals the conceptus remains in the lumen of the uterus and does not invade the stroma, attachment is a more appropriate term than the more frequently used term implantation. Attachment of the embryonic membranes to or within the epithelium of the uterus or endometrium does not occur instantaneously but proceeds gradually over a period of time. Attachment of the embryo may be said to have occurred when it becomes fixed in position and when it has established contact with the maternal organism. The trophoblast is the tissue of the embryo specialized for interaction with the uterus and giant cells of this tissue often have very invasive potentials and develop considerably during the course of attachment. The times from ovulation to attachment for the common farm animal species are given in Table 5.6. In the case of the horse, 3 weeks after fertilization specialized cells appear on the trophoblast and these enable the embryo to absorb uterine milk (histotrophe) which is a mixture of uterine secretion and tissue debris. These cells form a transient attachment to the endometrium and it is not until the 10th week after fertilization that the chorionic villi grow out into the folds of the uterine wall and initiate the final stages of true attachment. A diagrammatic representation of the process of attachment in sheep is given in Fig. 5.13.

The placenta is an organ of respiration, nutrition and excretion and therefore allows metabolic exchanges between the maternal and fetal components of which it is comprised. In addition, it is an endocrine gland secreting gonadotrophins and progesterone. Furthermore, the antigenic individuality of the embryo is protected by the placenta, and by its gonads, against immunological rejection by the mother.

The placenta grows at an extremely rapid rate in a linear manner and for a

Table 5.6. Stages of embryonic attachment in farm animals (adapted from McLaren, 1980).

	Cattle	Sheep	Pig	Horse
Beginning of maternal recognition of pregnancy (days)	16–17	12–13	10–12	14–16
Beginning of attachment (days after ovulation)	28–32	14–16	12–13	35–40
Completion of attachment (days after ovulation)	40–45	28–35	25–26	95–105

large part of the gestation period its size exceeds that of the embryo. Eventually, however, the exponential growth of the embryo exceeds the linear growth of the placenta in which, in the latter stages of the period, growth becomes more and more dependent on hypertrophic, rather than hyperplastic, processes. Although the placenta is an autonomous unit, it can be influenced by very nearly any factor that influences the size of the embryo, and, later, the fetus. For example, in multiparous animals the number of embryos/fetuses in the litter is inversely proportional to both fetal and placental size. Oestrogen has a marked inhibitory effect on placental growth and can also induce atrophy. This contrasts to the situation in the fetus where growth is stimulated by ovarian hormones. Generally speaking the size of the placenta is positively correlated with the size of the fetus.

5.5. Postgastrulation and Posttubulation Embryonic Development

The next stage of development after tubulation is characterized by the processes of torsion, in which the embryo becomes twisted and lies on its side, and of flexure, in which there is a bending of the embryo so that the ventral profile becomes concave and the dorsal profile becomes convex.

Up to this point development has been presented as very much a matter of cells, layers and tubes. From this point onwards, however, new tissues differentiate and diverse organs take shape. As a result, as time progresses the various parts of the body, the head, the neck, the trunk, the limbs and the tail, become recognizable. Within this overall pattern there is a gradient of activity in which the cranial structures differentiate earlier than those in the mid-body, which in turn exhibit an earlier development than those in the tail. The development of the pig embryo will be used here to demonstrate these points.

In the pig tubulation will be complete by about 15 days after fertilization. Visually, and following the pretorsion and flexure stages, the external appearance changes (Fig. 5.15). At 35 days the cerebral hemispheres have achieved a relatively high degree of development and the mid-brain is large and overhangs the developing cerebellum which is beginning to show signs of splitting into its two ultimate and distinct lobes. In the mouth the secondary palate is complete and the tongue has grown to a very distinguishable form. There is an elaborately coiled intestine present and lobulation of the lungs is apparent. The heart is by far the largest organ in the thoracic cavity and a rudimentary bladder is evident. Gonads have become either testes or ovaries according to the genetically determined sex of the embryo, and a double set of genital ducts is present. Cartilaginous models of many parts of the skeleton are present and the earliest sites of bone formation can be seen in the lower and upper jaws. Musculature of the limbs is apparent and the integument at this stage is generally transparent. Teeth are not yet present, only their germinal layers. Hair follicles and

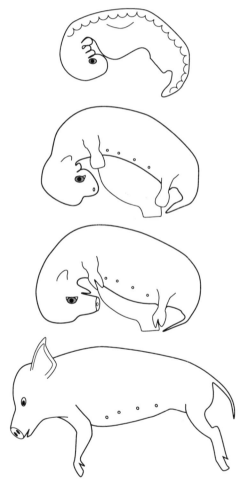

Fig. 5.15. Lateral views of external form of pig embryo, in descending order at about 18 days, 22 days, 35 days and at birth (not to scale) (based on drawings of Marrable, 1971).

rudimentary mammary glands are apparent. As can be seen from Fig. 5.15 with the first third of gestation having passed, the embryo already bears an immediately identifiable resemblance to the pig at, and immediately after, birth. Four limbs, a snout with a flattened nostrillar surface, a mouth with an opening extending halfway along the lower jaw and eyelids are evident. The shoulder, the leg and the foot, with digits, are all apparent and the thoracic and abdominal walls have become reinforced with muscle, the former more so than the latter. After this period, up to parturition, the organs, parts and tissues exhibit quite massive growth potentials giving, in the case of the pig, the external appearance indicated. This period of growth, in the fetal stage, will receive attention in chapter 6.

References

Austin, C.R. (1972) Fertilization. In: Austin, C.R. and Short, R.V. *Reproduction in Mammals. Book 1. Germ cells and fertilization*. Cambridge University Press, Cambridge pp. 103–133.

Garner, D.L. and Hafez, E.S.E. (1980) Spermatozoa. In: Hafez, E.S.E. (ed.) *Reproduction in Farm Animals*, 4th edn. Lea and Febiger, Philadelphia, pp. 167–188.

Hafez, E.S.E. (ed.) (1980) Functional anatomy of female reproduction. In: *Reproduction in Farm Animals*, 4th edn. Lea and Febiger, Philadelphia, pp. 30–62.

Hunter, R.H.F. (1982) *Reproduction in Farm Animals*. Longmans, London.

Jainudeen, M.R. and Hafez, E.S.E. (1974) Gestation, prenatal physiology and parturition. In: Hafez, E.S.E. (ed.) *Reproduction in Farm Animals*, 4th edn. Lea and Febiger, Philadephia, pp. 247–283.

Marrable, A.W. (1971) *The Embryonic Pig: a chronological account*. Pitman, London.

McLaren, A. (1972) The embryo. In: Austin, C.R. and Short, R.V. (eds) *Reproduction in Mammals. Book 2. Embryonic and fetal development*. Cambridge University Press, Cambridge, pp. 1–42.

McLaren, A. (1974) Fertilization, cleavage and implantation. In: Hafez, E.S.E. (ed.) *Reproduction in Farm Animals*, 3rd edn. Lea and Febiger, Philadelphia, pp. 143–165.

McLaren, A. (1980) Fertilization, cleavage and implantation. In: Hafez, E.S.E. (ed.) *Reproduction in Farm Animals*, 4th edn. Lea and Febiger, Philadelphia, pp. 226–246.

Prenatal and Postnatal Growth

6.1. Problems of Describing Growth

The dynamic changes which occur in the size, shape and proportions of an animal as it grows are so complex that any attempt at understanding requires the enormity of the problem to be lessened by introducing some simplifications.

A major question for the biologist and those engaged in animal production is how to understand growth well enough to make sensible comparisons and judgements between animals of different genotype and animals grown under different environmental conditions.

The first need is to try and understand some simple concepts about growth rate.

6.1.1. Growth in relation to time

When growth is related to time, a wide range of interesting principles is discovered. The general phenomenon is sometimes referred to as temporal growth. Animals cannot grow instantly. All the biochemical processes involved require time. For example, the acquisition of food takes time, the process of digestion takes time, the transcription of RNA takes time, and the building of new tissue takes time. In the wild, some animals grow very slowly, taking many years to reach maximum size. For example, elephants grow throughout the major part of their lives and may take 50 years to attain maximum size. Some birds, however, can complete their growth within a very few weeks, for example altricial birds such as kestrels which remain in the nest and are fed by their parents can be fledged when their weight exceeds that of the adult after only 2 months in the nest.

When the actual live weights of animals fed generously throughout life are plotted as a function of age or time, they produce a very characteristic growth curve. This is often termed a 'sigmoid' growth curve because of its resemblance to the letter S. It is has been described by many biologists as consisting of a self-

179

accelerating phase and a self-decelerating phase. This is in fact, as is usual in biology, an oversimplification. An alternative, but still an oversimplification, is to consider growth as being tripartite, with a self-accelerating phase, followed by a linear phase, and finally a decelerating phase which fades out as the animal reaches maturity.

The reasons for these different phases are complex. In fact each component of the body, such as a particular muscle or bone, has its own growth curve and live-weight change is the integral of all these.

The accelerating phase

In very simple terms, the self-accelerating phase can be understood by the fact that if growth were simply determined by the cells doubling at regular intervals, then the amount of growth in any period would be the square of that in the preceding one. For example, if in the first few hours of growth a single cell becomes two, and in a second similar period two cells become four, and so on, then the growth rate doubles in step with each doubling of the cells. For example, if the cells divide every 12h and achieve a doubling of mass in the same time, then we have the basis for an extraordinary calculation. Taking the starting weight of a gram, the absolute gain from day 1 to day 2 is 3g, from day 2 to day 3 is 12g and from day 7 to day 8 has become an astonishing 49kg which exceeds the highest growth rate of any mammal – that of the blue whale calf over the first 2 years from conception of 37.5kg per day. By day 11 the daily growth rate of our original 1g organism, if every cell is maintaining its doubling rate, has become over 3 tonnes. It is clear that these very rapid doubling times for the biomass may be achieved only for a very short time. This is true even for the ideal situation of bacteria dividing in a well-stirred broth, because eventually the end products of fermentation accumulate and the nutrients do not gain access so rapidly to the organisms at the centre of clumps. Problems of organization arise very quickly in multicellular organisms. Nutrients have to be transported to the cells on the inside of the mass and waste products transported to the outside either by diffusion between the cells or by a circulatory system. This necessitates the development of even more complex mechanisms to acquire and transport nutrients to where they are needed and the development of specialized organs of excretion. This slows down the doubling rate of the mass, although in absolute terms the rate of growth is increasing daily. As organisms get bigger they must necessarily become more complex.

The concept of percentage growth rate is important. It is quite possible in the early stages of growth *in utero* for the developing embryo to double in size over a 24h period. Inevitably, the doubling rate slows down and eventually the percentage growth rates fall. Percentage growth rates allow certain forms of comparison to be made if the animals or tissues are of different sizes. It becomes particularly important when comparing the relative growth rates of different tissues or parts within the same organisms. The concept can be explained by a numerical example. Two animals may gain at a rate of 200g per day. These two growth rates are in absolute terms the same. However if one animal weighs

100 kg and the other 200 kg, then in percentage terms they are quite different, the former is growing at twice the rate of the second, that is 0.2% per day compared with 0.1% per day.

Percentage growth rates are not necessarily the same for each component or tissue of the body and indeed this is how changes in form are brought about. This is known as differential growth and will be discussed in greater detail when we discuss proportionate or relative growth later in this chapter.

As growth progresses, two factors are in conflict. One is the accelerating force due to the increase in the number of replicating units, whilst the other counteracting force is the limitation of the greater complexity of the structures and the ability of the food supply to keep pace with the growth of the body. This often results in an extended linear phase of growth where the two forces are more or less in balance. This is illustrated in Fig. 6.1, in which growth curves for the major farm species are presented and also the rather peculiar growth curve of man.

The final phase of growth is the so-called self-decelerating phase as the animal approaches its mature weight. This is a phase in which there is an inbuilt restraint on further growth, which progressively reduces the proportion of intake which exceeds the maintenance requirement. This is probably a combination of many signals, but includes in this array the secretion from the hypothalamus of somatostatin. The most noticeable effect is a stabilization of the feed intake and a gradually diminishing increase in body weight until the intake equals the maintenance requirement. This asymptote can be regarded as the mature body weight, but it is not a stable value and varies considerably within individuals depending on the availability of feed, the demands of the reproductive cycle and in some cases the season of the year.

The pattern of growth of animals on virtually unrestricted food supply has been subjected to much mathematical analysis. It is not too remarkable that a process which has been framed within a set of biological constraints should be amenable to such treatment. However, before discussing the undoubted value of this approach, it should be pointed out that these rules are not inviolate and that though some of the solutions are extremely elegant, it is not the equation which drives growth. Equations which describe growth can be very helpful in terms of producing a workable simplification and allowing certain predictions to be made. They are however, in the last analysis, mainly pragmatic and descriptive.

Samuel Brody in his classic volume *Bioenergetics and Growth* (1945) made many valuable contributions in this area. He showed that very complex patterns of growth could be made much simpler by plotting the growth on semilogarithmic paper, that is the log of the weight against time. Using simple power equations a good description of growth could be achieved. He described the self-accelerating phase in terms of the following equation:

$$W = Ae^{kt} \tag{1}$$

where w = the natural logarithm of weight of the animal at time t, A = the

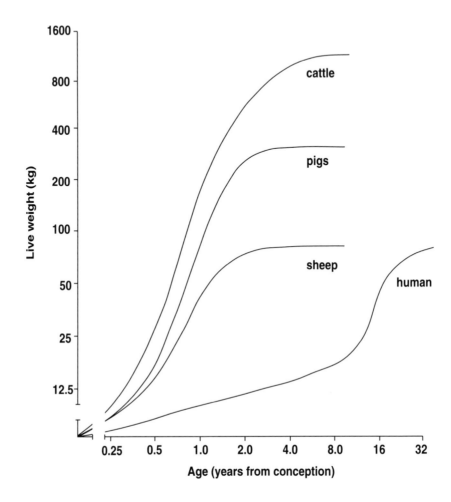

Fig. 6.1. Growth curves for major farm species and for man.

natural logarithm of W when $t = 0$ and k = a constant, representing the instantaneous, relative growth rate or, when multiplied by 100, the percentage growth rate.

The self-decelerating phase can be similarly expressed as:

$$W2 - W = A(1 - e^{-k(t - t^*)}) \tag{2}$$

where W = weight of the animal, A = asymptotic weight ('Mature weight'), k = an index of the decay or acceleration of the curve with time and t = time in days.

The changing shape of any growth curve can of course be fitted by using a series of constants, although the more constants that are used the more obscure becomes any relationship with anything recognizable as a biological factor. Over short periods of time a simple linear equation may suffice, and even over longer periods of time when the growth is obviously changing in rate a quadratic equation may give an extremely good fit to the data.

Many other equations have been fitted with success to data. Two in particular are worthy of mention, namely the Gompertz and von Bertalanffy equations. The Gompertz equation is attractive in that it allows both the accelerating and decelerating phases of growth to be incorporated in the same equation. Essentially, it predicts the weight of the animal from a constant raised to a power which is itself raised to a further power based on the time from the origin of the study, viz:

$$W = ae \exp(-be \exp-kt)$$

or

$$W = A - (\log A \, x \, (- \log kt)$$

It has been shown to give a good fit to data for prenatal growth, particularly in the case of lambs and deer calves (McDonald *et al.*, 1977; Adam *et al.*, 1988).

The von Bertalanffy equation (von Bertalanffy, 1960) is an attempt to model the sum of the anabolic factors and the catabolic factors (those that build up or break down tissues) in one equation. It has the basic form:

$$dw/dt = aW \exp m - bW \exp n$$

where the first term on the right-hand side represents the sum of the anabolic factors and the second the sum of the catabolic factors.

A very generalized equation which has extraordinary flexibility, and therefore a very great potential to fit most data very well, is the Richards equation (Richards, 1959, 1970).

It has the generalized form:

$$W = An - (An - Y^*n) e - kt$$

where W = body weight at age t, A = mature body weight, Y^* = weight at start, k = rate constant, n = weight exponent constant.

The value of n is in fact quite interesting. Bakker and Koops (1978) showed that according to the value chosen for n, the Richards equation transforms into any one of a number of the popular growth equations. For example, when it takes the value of 1 then it takes the form of the Brody growth equation, when it takes the value 0 it is has the Gompertz form, when its value is 1/3 then it is the von Bertalanffy equation and finally when its value is −1 then it gives a logistic equation which has also been widely used in growth analysis. This demonstrates very clearly the fact that growth equations are essentially describing the same thing and the best ones will belong to the same family, in this case the Richards family of equations.

Deviations from standard growth curves

Much as it is tidy to consider growth as progressing along a mathematically defined growth curve, several events occur which may deflect growth either of the whole animal or of some of its tissues or organs. Two examples can be given to illustrate this. Even when the nutritional background is stable, factors such as the advent of puberty or changes in daylength may affect some species. A classical example is the prepubertal growth spurt in humans. Note again that however devious the growth curve, it is always possible to fit some sort of mathematical expression which will describe it by using a large number of constants. Though this may be satisfying to a mathmetician, unless some biological meaning can ultimately be attached to the components of the equation, it does very little for the understanding of the underlying biology.

6.2. Describing Prenatal and Postnatal Growth

Once differentiation has occurred, describing growth can become extremely complex. Any description of the unfolding of the innate growth pattern of each tissue or part opens up one of the most vast areas of biology. It is unfortunate that specialization in education often separates agricultural and veterinary disciplines from general biological considerations at an early stage. Therefore, the purpose of much of this account will be kept at a very general level and an attempt will be made to highlight certain principles and concepts.

The physical effort required to dissect animals into constituent parts at different stages of growth is enormously taxing. Muscle-by-muscle dissection of the whole animal is a battle against time, because of the twin problems of moisture evaporation and bacterial contamination. To undertake this work requires a considerable team of skilled operators and recent constraints on funding have meant that much of the classical work in this area was undertaken over 30 years ago. This was an era when salaries were relatively low and when there were fewer counter-attractions for students of more glamorous fields of molecular biology. It was also an era which did not have the advantage of computerized data handling and rapid chemical analysis.

It is regrettable that many authors from this period, in their attempts to justify hundreds of hours of detailed dissection, months of chemical analysis and the computation of tens of thousands of numbers, succeeded only in the production of extensive tables and equations. Such dedication, though commendable, falls short of the scientific ideal, namely that while it is important to generate new facts, it is also important to develop usable ideas and insights accessible to those who are not specialists in the field. No apology is made for concentrating on some general principles which it is believed will help to maintain a respect for, and an interest in, the growth process as an inspiring example of biological diversity and adaptability, as well as a process harnessed by man for agricultural purposes.

6.3. Targets of Growth

The functional needs of the animal change as it develops and matures. Growth is not a uniform process, merely aimed at transforming an embryo into an adult, but a series of adaptations to the current and future needs of the animal. The problem with mathematical descriptions of growth is that although they describe simply and with reasonable precision the general process of live-weight growth to maturity, they describe less well the subtleties of growth and impact of changes in physiological need arising from, for example, changes in reproductive or nutritional status.

These problems can be easily illustrated by referring to a wider biological context. Consider first the problem of providing a mathematical basis for growth in the Ranidae (frogs). The targets of the growth genes of a frog are not aimed in one single conceptual direction. The first targets are related to the functional success of a tadpole living in an aqueous environment. Thus the development of a muscular tail is critical to the survival of the tadpole and it has its own growth curve to maturity. This however is only an intermediate target. The cells of the tail eventually become subject to a new set of genetic instructions and by apoptosis and cell deletion growth is reversed and the structure is rapidly absorbed. At about this stage, the many structures appropriate to the terrestrial life of the frog are initiated and the amino acids of the tail muscle are re-assembled into the new format of limb muscles and so on. An even more dramatic example of changing growth objectives is provided by the Lepidoptera (butterflies and moths), along with many insects which establish remarkably different growth formats within a single lifetime. The metamorphosis of the caterpillar through the chrysalis stage to the adult imago is a clear example of one set of genes being shut down and another array switched on as the genetic programme unfolds.

Returning to the vertebrates there are many spectacular examples of redirection of resources depending on the functional need at the time. The different morphology of birds at hatching is a striking example. Precocial birds (from *praecox* – early developing) such as the domestic hen, or game birds like partridge and pheasant, stride out of the egg fully capable of locomotion and able to forage for themselves. They are amazingly advanced (or precocious) in both physical and behavioural terms. They have a well-developed and versatile digestive tract and reach a high degree of independence at a very early age. Altricial birds (from *altrix* – a nurse) such as newly hatched raptors and pigeons are by contrast totally dependent on parental care and remain in the nest. This is reflected in their anatomical development. Most nature observers will be aware of the astonishing fragility of such chicks with their wide gaping beaks, naked bodies and disproportionately large abdomens. It has been demonstrated by Kirkwood and Prestcot (1984) that altricial birds have, in relative terms, both a very large gut and a very high feed intake capacity. They therefore grow extremely rapidly towards their mature size. As the time approaches for their departure from the nest, the proportions are transformed and there is rapid development of the pectoral muscles in preparation for flight.

A poetic and accurate description of the rapid changes in altricial birds is given by the famous British naturalist Gilbert White in his *Natural History of Selborne* (1789). In letter 20 of his correspondence with his philosophical friend, the Honourable Daines Barrington, he wrote of a further episode in his study of his favourite family of birds the hirundines (swifts, swallows, etc.). His enquiring mind responded to chicks taken from the nest of a swift in summer as follows:

> The squab young we brought down and placed on the grass plot, where they tumbled about, and were as helpless as a new-born child. While we contemplated their naked bodies, their unwieldy disproportioned abdomens and their heads, too heavy for their necks to support, we could not but wonder when we reflected that these shiftless beings in a little more then two weeks would be able to dash through the air almost with the inconceivable swiftness of a meteor; and perhaps, in their emigration must traverse vast continents and oceans as distant as the equator. So soon does nature advance small birds to their *helikia* (maturity), or state of perfection; while the progressive growth of men and large quadrupeds is slow and tedious.

White had a clear perception of the huge transformations necessary before the final manifestation of the genotype.

In an unconstrained environment where food is not limiting, the concept of ultimate targets at which growth is aimed is helpful. The targets can include size and ideal proportion of organs and parts. For this the Greek *helikia* is a convenient label. In mammals, the main targets could be considered as the proportions which are typical of the growing animal as it reaches adulthood. Thus in our own species it is no accident that the *helikia* regarded as an ideal is the configuration of the young adult, since this is the age at which pairing of male with female usually occurs. In some species, notably the ungulates, the young adult male is not sufficiently strong or skilled to establish his dominance amongst other males competing for control of a harem or for territory. In this case, the fullest expression of the *helikia* is in the older male where muscling is at its greatest, the competitive weaponry, such as antlers and horns, at their most imposing and fighting skills have been practised and honed to perfection.

In mathematical terms the *helikia* can be considered as the mature weight which is the asymptotic value of the upper part of the growth curve. It is often designated in equations as 'A'. It is widely used as a scaling factor to compare animals with characteristically differing size, and its fourth root ($A^{0.25}$) or the closely related function ($A^{0.27}$) has been shown by Taylor (1980) and others to allow comparisons to be made between species in terms of the rate at which they reach differing degrees of maturity.

6.4. Sequential Growth Targets

It is tempting to consider the *helikia* or adult form as the only growth target in mammals. However, it is possible to envisage growth at different stages as being directed at a number of intermediate growth targets, particularly if the growth process is considered in its widest sense as lasting from conception to death.

To illustrate the changing physiological objectives an illustrative list is provided below. Because the objectives are complex it is necessary to assign some rather crude labels to these stages.

1. The embryo as an effective parasite

The development of the blastocyst and the competition for uterine space is a clear growth target in postconceptual life. It is of paramount importance that the developing embryo has command of a sufficient area of uterine space for the supply of nutrients which will allow full development of the fetus.

2. The fetus as a competitor

Dziuk in Illinois has drawn attention to the intensely competitive environment within the uterus of the fetuses in polytocious species (Dziuk, 1992). He makes an analogy with the overproduction of fruits on a peach tree or of peas in a pod relative to the eventual ability of the parent to support the offspring. As the fetus develops it establishes not only its placenta but also priority for a circulatory system and liver. These early structures reflect the need to collect, process and distribute nutrients and oxygen to all the developing structures and to excrete waste products.

3. The fetus as a template for growth

Although the brain, limbs, bones, digestive tract and lungs have little function *in utero*, they will be required in fully operational form at birth. In anticipation of this, all key organs are inititiated and developed in a coordinated way but only so far as it will be appropriate at birth. Thus the ruminant is not born with a large functional rumen nor the ram lamb with horns. The form of the internalized offspring must also be compatible with the birth process. It is therefore inappropriate for fetuses to develop excessively large heads and shoulders prior to birth because of the complications that can arise in expelling the offspring through the birth canal.

The functional needs of the newborn lead to some charming adaptations. The baby kangaroo or joey is born with forelimbs which greatly exceed the size of the hindlimbs. This is because immediately after birth, the forelimbs are used to grasp the fur of the mother and allow it to travel 'arm over arm' to the pouch, where it can cling to the mother and establish itself on a nipple. The disparity in size of fore- and hindlimb is quickly reversed during growth to reflect the dominance of the hindlimb in juvenile and adult locomotion in this species.

4. Semi-independence at birth

The targets of growth *in utero* that eventually enable the neonate to operate independently of the dam are very remarkable. At birth, almost instantly in place are functioning lungs, a redirected circulatory system, fully operative suckling reflexes, limited communication skills and, in the case of the ungulates, limbs which are fully functional for locomotion within minutes.

5. Newly weaned juvenile

The transfer from total nutritional dependence on the dam, to the ability to forage and digest food obtained from the wider environment, represents a further set of growth targets. In the case of the ruminant, the change from abomasal digestion of milk to ruminal fermentation of roughage represents a major functional and anatomical change. In many species, the dentition changes at about this period, becoming more appropriate for the diet of the independent growing animal.

6. The growth phase

Each species must attain a certain optimal size before embarking on the responsibilities and challenges of reproduction. The period from weaning to puberty is usually the period of greatest absolute growth rate, and is characterized by the animal utilizing its nutritional resources to the full to attain a size which will enable it to reproduce successfully. It is this period which forms the main focus of animal production systems. In some species, this period is extended over several years and growing seasons, during which time the animal learns how to utilize its feed resources, defend itself from predators and develop the social skills of its species.

Humans and the apes have an extended period of prepubertal growth where the expression of the growth curve is apparently delayed. The explanation is thought to be associated with the survival advantage of an extended learning period during which the animal learns various complex skills including communication within its social group, familiarity with a territory, and strategies for hunting, defence and protection from the environment.

7. Puberty and the onset of reproductive capability

As the objective of increase in size is achieved and the animal approaches adulthood, then it becomes possible to superimpose the refinements which are appropriate to the mode of reproduction in the species. Many reproductive structures remain in infantile proportions until signals, which mark the onset of puberty, are initiated from the hypothalamus. Gonadotrophins switch on a new array of genes which modify growth to produce the secondary sex characteristics. In the male, they cause testicular enlargement and the development of the glands and structures associated with the genital tract. There may also be a change in musculature which reflects the increasing need of the male to compete with other males for the right to reproduce. The growth changes may be strictly functional, as in the case of the greatly enlarged muscles of the neck in the bull. Secondary sex characteristics can, as Berg and Butterfield (1976) have pointed out, be somewhat symbolic changes, as is the case with the North American bison. During puberty the male of this species greatly lengthens the dorsal processes of the neck, so giving an imposing profile designed to intimidate potential competitors.

In the female, the gonadotrophins released at puberty cause the ovaries to be activated. This has a considerable influence on growth. The dormant and infantile uterus enlarges and eventually changes to become receptive to the possibility of fertilized eggs. In some species, perhaps most notably humans, puberty is accompanied by rapid enlargement of the mammary glands.

8. Reproductive phase

The reproductive phase is characterized by repeated cycles of growth. The cycle may be entrained on an annual rhythm as in seasonally breeding species or operated independently of season as is mainly the case in pigs. The variations in relative size of organs and parts is not restricted to the organs directly associated with reproduction. The mammalian female is characterized by the storage of energy in adipose depots during pregnancy, and the progressive release of this energy in the lipid of the milk during lactation. Dairy cows gradually lose condition (or fat) during lactation as also do sows, particularly if the litter is large and the lactation extended. Some sea-going mammals are extreme examples of this phenomenon. In seals (Pinnipedia) which come ashore to breed, the whole of the lactation occurs on land but the dam does not feed at all during this period. The lactation is thus supported only by body reserves of fat and protein. Another extreme is the stag (the male of the Scottish red deer, *Cervus scotticus*) which usually does not feed during the whole of the rut (breeding season). The breeding life of both male and female is thus marked by considerable fluctuations in body weight and in the amount of fat reserves in the body.

9. Senescence and death

Death is part of the genetic programme. Each generation must make way for the next, and in most species the conclusion of the reproductive phase brings about the onset of senescence. Even the sagacity of the elder statesmen is of limited value to the species. Eventually inbuilt failure of one or more vital systems causes the death of the individual. This allows the new generations to develop without competition from the previous one. Elderly animals often separate from the social group, and there is weight loss and wasting which could be described as a form of negative growth.

6.5. Changes in Proportion During Growth

The growth cycle of each organ and tissue does not occur in synchrony. In general the shape of animals and the proportions of the tissues and parts alter considerably during growth in response to the current and future physiological needs.

There is a vast literature on the subject, of which only a tiny fraction will be detailed here. It should be acknowledged that every nation that has made a contribution to animal science has produced considerable experts in the field, usually in an earlier era. It is impossible within the limited scope of this book to give due justice to each of these, and the serious student is commended to the original papers and compendia on these studies, such as the monumental treatise of Samuel Brody – *Bioenergetics and Growth* (Brody, 1945).

The work of Sir John Hammond and the group of talented students which worked with him in Cambridge, England, during the period from 1930 to 1960, set the foundations for much current thinking concerning growth. This was the outcome of much painstaking work whereby the major farmed species were serially dissected from an early age until they reached marketable weight. The original accounts are worth much study, particularly as they are associated with extensive tables and appendices which provide primary data for all forms of subsequent analysis. This work points to the major issues in growth studies.

The approach was one of extreme beauty and has been a model for many subsequent investigations of growth. The studies were conducted in two stages:

1. Serial dissection of animals kept under typical conditions of environment and nutrition.
2. Attempts to examine the deviations from the basic growth pattern consequent upon perturbations of the nutritional state during life.

The major studies were carried out by:
Sir John Hammond (1932) – sheep.
Walton and Hammond (1938) – horse.
McMeekan (1940 a,b,c) – pig.
Wallace (1948) – sheep (prenatal).
Pállson and Vergés (1952) – sheep (postnatal).
   and also derivative studies by:
Pomeroy (1955)
Wilson (1952, 1960).

This collection of studies has been emulated but not surpassed by other workers in more recent years, for example the work of Butterfield in Australia (e.g. Butterfield and Berg 1966 a,b), and the workers in Alberta, Canada (e.g. Richmond and Berg, 1971 a,b,c).

Hammond was primarily interested in the way in which the bone and musculature developed in relation to the meat-eating qualities of farm animals.

He was among the first to point out the sequence of events whereby during growth particular tissues and parts underwent their growth cycle in a particular sequence, so giving rise to an order of maturity. The tissue sequence was described as: 1, nervous tissue, 2, bone, 3, muscle and 4, fat (Fig. 6.2).

Hammond also noted that there was a strong tendency for the extremities to complete their growth cycle first, followed by development of the proximal and axial parts. The phenomenon is readily seen in the extreme 'legginess' of ungulates at birth and the relatively large head. Hammond and his group also observed that growth in the limbs followed the sequence of metacarpals and metatarsals, then radius–ulna and tibia–fibula, then humerus and femur, and finally scapula and pelvis. This he described as 'waves of growth starting at the extremities' which then worked their way towards the axial skeleton and the loin. This was called 'centripetal' growth or centre-seeking growth. The changing form in human growth is illustrated in Fig. 6.3.

6.5.1. The first controversy: live weight as a determining variable

In parallel with Hammond's work, the zoologist Julian Huxley published the view that the proportions of the animal were determined by the overall weight. This theory, which became known as growth allometry, was based on his observations of the growth of the fighting claw of the male fiddler crab which becomes disproportionately large as the animal matures. Huxley found that if the natural logarithm of the claw weight was plotted against the natural logarithm of the weight of the body minus the claw, the resulting relationship was virtually a straight line:

$$\text{Log } Y = \log a + b \log X$$

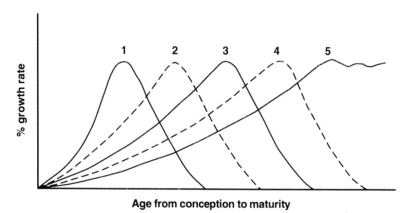

Fig. 6.2. Waves of growth: 1 = nervous tissue; 2 = bone; 3 = muscle; 4 = fat; 5 = daily feed intake. When all the tissues reach mature size daily intake may have declined from its maximum value and often fluctuates on a seasonal basis.

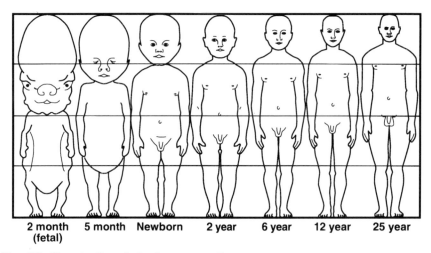

2 month   5 month   Newborn   2 year   6 year   12 year   25 year
(fetal)

Fig. 6.3. Changing form during human growth.

where Y = claw weight, a = the value of Y when $\log X = 0$ and X = weight of body − claw.

Such a relationship implies a constant ratio between the percentage growth rates. Since the claw was growing relatively faster than the rest of the body, the value of b was greater than 1. Indeed similar relationships can be constructed to compare the growth of any part of the body relative to the whole. If the value of b is greater than 1, then the part is growing more rapidly than the whole and is said to be later maturing. If the value of b is less than 1, then the part is growing more slowly than the whole and in that context is said to be 'early-maturing'. If the value of b is unity then the part and the whole increase in percentage terms at the same rate and remain in a constant proportion to each other.

The apparent dependence of proportions and the size of parts on body weight was contrary to Hammond's experience of farm animals with different nutritional experience. It was well known in farming circles that to obtain a satisfactory fat cover on slaughtered beasts, the animal had to be fed generously particularly for a period immediately prior to slaughter. However, animals suffering from undernutrition, as for example those kept outside during the winter, had a very different appearance to those which had been stall fed indoors. The outwintered animals were found to have a lean and rangy appearance. Hammond resolved to undertake a series of experiments in which the nutrition throughout life was systematically altered to demonstrate the fact that for farm animals at least, a system of allometry based on live weight would not adequately describe the growth changes which occurred.

The format he chose has become a model for hundreds of subsequent nutritional experiments. Essentially he set up experiments with two periods during which he fed animals either a high or low plane of nutrition to give a two by two factorial design:

	Period 1	Period 2
	High	High
Plane of nutrition	High	Low
	Low	High
	Low	Low

From this series of experiments he quickly drew the conclusion that when nutrition was limiting then the tissues and parts had a different priority for available nutrients, based somewhat on the sequence in which they developed during growth, and also on their functional priority. This he developed into a very elegant and simple diagram (Fig. 6.4)

The experiments were a dramatic confirmation that the proportion of tissues was not determined by the live weight of the animal but by other factors related to the functional requirements of the animal under adverse conditions.

The results led to the formulation of five major principles of the Hammond School, which have been regarded by adherents as largely axiomatic ever since. These were set out by Pállson (1955) and they are given below in a slightly abbreviated and rearranged form but with no attempt to change the meaning.

1. Severe undernutrition of the dam has no retarding effects on the development of the fetus until the later stages of fetal life.

2. Restricted nutrition from late fetal life has an increasing retarding effect on the different parts and tissues in the direct order of maturity, with the latest maturing parts being the most affected.

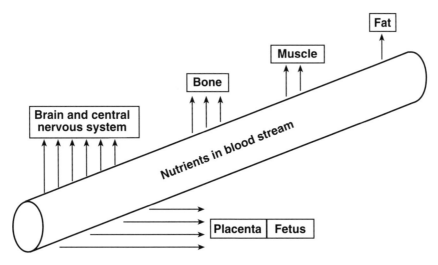

Fig. 6.4. Partition of nutrients. The 'priority' for nutrients from the bloodstream is indicated by the direction and number of arrows for each tissue. As the plane of nutrition falls, so an arrow is subtracted from each location. During starvation, the 'fat' and 'muscle' arrows may reverse in direction.

3. A submaintenance diet causes the mobilization of nutrients from tissues in the reverse order of their maturity.

4. Tissues retarded by restricted nutrition exhibit great powers of recovery when a high level of nutrition is restored.

5. During late fetal life to maturity any part, organ or tissue of the animal's body is proportionately the most retarded in development by restricted nutrition at the age when it has its highest natural growth intensity.

6.5.2. The second controversy: should fat be included as part of the independent variable?

Of the five propositions of the Hammond School above, much support can be found for the first four, but the last one has been the subject of much controversy. The reason why it appears to be so problematical, is that it seems to fly in the face of other biological principles. These would suggest that the balance of components within the functioning units of the animal must be maintained in optimal proportions, if the animal is to survive and operate effectively.

Why should a common occurrence in the life of the species such as undernutrition be capable of perturbing the functional harmony between tissues and parts? This dilemma brought contradictions which were apparent from the data of the Hammond group. One such problem was the difficulty of reconciling the fact that the head (an early-maturing part) usually appeared to have its greatest proportionate size in the group exposed to the Low–Low treatment. This treatment subjected the animal to a low plane of nutrition from an early age and according to the theory, might have been thought to differentially impair the growth of the head. Even Wallace, one of the members of the group, observed that it appeared to him that the proportions of the skeleton of the pigs on all the treatments appeared normal for the size of the total skeleton independently of the nutritional treatments. This remark, undeveloped in Wallace's paper, held the key to a better understanding of what the results represented in adverse nutritional circumstances.

The major problem about taking live weight as the reference for slaughtering the animals is that it contains a highly variable component, namely the adipose depots. The treatments had a profound effect on the proportion of fat in the body (Table 6.1). Although the pigs in Table 6.1 were slaughtered at the same live weight (about 91 kg), the effect of the large variations in lipid content of the carcass was to cause a comparison to be made of pigs with vastly different lean body mass. Indeed the lean body mass of the Low–High group was less than 80% of that of the Low–Low group. The animals were therefore smaller, and in physiological terms effectively younger and less mature.

Could the systematic differences in the proportions of tissues within the lean body be largely accounted for by the differences in lean body mass and therein relative maturity? This problem was addressed by Elsley *et al.* (1964) and Fowler (1968) who reanalysed much of the data of the Hammond School

Table 6.1. Proportions of tissues in pigs slaughtered at 91 kg (after McMeekan, 1940a,b,c).

	Plane of nutrition			
	High–High	High–Low	Low–High	Low–Low
Lipid in carcass (%)	38	33	44	28
Lean in carcass (%)	62	67	56	72
Mass of lean as % of Low-Low	86	93	78	100

using regression analyses of the logarithms of the mass of the parts on the mass of the whole. In these allometric equations fat was excluded from the independent and dependent variables. The results proved beyond doubt that most of the apparent variation resulting from the effects of nutritional treatments could be accounted for by differences in the lean body mass. The functional integrity of the body was therefore maintained.

It should not be concluded from the above that nutritional changes cannot affect the proportions of the lean body at any given weight of the lean body. Several adaptations are possible which have a functional significance. For example, an animal which is enduring prolonged nutritional deprivation cannot afford the luxury of a metabolically expensive and active large gut. During prolonged inanition, the gut becomes proportionately smaller as do the associated organs such as the liver. In accordance, however, with the fourth Hammond principle, rehabilitation of the animal results in a rapid recovery of the gut to take advantage of the improved nutritional status.

A classical example of the above was again undertaken with pigs at Cambridge, but by a different group under the direction of Professor McCance. Pigs were maintained at juvenile weight of only 5 kg by chronic undernutrition for a year. Their appearance at the end of this period was surprisingly normal (Fig. 6.5). Several minor problems occurred such as impaction of the teeth which continued to grow slowly, and the skin became very wrinkled and hairy. The adipose tissue disappeared virtually completely. Although the jaw became lengthened to accommodate the teeth, the overall appearance of the animal in terms of the proportions of tissues and the disposition of the limbs remained extraordinarily normal. When these animals were restored to a normal diet they grew out perfectly normally and attained in most cases an adult size within 90% of the mass of the controls and appeared to be reproductively adequate (Lister and McCance 1967). This extraordinary resilience is a great testimony of the ability of an individual within a species to protect its vital functions in an adverse environment

It can therefore be proposed that a further principle governing the responses of animals to nutritional changes is as follows:

That the animal tends to adjust to environmental and nutritional changes in such a way that the vital functional relationships between essential body

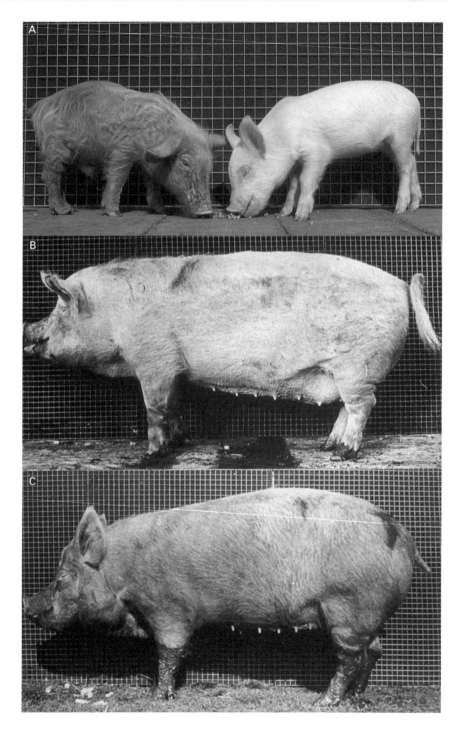

Fig. 6.5. (A) A one-year-old severely undernourished pig (left), a normal pig of the same weight (right); (B) the rehabilitated adult pig; and (C) the normal adult pig. Reproduced by kind permission of Dr David Lister.

components is preserved, or that the proportions are modified to a form which will give the animal its best chance of survival and successful reproduction.

6.6. Functional Units

Different components of the body have specialized functions, although they act in concert as far as the whole organism is concerned. Not all components, however, are equally vital to life and there is scope for some flexibility in the proportions of one to the other. Evaluation of many sets of data show that those tissues or parts which have a similar function or which are parts of a particular system such as the respiratory, digestive or circulatory systems tend to maintain a relatively strict proportionality within the system. These functional entities tend to operate as a unit so that they maintain their functional integrity. An illustration of the types of groupings which could be considered as functional units is given in Fig. 6.6.

The unit, however, may be scaled up or down in relative terms to accommodate the total functional need of the animal at the time. Ideas of this nature have been suggested by many authors over a considerable time. For example, Gutman and Gutman (1965) suggested that when two features are close to each other either structurally or operationally, they will be closer to each other in the sense of statistical correlations than features which do not share a similarity of function.

An interesting illustration of this is the way in which structures of containment such as the abdominal wall or the rib cage respond to changes in the size of the organs which they contain. Lodge and Heap (1967) were among the first to demonstrate that during pregnancy the muscle of the abdominal wall increased in mass relative to other muscles, but after parturition rapidly returned to proportions typical of the unbred animal. In a reanalysis of the 1952

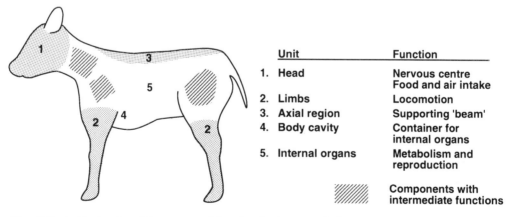

Unit	Function
1. Head	Nervous centre Food and air intake
2. Limbs	Locomotion
3. Axial region	Supporting 'beam'
4. Body cavity	Container for internal organs
5. Internal organs	Metabolism and reproduction
	Components with intermediate functions

Fig. 6.6. An illustration of the concept of functional units as applied to a representative farm animal.

Table 6.2. Comparison of relative proportions of the mass of organs contained in the rib cage of lambs subjected to different planes of nutrition compared with the relative proportions of the skeletal component of the rib cage (original data from Pállson and Vergés, 1952).

	Plane of nutrition			
	High–High	Low–High	High–Low	Low–Low
Organs of thorax and anterior digestive tract	100	97	101	115
Ribs + sternum	100	102	102	122

data of Pállson and Vergés, Fowler (1968) showed that responses to treatment, in terms of the organs of the thorax and anterior alimentary tract, were reflected in the relative weight of the skeletal tissues of the ribs and sternum (Table 6.2).

A further instructive example of the way in which function can be used to interpret changes in the proportion of tissues and organs is derived from work on the effect of daily injections of recombinant porcine somatotrophin by McKeith *et al.* (1989). The basic data showed that one effect of treatment was to increase the relative weight of components of the offal whilst at the same time reducing the amount of lipid in the carcass. An interesting question is whether the effects on the organs were independent of the weight of the lipid-free body. Some comparisons are shown in Table 6.3.

From the results in Table 6.3 it can be seen that whilst the stomach and heart tend to increase in proportion to the changes in lipid-free mass, those organs concerned most directly with the total metabolism, that is the liver and the kidney, increase proportionately more than the lipid-free mass, suggesting that

Table 6.3. Effects of treatment of pigs growing from 57 to 103 kg live weight with injections of porcine somatotrophin at either 0, 3 or 6 mg per day on lipid-free carcass and on organ weights. Results are expressed as proportions of the '0' dose group scaled to an arbitrary value of 100.

	Somatotrophin dose (mg day^{-1})		
	0	3	6
Fat-free carcass	100	111	118
Stomach	100	105	109
Heart	100	115	124
Liver	100	120	132
Kidney	100	124	137

one effect of somatotrophin is to differentially increase the overall rate of metabolism.

6.7. Conclusions

The choice of a basis for comparison for proportionate growth is intimately related to the type of question being asked. The baseline or independent variable is usually more informative if highly variable tissues such as fatty tissue and abdominal organs are excluded. If more detailed anatomical investigations are sought, then consideration should be given to using a very limited baseline along the lines of a functionally integrated unit. The complexity of growth makes rules for its analysis potentially dangerous. The touchstone of success is the ability to imbue a mass of data with clear simplifications which illuminate the important issues without invoking an impenetrable set of equations and without reducing the growth process to a naive list of masses and percentages.

References

Adam, C.L., McDonald, I., Moir, C.E. and Pennie, K. (1988) *Animal Production* 46, 131–138.

Bakker, H. and Koops, W.J. (1978) In: DeBoer, H. and Martin, J. (eds) *Patterns of Growth and Development in Cattle*. Martinus Nijhoff, The Hague, p. 705.

Berg, R.T. and Butterfield, R.M. (1976) *New Concepts in Cattle Growth*. University of Sydney Press, Sydney.

Brody, S. (1945) *Bioenergetics and Growth*. Reinhold, New York.

Butterfield, R.M. and Berg, R.T. (1966a) *Research in Veterinary Science* 7, 326.

Butterfield, R.M. and Berg, R.T. (1966b) *Research in Veterinary Science* 7, 389.

Dziuk, P. (1992) *Perspectives in Biology and Medicine* 35(3), 357–360.

Elsley, F.W.H., McDonald, I. and Fowler, V.R. (1964) *Animal Production* 6, 141.

Fowler, V.R. (1968) In: Lodge, G.A. and Lamming, G.E. (eds) *Growth and Development of Mammals*. Butterworths, London, p. 195.

Guttman, R. and Guttman, E. (1965) *Growth* 29, 219.

Hammond, J. (1932) *Growth and Development of Mutton Qualities in the Sheep*. Oliver and Boyd, Edinburgh.

Kirkwood, J.K. and Prestcot, N.J. (1984) *Livestock Production Science* 11, 461–474.

Lister, D. and McCance, R.A. (1967) *British Journal of Nutrition* 21, 787.

Lodge, G.A. and Heap, F.C. (1967) *Animal Production* 9, 237.

McDonald, I., Wenham, G. and Robinson, J.J. (1977) *Journal of Agricultural Science, Cambridge* 89, 373–391.

McKeith, F.K., Bechtel, P.J. and Novakofski, J. (1989) In: Vander Wal, P., Nieuwhof, G.J. and Politiek, R.D. (eds) *Biotechnology for Control of Growth and Product Quality in Swine. Implications and acceptability*. Pudoc, Wageningen, pp. 101–110.

McMeekan, C.P. (1940a) *Journal of Agricultural Science, Cambridge* 30, 276.

McMeekan, C.P. (1940b) *Journal of Agricultural Science, Cambridge* 30, 387.

McMeekan, C.P. (1940c) *Journal of Agricultural Science, Cambridge* 30, 511.

McMeekan, C.P. (1941) *Journal of Agricultural Science, Cambridge* 31, 1.

Pállson, H. (1955) *Progress in the Physiology of Farm Animals* vol. 2, Butterworth, London, p. 430.

Pállson, H. and Vergés, J.B. (1952) *Journal of Agricultural Science, Cambridge* 42, 93.

Pomeroy, R.W. (1955) *Progress in the Physiology of Farm Animals* 2, 395.

Richards, F.J. (1959) *Journal of Experimental Botany* 10, 290.

Richards, F.J. (1970) The quantitative analysis of growth. In: Steward, F.C. (ed.) *Plant Physiology*. Academic, New York.

Richmond, R.J. and Berg, R.T. (1971a) *Canadian Journal of Animal Science* 51, 31.

Richmond, R.J. and Berg, R.T. (1971b) *Canadian Journal of Animal Science* 51, 41.

Richmond, R.J. and Berg, R.T. (1971c) *Canadian Journal of Animal Science* 51, 523.

Taylor, S.C. (1980) *Animal Production* 31, 223.

von Bertalanffy, L. (1960) In: Nowinski, W.W. (ed.) *Fundamental Aspects of Normal and Malignant Growth*. Elsevier, Amsterdam, p. 137.

Wallace, (1948) *Journal of Agricultural Science, Cambridge* 38, 93, 243, 367.

Walton, A. and Hammond, J. (1938) *Proceedings of the Royal Society, London* 125, 311.

White, G. (1789) *A Natural History of Selborne*. (New edn published by Century Hutchinson Ltd, London, 1988.)

Wilson, P.N. (1952) *Journal of Agricultural Science, Cambridge* 45, 67.

Wilson, P.N. (1960) *Journal of Agricultural Science, Cambridge* 54, 105.

Efficiency and Growth

'All flesh is grass'. This biblical insight is the progenitor of many interesting questions. How much grass equals one cow, or how much barley equals one pig? One definition of growth is the process by which feed materials in the environment are incorporated into animal tissue. The provision of food for the animal is the major cost of most animal production systems. In the commercial world of animal production, 'increasing efficiency' has become a key strategy. It is the major goal of genetic improvement and of modern nutrition.

The purpose of this chapter is to illustrate some of the main concepts and give examples of useful approaches to understanding efficiency in relation to growth. The ultimate limitations to the drive for efficiency include not only the obvious laws of thermodynamics and of biochemistry, but also the limits set by the fundamental relationships between form and function.

7.1. Numerical Concepts of Efficiency

Efficiency is usually expressed as a ratio of output/input. For example, the recovery of energy in the animal body from the energy supplied in the food could be expressed as:

$$\frac{\text{Total energy in body gain}}{\text{Total energy in feed}}$$

This type of ratio is often converted to a percentage:

$$\frac{(\text{Total energy in body gain}) \times 100}{\text{Total energy in feed}}$$

There are exceptions to this general form which have become established by custom. For example, the reciprocal of efficiency is sometimes used, particularly in relation to the weight of feed required per unit of live-weight gain:

$$\frac{\text{Amount of feed consumed (kg unit time}^{-1})}{\text{Live-weight gain (kg unit time}^{-1})}$$

This format would normally be described as 'feed conversion ratio' or as 'feed conversion'. Unfortunately, it is sometimes referred to wrongly as 'feed efficiency', and this can cause confusion.

The spectrum of possible expressions of efficiency is very wide. It extends from purely economic considerations, such as monetary return / total monetary expenditure at one end to detailed efficiencies of biological processes at the other. For example, one may be concerned with the efficiency with which a dietary input of protein is recovered in the tissues of the whole animal:

$$\frac{\text{Gain of weight of protein in body tissues}}{\text{Weight of protein provided in feed}}$$

or the efficiency with which a limiting nutritional resource such as the amino acid lysine is recovered as lysine in the edible part of the carcass:

$$\frac{\text{Gain in muscle lysine}}{\text{Weight of lysine in diet}}$$

These illustrations show that the range of possible relationships can encompass efficiency at the macroeconomic and agricultural levels and can be extended in exquisite detail to cellular and biochemical levels.

7.2. Energy as a Baseline for Feed Input

Although many nutrients are required for growth, food energy is usually chosen as the base requirement and other nutrients are expressed in relation to it. Life itself is an energy-consuming process, a controlled combustion, and it is this concept which is encapsulated in the title of Max Kleiber's famous book *The Fire of Life* (1961). The carbohydrates, proteins and fats of food all act as fuel for the life processes of the animal. Describing them in terms of their energy-yielding potential on combustion is a means of exchanging each to a common currency.

Although feed materials are the 'fuel of life', energy is not strictly a nutrient. The energy is only released from the food by the complex process of metabolism. All the organic constituents of normal diets are susceptible to oxidation. Some molecules are not oxidized immediately but are reconstituted into new molecular structures which become incorporated into the animal's tissues. This is another way of viewing growth and can be described as 'chemical growth'.

The energy produced by physiological oxidation is harnessed by the animal in two main ways. First it is used for *work*, as for example powering the movement of its various muscles such as the heart pumping blood, the diaphragm and intercostal muscles cooperating in respiration, and the muscles of the limbs providing locomotion. Secondly, the energy is used to maintain body temperature which is effectively the generation of *heat*. If the animal has no opportunity to transfer work to the environment by for example walking up hill itself, then eventually all the work it performs within its own system is transferred to the environment as heat in some form. It is immaterial whether the total oxidation of the food chemicals occurs in the laboratory or in the animal, because the total energy yield from all the steps is the same in each case. This is of course consistent with the first law of thermodynamics which requires that all the energy transfers between a system and its surroundings be accounted for by the sum of the heat and work transferred between them.

7.3. Units of Energy

The various units used in science to describe amounts of energy, reflect the fact that energy can be measured both in terms of work or heat. Some definitions and derivations are given below.

7.3.1. The joule

The standard SI (*Système International*) unit of energy is the joule. It is defined as a force of 1 newton acting over 1 metre in the direction of action of the force. One newton is the force which when acting on a mass of 1 kilogram increases its velocity by 1 metre per second every second along the direction that it acts. In mathematical shorthand:

$$1 \text{ joule} = 1 \text{ kg } (m^{-2}s^{-2})$$

The non-physicist may have some difficulty in relating to a unit with this definition. For those with some understanding of electricity it may help to state the relationship between the joule and an electrical unit of energy:

$$1 \text{ joule} = 1 \text{ watt flowing for 1 second.}$$

7.3.2. The calorie

Historically the unit of heat used to describe the energy-yielding capability of foods (and fuels) on complete combustion was the calorie. The definition of the calorie is the amount of heat required to raise 1 gram of pure water from 14.5 to 15.5°C. The calorie has a constant relationship to the joule:

$$1 \text{ calorie} = 4.184 \text{ joules.}$$

The above definitions allow interconversion of all the units of energy but confusion can arise because of the long-term association of the word calorie with the energy in human food. The calorie defined above is sometimes called a 'small' calorie to distinguish it from Calorie with a capital C, which is equivalent to 1000 small calories or 1 kilocalorie. The calorie system is still widely used by the human food industry and often it is the large Calorie which is used, but unfortunately not always with the intended capital C. Scientific papers in Europe almost always report in the SI system using the joule. The feed industry and animal science groups of the USA still maintain their use of the calorie system but use kilocalorie rather than the large Calorie. Table 7.1 provides interconversion factors for energy units.

To show some of the interconversions listed in Table 7.1 in action, consider:

Heat output per day of a 70 kg human adult male =

10,125 kJ = 10.1 MJ = 2420 kcal = 2.42 Mcal =
117 W for 24 h = 2.81 kW h^{-1}

7.4. The Gross Energy of a Feed

The gross energy of a feed material can be considered as the heat generated when a unit mass of feed is completely combusted in oxygen to yield water and carbon dioxide under standardized conditions of temperature and pressure. The heat of combustion is measured in the laboratory by use of a combustion calorimeter. The most common form is often referred to as a 'bomb' calorimeter because the combustion chamber is a robust steel cylinder somewhat resembling a bomb. Essentially, a weighed amount of the test material is combusted in an atmosphere of pure oxygen and the heat generated collected as a temperature rise in an insulated water jacket surrounding the bomb. The apparatus is illustrated in Fig. 7.1

In chemical terms the combustion is regarded as a change of enthalpy in the combusted material and some textbooks refer to this as the enthalpy of

Table 7.1. Interconversion factors of energy units.

1 joule	= 0.239 calories
1 joule	= 1 watt flowing for 1 second
1 calorie	= 4.184 joules
	= 4.184 watts flowing for 1 second
1 kilocalorie	= 4.184 kilojoules
1 kilocalorie	= 1 'large Calorie' (not recommended)
1 megacalorie	= 4.184 megajoules
1 kilowatt h	= 3.6 megajoules

combustion. In nutrition science, the growth potential energy of a food is its heat of combustion. The fuel-like nature of foods is easily understood by reference to kitchen emergencies, such as chip-pan fires and the burning of toast. Domestic sugar (sucrose) makes an effective, though expensive, substitute for firelighters.

To illustrate some of the principles it is simplest to take a purified nutrient such as glucose:

Glucose plus oxygen → carbon dioxide plus water plus energy

The molecular format for this reaction is:

$$C_6H_{12}O_6 + 6O_2 \rightarrow (6CO_2) + 6H_2O + energy$$

In some cases the reaction is considered in terms of 'molar' proportions whereby the atomic weights of each of the atoms of each molecule are summated and converted to a weight. The method is shown below. Assuming the atomic weights of carbon, hydrogen and oxygen to be 12, 1 and 16, and the reference weight to be 1 g per atom of hydrogen, then:

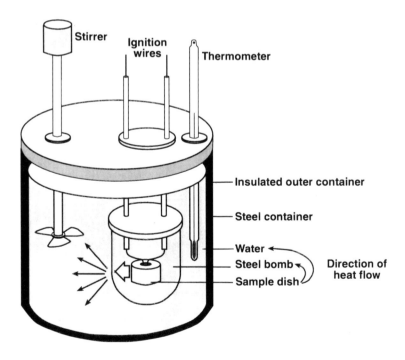

Fig. 7.1. Bomb calorimeter.

$[(C_6) 6 \times 12] + [(H_{12}) 12 \times 1] [(O_6) 6 \times 16] = 180\,g$ glucose
$+ [(6O_2) 12 \times 16] = 192\,g$ oxygen

yields

$[(6CO_2) 6 \times 12 + 12 \times 16] = 264\,g$ carbon dioxide
$+ [(6H_2O) 12 \times 1 + 6 \times 16] = 108\,g$ water
$+$ energy $(2806\,kJ)$

Expressed linearly this gives:

$180\,g$ glucose $+ 192\,g$ oxygen $\rightarrow 264\,g$ carbon dioxide $+ 108\,g$ water $+ 2806\,kJ$

The value 2806 refers to the kilojoules of heat liberated per gram molecule of glucose burned, that is 180g glucose This is equivalent to a gross energy density of 15.6kJ per g or 15.6 megajoules (MJ) per kg of anhydrous pure glucose. In utilizing published values of the heats of combustion, it is essential to be aware that they can be given either per unit weight of material or in the case of pure chemicals per gram molecule.

The purified constituents of food have characteristic energy-yielding capabilities when completely combusted. The main ones are given in Table 7.2, and a consideration of these values indicates immediately why both animals and plants use lipids as a major means of energy storage.

Table 7.2. Gross energy yields resulting from the complete combustion of dry purified food materials expressed as MJ kg^{-1}.

Material	MJ kg^{-1}
Glucose	15.6
Sucrose	16.5
Starch	17.5
Cellulose	17.5
Vegetable oil	38.9
Animal fat	39.4
Protein	23.6

7.5 Definitions of Feed Energy in Animal Systems

The animal is not efficient at transforming feed energy into its own body energy. There are several components to this inefficiency.

The initial energy of the feed or its 'gross energy' (GE) is simply the heat of combustion of a unit mass of the original material. Not all the substances contained in a feed enter into the metabolic processes of the animal and the visible evidence of this is the production of faeces. The energy lost in the form of faeces can be subtracted from the original gross energy of the corresponding feed. This is called 'apparent digestible energy'. The word apparent signifies that faecal matter is not strictly undigested material or its derivatives alone but contains substances which were once 'part' of the animal such as sloughed off cells from the wall of the gastrointestinal tract and the residual of secretions into the tract such as mucus. Therefore:

Apparent digestible energy (DE_a) = Gross energy of feed
− gross energy of faeces

Not all the apparently absorbed energy is useful to the animal. Energy is lost in the process of fermentation, particularly in ruminants through the evolution of combustible gases such as methane and hydrogen. A further loss of dietary energy occurs in the energy component of urine, due mainly to its content of urea which is an end product of protein breakdown or catabolism in the mammal. This is the main means by which the nitrogen from excess amino acids is eliminated. In birds the end product of amino acid breakdown is uric acid. When these losses of energy are subtracted from the apparent digestible energy, the balance is called metabolizable energy (ME), so that:

$ME = GE - DE_a$ − urinary energy − energy of combustible gases

Metabolizable energy is effectively available to the animal for its metabolism and is drawn on to produce either heat, work or growth. The list of products can be extended to offspring in reproducing females, wool and fibre, milk in lactation, and in the case of laying hens, eggs.

7.6. The Partition of Metabolizable Energy in the Growing Animal

Are some species more efficient than others? Are some individuals within a species more efficient than others? What effect does rate of growth have on efficiency? What is the optimum stage to slaughter the animal for maximum efficiency?

To dissect out some of the answers to these questions it is important to understand the interactions between growth and feed intake and also the nature of the energetic costs and overheads which are incurred during growth.

The partition of energy between the various functions is central to the

discussion of general aspects of efficiency in mammals. From the point of view
of efficiency of growth, it is clearly desirable that the maximum amount of the
dietary molecules are converted into the molecules of growth and the minimum
oxidized to carbon dioxide and water. One way of expressing this efficiency is
in terms of the usable energy supplied to the animal and the amount of that
energy retained by the animal or lost as heat:

$$ME = R + H$$

where ME = intake of metabolizable energy, R = energy retained as body
tissues, H = total heat loss from the animal

   The recovery of metabolizable energy from the diet in terms of the energy
of the tissues laid down is at best about 40% and in the majority of cases far less.
This figure is reduced much further if only the energy recovered in the edible
parts of the animal is considered. The components of the heat loss are
biologically of extreme interest and considerable scientific effort has been
expended in attempting a logical partition of the contribution of different
metabolic processes. The major factors which will be considered are:

1. The heat production which is the corollary of the animal staying alive.
2. The heat production associated with the deposition of protein or fat or with
positive energy retention in the animal.

   The principal factors affecting the efficiency of energy utilization for
growth are discussed below.

7.7. Maintenance and Basal Metabolism

The minimum heat production from a healthy animal is achieved when it is not
given food for some time (fasted) and is kept in a thermoneutral environment
with a minimum of activity. Cognate measurements are variously described as
basal metabolism (mainly used in human energy studies), or fasting metabolism,
or minimal metabolism, or postabsorptive metabolism. The experimental
conditions are relatively easy to achieve within the confines of an animal
calorimeter. Because of this measurements have been undertaken with a wide
range of species.

   The nutritional concept of maintenance though related to minimal metabolism is not the same, because the animal is not fasting but eating. The
metabolizable energy for maintenance (ME_m) is defined as the rate of heat
production of an animal kept in a thermoneutral environment, when the rate of
intake of metabolizable energy in feed exactly balances the rate of heat loss. An
animal which is fed so that it is stable in weight and also in chemical
composition over a period of time is in a state of maintenance. The maintenance
heat production (ME_m) is always higher than basal metabolism because the
process of eating, digesting and metabolizing food requires energy and this

eventually emerges from the animal as heat. The heat production of an animal on a maintenance diet is a summation of heat production of many processes. The major components are listed in Table 7.3. Maintenance heat production combines factors 1 and 2 listed in the table, whereas fasting metabolism excludes the factors associated with 1.

Basal metabolism and maintenance requirements are usually expressed on a daily basis. For example, maintenance energy supplied in the diet would be expressed as megajoules of metabolizable energy per day, or ME_m (MJ day^{-1}).

Maintenance and basal metabolism are also a function of the mass of the animal. Extensive studies on a wide range of species of different adult size and also on farm animals at different stages of growth show that daily heat production does not increase in direct proportion to live weight. Mathematical analysis of such data shows that the relationship is logarithmic, whereby the log of heat production on the log of body mass produces a linear relationship of the form:

$$\log \text{heat production} = \log a + b \log M$$

where b is a constant of 0.75 and M is body mass.

or

$$ME_m = M^{0.75} .$$

The exponent or scaling factor for weight, the power 0.75, is remarkably constant across a wide rang of studies. Just why the heat production of the animal per unit of weight declines with increasing size is not entirely understood. The simplest explanation is that the rate of heat loss from the animal is a function of its surface area and so varies as the 2/3 power of weight, i.e. $M^{0.67}$. There must also be a linear component perhaps relating to the absolute protein mass of the body and the interaction of these two components may account for the numerical raising of the exponent to the 3/4 power. Whatever the true explanation is, the relationship with weight raised to the 3/4

Table 7.3. Some major contributors to heat production in an animal receiving a maintenance supply of dietary energy (ME_m).

---

1. *Factors relating to the processing of the diet by the animal:*
   - Work done in location, prehension and mastication of feed
   - Work done by movement of digestive tract
   - Heat of fermentation of certain dietary constituents
   - Heat increment associated with the metabolic processing of nutrients
2. *Factors mainly associated with non-food-related activities:*
   - Maintaining body temperature
   - Work of circulation, respiration, maintenance of posture, standing and locomotion
   - Energy cost of basic metabolic processes including tissue turnover

---

power applies over the widest possible weight difference from shrews to elephants and allegedly to whales, although it is difficult to imagine accurate measurements with large marine mammals.

A general relationship with a wide application over all eutherian (placental mammals) relates to the minimum metabolism $H_{(min)}$, which usually has a value which is 70–80% that of ME_m.

$$H_{(min)} = 300 \text{ kJ } M(\text{kg})^{0.75}$$

Pigs are among the most amenable animals on which to conduct accurate measures and a survey by the working party of the Agricultural Research Council on the Energy Requirements of Pigs of all reported experiments up to 1981, suggested that for pigs ranging from 5 to 200 kg the following equation predicted the daily ME requirement for maintenance (ME_m).

$$ME_m = 458 \text{ kJ } M(\text{kg})^{0.75}$$

From such information it is possible to identify how much food will be required to meet the maintenance requirement of the animal at any weight. An illustration of this is given in Table 7.4.

From the point of view of the efficiency of growth, feed used for maintenance, though necessary, is an overhead cost. Since it is virtually a function of mass of the animal and time, it is pertinent to ask what scope there is for reducing this overhead to benefit the economy of growth? The answer is that since the maintenance requirement on any given day is constant, then the means for reducing the cost to a minimum is to aim for the maximum growth on that day so spreading the overhead cost over as many grams of gain as possible.

Increasing growth rate is an axiomatic means of increasing the efficiency of growth because it diminishes the overhead cost of maintenance.

Table 7.4. Maintenance requirement of pigs of different weights.

Live weight (kg)	Metabolic weight (kg)	ME_m (MJ day^{-1})	kg day^{-1} of standard feed (12 MJ ME kg^{-1})
5	3.44	1.58	0.13
10	5.62	2.57	0.21
20	9.45	4.33	0.36
40	15.91	7.29	0.56
80	26.75	12.25	1.02
160	44.99	20.61	1.72
320	75.66	34.65	2.89

7.8. The Utilization of Dietary Energy above Maintenance

When the daily feed intake exceeds that required for maintenance (ME_m), then there is a balance of energy available for growth. The energy retained by the animal in the growing tissues is less than the excess energy over maintenance because there is an energy 'cost' of growth. This efficiency, or inefficiency, of deposition is due to the cumulative inefficiency of all the biochemical reactions involved in the growth of a tissue plus the heat increment of the additional food (Fig. 7.2).

Understanding the energetics of growth can be simplified in one of two ways:

1. By considering energy retained in tissue as a uniform concept and defining the retained energy as ME_R and the efficiency of this process as k_R.
2. By separating the main chemical entities of growth into two components and defining them as energy retained as fat or lipid (F) with an efficiency of deposition of k_f and energy retained as protein (P) with an efficiency k_p.

The so-called metabolizable energy system for describing the energy

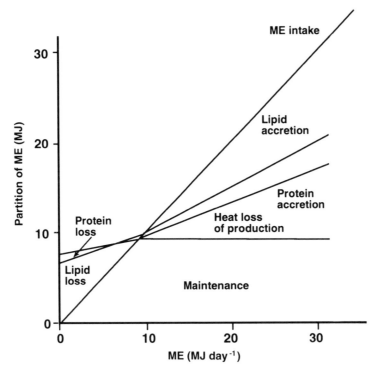

Fig. 7.2. Partition of energy in a pig of 60 kg live weight for different rates of daily intake of ME.

requirements of ruminants (Blaxter, 1967) is based on **1** above and identifies three components:

$$ME_{I} = F/k_{m} + R/k_{R}$$

where ME_{I} = intake of ME (MJ day^{-1}), F= fasting metabolism (MJ day^{-1}), k_{m} = efficiency of utilization of ME for F, R = retention of energy in body tissues, k = efficiency of utilization of ME for R.

If animals grew with a constant ratio of protein to lipid then there would be no difficulty in accepting method **1** as adequate. However, the complication arises when different animals or the same animals deposit different proportions of protein and fat because the energy cost of deposition differs. A greater amount of heat is generated per megajoule of energy stored as protein than for each megajoule stored as fat. This is because the deposition of protein in muscle is considerably more complex in operational terms than the deposition of fat in adipose tissue. In addition, protein undergoes continuous synthesis and degradation (protein turnover) and during active accretion it appears that there is an increase in energy 'wasting' turnover. Some efficiencies of energy deposition in tissues are given in Table 7.5.

This information provides the basis of a model for growth first proposed by the great Polish physiologist Kielanowski (1965). His model was generated from a large-scale experiment in which pigs were systematically fed different amounts of feed to ensure different growth rates, different body compositions and different days to slaughter. The whole empty body was then subjected to chemical analysis. His data allowed parameters to be fitted to the following multifactorial equation:

$$ME_{1} = (ME_{m} \times d) + (1/k_{p})P + (1/k_{f})F$$

where: ME_{I} = total metabolizable energy in feed over experimental period (MJ), ME_{m} = mean daily requirement for maintenance (MJ), d = days of experiment, P = energy retained as protein over experiment, k_{p} = efficiency of utilization of ME above maintenance for energy deposited as protein, F = energy retained as

Table 7.5. Efficiencies of energy deposition in tissues for different species.

Tissue	Symbol	Energy retained / ME above maintenance	
		Pigs	Ruminants
Lipid	k_{f}	0.74	—
Protein	k_{p}	0.54	—
All growth	k_{g}	0.70	0.32–0.55
Milk	k_{c}	0.65	0.56–0.66

fat over experiment, k_f = efficiency of utilization of *ME* above maintenance for energy deposited as fat.

Since this piece of pioneering work, the components of the model have been refined but the concept has been used in many academic and commercial models of animal growth. These models are now used extensively to predict the effects of dietary or genetic change on growth and efficiency.

7.9. Growth Rate, Feed Intake and Efficiency

Max Kleiber (1961) was among the first to recognize the intriguing nature of the interrelationship between intake and efficiency. He initiated a query which has become known as 'Kleiber's Conundrum', which in essence asks which is the more efficient way of converting 1 tonne of hay into meat, between allowing it to be eaten by a steer weighing 591 kg or by 300 rabbits of the same combined weight?

His method of calculation was instructive because he divided the daily intake of metabolizable energy of either a steer or a rabbit by its daily fasting heat loss to arrive at a value which he called the relative food capacity. By this means he showed that the relative food capacity of the steer and the rabbit are about equal, with an approximate value of 5. From this he concluded that the efficiency is in fact equal. The calculation ignores many other contributing factors to efficiency, including the energy concentration of the tissue and the stage of growth of the two species, but it is an amusing and informative exercise. Kleiber's calculations for these and others species are given in Table 7.6.

Although the data of Kleiber provide a good explanation of why the mass of animal is irrelevant to its potential efficiency, the weights of animal he chose are not really typical of animals at the mid-point of their economic growing period. More typical values drawn from information provided by the Agricultural Research Council (1980, 1981) are given in Table 7.7. Since fasting heat loss is already scaled to metabolic body weight across species, it is unnecessary for

Table 7.6. Calculations of the relative food capacity of different species (after Kleiber, 1961).

1	2	3	4	5
		Daily *ME**	Daily fasting	Relative food
	Liveweight	intake	heat loss	capacity
Species	(kg)	(MJ kg^{-1} $M^{0.75}$)	(MJ kg^{-1} $M^{0.75}$)	(col. 3/col. 4)
Chicken	0.08	1.50	0.33	4.4
Rabbit	2.4	1.85	0.21	5.1
Sheep	50	1.28	0.29	4.4
Pig	130	1.51	0.27	5.7
Cattle	435	1.72	0.35	4.9

*Metabolizable energy.

the sake of the comparison of intake capacity to include it in the calculation.

The results in Table 7.7 provide an interesting insight into perceived impressions of feed capacity. In terms of intake the non-ruminant pig clearly has a higher capacity to ingest metabolizable energy per unit of metabolic body weight than do the ruminant species. What is more surprising is that this is also true of dry matter intake. The explanation is of course that the ruminant is usually thought of as a consumer of roughage materials which have a high moisture content. Pigs on the other hand (and chickens) under normal farming practice eat cereal-based diets. A pig confronted only with the aerial parts of grass will struggle to meet even its basic maintenance requirement because of its inability to handle cellulose efficiently.

The effect on growth of increasing feed intake and the consequences for efficiency can be investigated using as a basis the information given above (Tables 7.8 and 7.9). The range of intakes given in Table 7.8 are consistent with the range encountered in field data and show clearly how, if all other things remain equal, that increases in feed intake have a profound effect on the efficiency of feed utilization.

Table 7.7. Calculations of relative feed intake of pigs, cattle and sheep when considered at the mid-point of their productive growth.

Species	Liveweight (kg)	Daily intake of feed dry matter (g kg^{-1} $M^{0.75}$)	Daily ME intake (MJ kg^{-1} $M^{0.75}$)
Cattle	300	89	0.98
Sheep	40	87	0.96
Pigs	60	125	1.71

Table 7.8. Effect of growth from increasing feed intake on the efficiency of food utilization. Modelled results for a pig over 20–90-kg-live-weight range.

% increase of feed intake over twice maintenance	1 ME intake (kJ kg^{-1} $M^{0.75}$)	2 Feed (g kg^{-1} $M^{0.75}$)	3 ME above maintenance (kJ kg^{-1} $M^{0.75}$)	4 Live-weight gain (g kg^{-1} $M^{0.75}$)*	Feed/gain ratio (col. 3/col. 5)
0	880	73	440	23.8	3.07
+20	1056	88	616	33.4	2.63
+40	1232	102	792	42.9	2.38
+60	1408	117	968	52.4	2.23
+80	1584	132	1144	62.0	2.12
+100	1760	147	1320	71.5	2.05

*Energy density of gain assumed to be 12 kJ g^{-1} and efficiency of energy retention in gain (k_g) 0.65.

Table 7.9. Effect on the feed cost of 1 kg body-weight gain of increases in the concentration of lipid in that gain.

g lipid in 1-kg gain	Energy required above maintenance for components of body gain: Lipid (MJ *ME*)	Protein: g	Protein: MJ *ME*	Total	Feed equivalent of *ME* (g kg gain^{-1})
0	0.0	213	9.0	9.0	750
50	2.8	202	8.6	11.4	950
100	5.5	192	8.1	13.6	1133
200	11.0	170	7.2	18.2	1517
400	22.0	128	5.4	27.4	2283

7.10. The Effect of Choice of Slaughter Weight on Efficiency

The gross efficiency of lean tissue production changes through the growth period. The lean tissue of the newly born animal has a high cost per kilogram because it is loaded with the 'maternal overhead'. Thus the cost of keeping the dam through the reproductive cycle must be shared between the viable offspring. As the animal grows the maternal overhead per kilogram of lean tissue or carcass diminishes. Eventually a new factor enters the equation which can be labelled as the cost of increasing maturity. The signs of this are:

1. Increasing fat associated with each kilogram of lean.
2. A steadily diminishing growth rate of the lean.
3. A stabilization of intake as it approaches its asymptote.

Associated with **1** there is the increased energy cost of the gain, and with **2** and **3** there is an increased maintenance overhead on each kilogram of lean. This is illustrated for the least complicated species, the pig, in Table 7.10.

A similar approach may be used to investigate the relative effects of changing the components of cost, if indeed such changes were possible. The figures given in Table 7.10 are based on the assumption that each component is changed independently of the others.

The values in Table 7.11, although calculated according to the best available information, are somewhat arbitrary. They show the relative priorities clearly. The effects of changes in 1 and 2, although of considerable interest from a biochemical point of view, make relatively little impact on overall efficiency. Although an extra two piglets born (change 3) is highly desirable, it is surprising how moderate is the impact on efficiency. Nevertheless if a quantum leap can be made, as would be the case with moving to prolific sheep or to twinning in cattle, then there are major possibilities. However if only 10% of females double

Table 7.10. Overall cost in terms of metabolizable energy (*ME*) of producing 1 kg of lean tissue in the carcass of the growing pig.

Live weight (kg)	Cumulative *ME* intake (MJ)	Weight of lean tissue (kg)	*ME* cost per kg of lean tissue (MJ)
Birth*	714	0.5	1428
10	832	4.3	193
20	1000	8.8	113
30	1240	12.9	96
40	1492	16.8	89
50	1756	20.5	86
60	2044	24.0	85
70	2356	28.0	84
80	2704	31.6	86
90	3100	34.2	91

*Maternal costs based on annual production of a sow producing 21 piglets and eating 1.25 tonne of standard feed.

their output of offspring per year then, in principle, the figures would be somewhat similar to those given for the pig. Change 4 shows the absolute importance of maximizing growth rate and change 5 the benefits of minimizing maintenance requirements by providing an environment within the thermoneutral zone.

The importance of a reduction in lipid associated with lean gain has been

Table 7.11. The reductions in cost of *ME* per kg of lean in a pig of 90 kg live weight (32.4 kg lean in whole animal) resulting from a change in a favourable direction of some major components of efficiency.

Component	Reduction of MJ *ME* required per kg lean tissue
1. 10% reduction in heat loss associated with lipid deposition	0.81
2. 10% reduction in heat loss associated with protein deposition	1.02
3. 10% increase in piglets per sow per year assuming, initially, 21 piglets per year	1.89
4. 10% reduction in days to slaughter	2.50
5. 10% reduction in heat production associated with maintenance	2.50
6. 10% reduction in lipid gain	2.82
7. 10% reduction in slaughter weight	4.57

mentioned earlier in this chapter. Perhaps the most surprising outcome is the critical nature of slaughter weight. There are many factors which have an impact on this, but the fact that the biological optimum may be some way off accepted practice needs consideration. The relative cost of diets and housing at different stages of growth must be taken into account.

In practice, two influences on ideal slaughter weight have tended to operate in rather opposing directions:

1. In the case of ruminants, increasing use of concentrated feeds has resulted in animals being fatter at a lighter weight. Because of a longstanding belief that some covering of fat is desirable, the optimal weight for slaughter from the butchering perspective is lighter than it would be otherwise.
2. In contrast to the above, genetic selection for efficiency and growth rate has tended to increase mature size. As a consequence, these 'advanced' breeds and strains are regarded as too lean at traditional slaughter weights and are also likely to be slaughtered at weights which are below the 'efficient' optimum.

Optimizing slaughter weight is therefore a compromise between biological efficiency and market preferences. This interface has become increasingly difficult and it requires a considerable degree of understanding for new production ideas to translate into the downstream economics of meat production.

7.11. Once-bred Gilts and Once-bred Heifers

Reference to this in a different context occurs in chapter 9.

The maternal overhead on meat production is very considerable if the sole purpose of the mother is to produce offspring for slaughter. In the case of the dairy cow, the maternal cost is offset by milk production and in the case of the fibre-producing mammals some abatement of the cost is provided by the sale of wool and mohair. However, in pigs and dedicated beef breeds the full impact of maternal cost is borne by the sale of the offspring for meat. Since most female animals are capable of reproduction at a stage of maturity where there is still considerable growth potential in the mother, approaches have been developed which seek to spread the cost of maintenance during pregnancy over productive maternal growth as well. When the mother has delivered her offspring or litter and sustained it over the initial period of dependency on colostrum, then she is slaughtered and replaced by another female on the verge of reproductive capability. This approach of 'kill the mother save the calf or piglet' gives two products for the price of one maintenance. The two products are maternal meat and offspring. It is an imaginative way of capitalizing on insights into biological efficiency.

There are of course some practical difficulties in that not every calf born will be female and not every litter of pigs will have a suitable genotype for a replacement female animal. There are a number of ways in which such difficulties can be overcome. One is to make the system a partial one, so that conventional and once-bred female systems work side by side. Another possibility is that it may soon become possible to determine the sex of offspring so that such a system could be virtually self-replenishing.

7.12. Efficiency, Slaughter Weight and Marketing

Efficiency of production, although a major interest to the farmer, is of little direct concern to the butcher as he strives to meet consumer concerns. The consumer's preferences for a particular size of joint such as a leg of lamb, a pork chop or a slice of bacon may determine whether or not meat is purchased from the retailer. Similarly the fat to lean ratio has historically influenced the trade in determining the preferred size of carcass to purchase. Butchering techniques and processing equipment are all geared to these traditions so that deviations from standardized slaughter weights are penalized. If the genetics of animals and the nutrition were also stable, one would expect the slaughter weight to closely approximate to the optimum in terms of both efficiency and marketing.

References

Agricultural Research Council (1980) *The Nutrient Requirements of Ruminant Livestock*. CAB International, Wallingford.

Agricultural Research Council (1981) *The Nutrient Requirements of Farm Livestock, No. 3: Pigs*. Agricultural Research Council, London.

Blaxter, K.L. (1967) *The Energy Metabolism of Ruminants*. Hutchinson, London.

Kielanowski, J. (1966) *Animal Production* 8, 121.

Kleiber, M. (1961) *The Fire of Life*. Wiley, New York.

Compensatory Growth

8.1. Introduction

Animals in the wild, particularly ruminants but also other herbivores, experience periods of alternating food abundance and poverty. Even under domestication the derivatives of these animals and others which humans have chosen to meet their needs do not always have sufficient food available at particular times to allow a full expression of their genetic potential for growth. In such cases a smooth progression along the sigmoid-shaped growth curve, predetermined for the individual by its genetic template, is disrupted. When this occurs and growth falls below genetic potential, it has been shown in many experiments that when food supplies again become abundant, growth rates accelerate and exceed those achieved by comparable animals fed well and continuously. This phenomenon is known as 'compensatory growth' and is a term which may be regarded as synonymous with the often-used alternatives of 'catch-up growth', 're-bound growth' and 'rehabilitative growth'. Whilst there are exceptions to this generalized picture, to be discussed later, this apparent tendency of animals to regain the position lost on their growth curves by exhibiting enhanced growth rates is both fascinating biologically and important economically. In terms of biology it is intriguing that nature has endowed animals with such an apparent ability to contend with fluctuations in food supply by 'storing' growth potential. Economically this ability allows owners of herbivorous animals, in particular, to plan feeding schedules so that maximum use of herbage grazed *in situ* can be realized whilst economizing on supplementary feeding during periods when natural food supplies are in deficit, for example in winter periods in temperate climates and in dry periods in other climates where rain and dry seasons alternate.

During the last 30 years six extensive reviews of experiments made on various aspects of compensatory growth in domesticated animals have been published (Wilson and Osbourn, 1960; Allden, 1970; O'Donovan, 1984; Ryan, 1990; Berge, 1991; Hogg, 1991). The first of these cast its net wide and considered work conducted on mammals, including non-herbivores, and birds.

The other five addressed a narrower field: cattle and sheep. This is perhaps understandable in view of the greater number of experiments made and because of the potentially greater economic importance of compensatory growth to these species. Additionally, however, work with mammals other than herbivores, and with birds, does not indicate in general that compensatory growth is exhibited to the same extent, if at all, and some possible reasons for this will emerge later. In ruminants and especially in cattle, compensatory growth tends to be greater when there is a change in diet type as well as in the amount of food offered. As a consequence of all of these factors, this consideration of compensatory growth will be orientated predominantly towards cattle grazed at pasture following periods of growth restriction, but will draw on evidence from other mammalian species, both herbivores and non-herbivores, and from birds, to illustrate where necessary particularly salient points. No consideration will be given here as to whether or not animals which have had their growth interrupted reach a normal mature size, since this is dealt with elsewhere (see chapter 6). The main concern will be a consideration of compensatory growth *per se*, the mechanisms which might be involved, which factors might affect it and what it is actually reflecting in the animal. As will become all too evident, there are considerable problems in interpreting data that have been published, and because of this a final section will attempt to focus attention on the pitfalls that await the unwary.

8.2. Factors Affecting Compensatory Growth

8.2.1. General factors

Wilson and Osbourn (1960) in their review of the literature, identified six factors which could affect compensatory growth (and ultimate compensation):

1. The nature of the restricted diet.
2. The degree of severity of undernutrition.
3. The duration of the period of undernutrition.
4. The stage of development of the body at the commencement of under-nutrition.
5. The relative rate of maturity of the animals concerned (whether species or breeds within species).
6. The so-called pattern of realimentation, that is whether or not sufficient food is available at all times after the growth restriction period.

For this discussion, the factors mentioned above can be re-arranged to place intrinsic features of the animal at the start of the period of undernutrition into one grouping and the dietary factors into another. The re-arrangement is shown in Table 8.1.

Table 8.1. Factors affecting compensatory growth.

Animal factors

1. The degree of maturity at the start of undernutrition: that is the proportion of expected normal mature mass already achieved
2. The proportion of body weight attributable to adipose depots at the start of undernutrition
3. The genotype
4. Changes in metabolic rate

Nutritional factors

1. The severity of the undernutrition, that is what fraction or multiple of the maintenance energy required is eaten on a mean daily basis
2. The duration of the period of undernutrition
3. The nutrient density of the food during undernutrition
4. Food intake during rehabilitation

8.2.2. Animal factors

Degree of maturity at start of undernourishment

In cattle there is only slight evidence to support the contention that the distance the animal is from its mature size at the start of any growth restriction has a part to play in affecting compensatory growth. There is some, but by no means conclusive, evidence to suggest that the younger the animal the more difficult it might be for it to exhibit compensatory growth. Berge (1991) examined the data from a large number of experiments in which growth restrictions had been experienced in beef cattle at various ages from birth up to 25 months of age, and whilst there are positive indications of a trend in this direction in this data the lack of total conviction in an age effect is apparent (Fig. 8.1). Other data do not allow firmer conclusions to be reached and one of the major problems is an inability to disentangle the often confounding effects of severity and length of the period of growth restriction experienced, attention to which is given later.

Adipose depots at the start of undernutrition

It is not illogical to assume that an animal with extensively developed adipose tissue, full of readily mobilizable lipid, may be able to withstand better, and for a longer period of time, restricted nutrition compared with an animal not thus endowed. Subsequently the degree of compensatory growth exhibited, if any, when nutrition improves, may depend on how severely such reserves and other tissues, such as muscle tissue, have been depleted and how far the chronological/physiological time axis has been distorted. Previous nutritional level and accompanying growth in relation to other factors such as weight relative to age, sex, severity and duration of the growth restriction and genotype, which are all

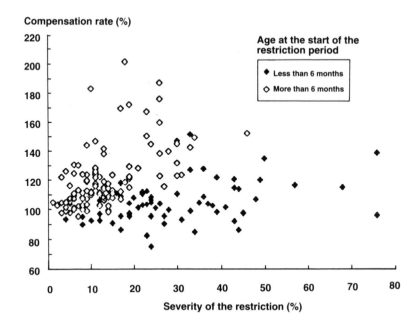

Fig. 8.1. Interaction between severity of food restriction and the age at the start of the restriction period on the subsequent compensatory growth in cattle. The severity of food restriction is expressed as the present live-weight difference between restricted and control animals at the end of the restriction period. The compensation rate is expressed as the ratio of the daily live-weight gain of previously restricted calves on the daily live-weight gain of control calves × 100. Data are from 74 experiments, each of the 177 points corresponding to a restricted group mean value (reproduced from Berge (1991) by kind permission of the copyright holder: Elsevier Science Publishers B.V., Amsterdam, The Netherlands).

discussed elsewhere in this chapter, can all interact in a most complex manner to fashion the ultimate response achieved within the period of time that the animal has to refashion its growth curve. For these reasons it is impossible to draw even the most tentative of overall conclusions and the reader is referred to chapter 6 where aspects of growth/nutrition interactions are considered.

Genotype

There is sparse information to indicate whether or not different genotypes within a species have different abilities to exhibit compensatory growth and that which is available does not allow any conclusions to be reached. For example, it would not seem unreasonable to assume in the case of cattle, and within the general framework of the points raised in section 8.2.2., that in two animals of similar live weight but of different maturation rates, and therefore of different body compositions (assuming equal and near optimal growth before restriction), the earlier maturing animal because it would have better deposits of lipid to draw on might therefore be in a better position to withstand nutritional restriction and to have a greater propensity for compensatory responses subsequently. No evidence can be found to support this hypothesis. Indeed the sparse information available in the literature on compensatory responses in different breeds of cattle is very equivocal (Lush *et al.*, 1930; Steensberg and Ostergaard, 1945a, b; Joubert, 1954; Brookes and Hodges, 1959; Bond *et al.*, 1972; Meadowcroft and Yule, 1976). The confounding animal factors referred to above again precipitate an impervious layer of complexity around all results that have so far been presented in the scientific literature. Again the reader is referred to chapter 6 for a basic understanding of how body composition is influenced by differences in maturation rates of different tissues and the interactions with sex, genotype and nutrition which can take place and which finally determine growth rate, size and body composition.

Changes in metabolic rate and nutrient utilization

Heart rate can be taken as a reasonable indicator of metabolic rate, and work with the horse (Ellis and Lawrence, 1978b) provides evidence of an immediate response in heart rate to an increased availability of food following a very marked decrease in heart rate during the winter period of food restriction (Fig. 8.2). These horses exhibited increased food intakes in the summer grazing period in parallel with the marked change in heart rate on experiencing an enhanced food supply at pasture, but the food intakes during the restricted winter period decreased as the period progressed, in parallel with the reduced heart rates (see also section 8.2.3). This apparent decrease in maintenance requirement due to a depressed metabolic rate suggests an animal becoming more and more efficient in utilizing a reduced food intake, but clearly if such a reduced maintenance need is induced via this route then it is not carried over into the period when food is freely available. The concept of a reduced maintenance requirement in animals recovering from periods of growth restriction, because of carry-over effects of lowered metabolic rates allowing more food to be available for growth purposes, therefore finds no support from these data. Confirmatory support for this is found in the work of Thomson *et al.* (1979), who measured fasting heat production during undernutrition and subsequently when food again became plentiful.

Some indirect calorimetric work points to a further twist to this spiral in showing that compensatory growth involves a shift in efficiency of use of metabolizable energy above and, for maintenance, a shift which stabilizes about 3 weeks after changing to a plentiful supply of food (Carstens *et al.*, 1987). This led this group to conclude that lower net energy requirements for growth and changes in gut fill account for a high proportion of compensatory live-weight-gain responses (Carstens *et al.*, 1988).

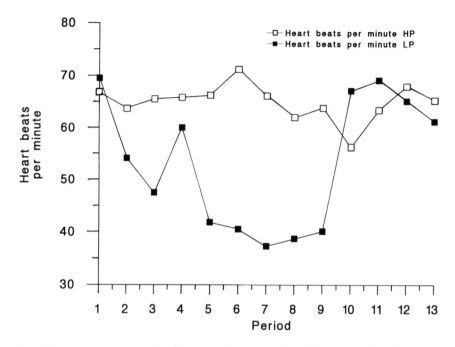

Fig. 8.2. Heart rates of filly foals during a winter period of 147 days (periods 1–9) on either a high (HP) or a low (LP) plane of nutrition (daily live-weight changes (kg) +0.4 and −0.01 respectively) and during a subsequent period of grazing common to all animals (periods 9–13) (based on the data given in Ellis and Lawrence, 1978b).

8.2.3. Nutritional factors

Duration and severity of undernutrition

It is nearly always difficult to disentangle the effects of severity on the one hand from duration on the other. Wilson and Osbourn (1960) concluded from an examination of a number of data sets that the nature of periods of growth restriction could be classified simply into three categories:

1. Severe restriction, resulting in loss in live weight.
2. Restriction, resulting in maintenance of constant live weight.
3. Mild restriction allowing small but subnormal increases in live weight.

They concluded that the degree of ultimate recovery, in which interest is not primarily held here, increased as the restriction became progressively less severe but that the degree of compensatory growth exhibited during the early part of the recovery period, in which interest is primarily held here, increased as the restriction became progressively more severe. Even here, however, it is impossible to separate the effect of severity from that of duration, both in the inhibitory and recovery periods. Berge's (1991) compilation of a large number of data sets gives some indication that cattle restricted before 6 months of age show limited compensatory growth subsequently, almost independently of the severity of the restriction, while cattle restricted at ages beyond 6 months exhibit compensatory growth proportional to the degree of restriction, that is the more severe the restriction the greater the compensatory response (Fig. 8.1). Therefore the proposals of Wilson and Osbourn, based on data sets from different species, find support from these studies of sets of cattle data, although an age threshold would appear to be important in limiting any generalizations made. However it would appear that growth restriction periods which are very long, for example several years, may inhibit the capabilities to exhibit compensatory growth (e.g. Hogan, 1929) (see also chapter 6). Some idea of the quantitative relationship between severity of growth restriction in cattle and subsequent compensatory growth is possible by examining the data presented in Fig. 8.3 where some of the data sets given in detail by O'Donovan (1984) are presented. In this figure the negative correlation between restricted and compensatory live-weight gains is evident. However, interestingly, it can be seen in several of the data sets that where live-weight gains in the winter were relatively high, subsequent pasture live-weight gains were depressed and not enhanced.

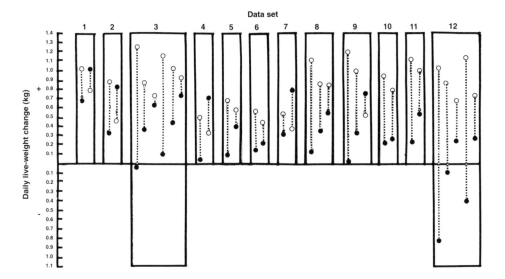

Fig. 8.3. Relationships between weight changes of cattle during restriction and subsequent compensation at pasture based on 12 data sets detailed in Table 2 of O'Donovan (1984). ● = winter; ○, = summer.

Nutrient density of food during undernourishment

> Germane to the consideration here of ruminant animals there is little evidence from experiments made to show that a protein deficiency is more important than an energy deficiency in affecting subsequent compensatory growth. Under practical conditions animals that are retarded in growth usually experience this retardation because of a deficiency in quantity of a diet which is overall reasonably well balanced with all nutrients. Therefore as this is largely synonymous with energy intake, and as on a worldwide basis energy is regarded to be of primary importance in relation to all naturally occurring dietary deficiencies (Huffman and Duncan, 1944), it is not too illogical to regard all other deficiencies as of secondary importance whilst at the same time accepting that deficiencies of specific vitamins and/or minerals could modify the picture but are outside the scope of this study. This is why in this case the prime consideration with cattle is that of compensatory growth in response to growth restriction previously experienced from a shortage of energy induced by intakes of reasonably well-balanced diets.

Food intake during rehabilitation

> Of all the factors considered to have a possible influence on compensatory live-weight gains, an increased appetite resulting in an increased food intake is generally thought to be the most important in the majority of circumstances. The effect can be very nearly all important: for example in pigs, Ratcliffe and Fowler (1980) considered that the compensatory growth which they observed in early life was due mainly, but not totally, to an increased food intake per unit of live weight. Other factors, such as changes in metabolic rate (see above) and effects of differences in maintenance requirements, can be tied into this factor to a greater or lesser extent.
>
> But why should there be an increased food intake? As will be discussed later, evidence of compensatory growth in the gastrointestinal tract is equivocal. If evidence were unequivocal then an understanding of a greater intake when food again becomes plentiful would be at least partially possible if the result were a larger digestive capacity in relation to body size. On the other hand, it has to be appreciated that the intake of herbage by grazing animals will reflect not only the space available in the gastrointestinal tract but will also be, in the first instance, a function of bite size of the animal and the time it spends grazing. In this particular context the findings of Ferrer Cazcarra and Petit (1995) suggest that increased intakes of herbage in relation to body size in animals previously retarded in growth in winter periods are related not to bite size, which is linearly related to live weight, but to a willingness to graze for longer periods of time and to increase this grazing time at a faster rate compared with animals fed better to grow faster previously.
>
> Enhanced food intakes have been recorded in horses placed on pasture following a winter period of food restriction (Table 8.2). In some studies with cattle, sward height (Steen, 1994) and stocking density (Wilkinson and Prescott,

Table 8.2. Weekly dry matter intakes (kg) of herbage grazed *in situ* by filly foals after subjection to high (HP) and low (LP) planes of nutrition during a winter period of 147 days (see Fig. 8.2 for heart rates) (Ellis and Lawrence, 1978b).

	Week after turning to pasture					
	1	2	3	4	5	6
Live weight (kg)						
HP	143	150	153	150	147	148
LP	112	122	127	127	130	136
Total intake						
HP	23.00	22.70	27.36	31.07	32.13	33.49
LP	26.09	27.76	36.66	38.89	40.10	36.06
Intake per unit live weight						
HP	0.161	0.151	0.179	0.207	0.218	0.226
LP	0.233	0.227	0.289	0.306	0.308	0.265
Intake per unit live weight $^{0.75}$						
HP	0.557	0.529	0.629	0.724	0.761	0.790
LP	0.758	0.756	0.970	1.029	1.041	0.906

1970; Drennan *et al.*, 1982) have been shown to affect intakes of previously retarded animals when changed to pasture grazing after winter feeding. Nevertheless, in spite of this fact, and as might logically be expected in a species with capacious total gastrointestinal tracts (contributed to significantly by the rumen) relative to live weight, in cattle enhanced food intakes after food restriction have been found in several experiments. The results of Wright *et al.* (1989) in Table 8.3 are of value in illustrating this particular point but importantly, in relation to point 6 raised by Wilson and Osbourn (1960) (see page 220), they show how consistent the comparative compensatory responses can be when food supplies vary in the recovery period. However, this is not to be confused with responses to fluctuating food supplies within an overall period of time in which growth rates of previously differently treated animals are being compared. As in the case of the horse data discussed previously, there is evidence of such an enhanced intake being preceded by a gradual reduction in intake as the winter period of restriction progressed (Lawrence and Pearce, 1964a). As a consequence, the summer grazing period could commence with animals containing, in comparison with their previously better fed contemporaries, a smaller gut fill. The smaller gut fill resulting from the previous restriction, which may or may not be accompanied by a smaller relative gut size, could contribute initially to a greater food intake and it is therefore likely that it could contribute too to the sharply enhanced live-weight increases

Table 8.3. Organic matter (OM) and digestible organic matter (DOM) intakes* at summer grazing in cattle (average age 240 days initially), previously given three planes of winter nutrition (H = high; M = medium; L = low) during a winter period of 182 days, and given access to sown pasture (S), hill reseed (R) or unimproved hill pasture (H) (Wright et al., 1989).

| | Summer pasture type[+] and winter nutrition | | | | | | | | |
| | S | | | R | | | H | | |
	L	M	H	L	M	H	L	M	H
Winter period									
Weight at start (kg)	212	218	206	212	218	206	212	218	206
Daily growth (kg)	0.50	0.75	0.96	0.50	0.75	0.96	0.50	0.75	0.96
Weight at turnout to pasture (kg)	303	355	391	303	355	381	303	355	381
Summer period									
Weight change in first 11 days (kg)	−43.5	−48.3	−57.5	−45.8	−47.7	−47.7	−44.2	−47.0	−56.9
Weight at end of summer (kg)	384	405	424	403	421	409	360	384	387
Daily growth (kg day^{-1})	1.07	0.86	0.71	1.16	0.94	0.72	0.78	0.54	0.51
Daily intakes									
OM (kg)	6.57	6.78	6.55	6.88	6.76	6.18	7.26	6.85	7.51
OM (g kg live weight^{-1})	19.1	18.3	16.2	19.8	17.9	16.4	22.6	19.0	20.5
DOM (kg)	5.32	5.40	5.23	5.15	5.06	4.62	4.74	4.46	4.89
DOM (g kg live weight^{-1})	15.5	14.6	18.0	14.8	13.4	122	14.8	12.4	13.4

*Authors state that intakes were determined immediately after turnout to pasture and at two later stages and that the overall means are presented because within-period intakes showed a similar pattern.

[+]For S and R pastures sward height maintained between 6 and 8 cm by addition and removal of non-experimental stock. For H pasture no similar grazing control. For first 11 days after turnout all animals retained on ryegrass pasture.

encountered in the initial part of the grazing period (e.g. see Table 8.4).

A final and particularly important point about food intake and its connection with live-weight gain is that retarded animals at the beginning of the grazing period will be smaller in size and will therefore have a smaller maintenance requirement. As a consequence an enhanced food intake in relation to size will imply that proportionately more food will be available for growth purposes.

Table 8.4. Daily weight changes (kg) in cattle during the first month of a grazing period and in each of the following 4 months following high, medium and low planes of nutrition in a preceding winter period of 168 days (for other data of this experiment see Tables 8.7 and 8.8) (Lawrence and Pearce, 1964a).

	Winter plane of nutrition		
	High	Medium	Low
Daily growth (kg)	0.73	0.33	0.01
Weight at beginning of summer grazing (kg)	155	125	101
Daily weight change			
First 4 days	-3.21 (12-)	-1.08 (11-)	+0.27 (6-)
	(0+)	(1+)	(6+)
Second 4 days	+0.48 (4-)	+2.28 (0-)	+3.17 (0-)
	(8+)	(12+)	(12+)
Third 4 days		no weights recorded	
Fourth 4 days	+1.06 (0-)	+1.68 (0-)	+1.68 (1-)
	(12+)	(12+)	(11+)
Fifth 4 days	+0.65 (2-)	+1.52 (1-)	+2.15 (0-)
	(10+)	(11+)	(12+)
Sixth 4 days	+1.46 (0-)	+1.90 (1-)	+2.55 (0-)
	(12+)	(11+)	(12+)
Seventh 4 days	+0.65 (3-)	+1.14 (0-)	+1.40 (0-)
	(9+)	(12+)	(12+)
1st month	+0.27	+1.28	+1.84
2nd month	+0.93	+1.40	+1.51
3rd month	+0.58	+0.96	+1.17
4th month	+0.22	+0.27	+0.41
5th month	+0.85	+0.97	+1.05

Numbers in parentheses refer to numbers of animals within a treatment group showing either positive (+) or negative (-) changes in live weight.

8.3. Components of Compensatory Growth

8.3.1. General

Compensatory growth is often described in the literature without any attempt being made to define the word 'growth' and there is often no indication of the changes taking place in the composition of the body of the animal for any given unit of live-weight change. In economic terms this is usually of greater importance to the animal owner than simple live-weight change, because ratios of carcass weight to live weight, and carcass composition, can play an important role in determining the profitability of the production system chosen to embrace the concept of compensatory growth. Changes in its body (carcass) tissues and organs are biologically important if the animal is to regain its normal functional integrity and re-establish its ability to withstand further periods of nutritional deprivation. In terms of both carcass and non-carcass components much may depend on the point on the growth curve at which growth retardation is experienced in relation to the earliness or lateness of maturity of the tissue or organ considered, as well as the severity of the deprivation and the time after deprivation at which measurements are taken. However some guidance is possible by consideration of the conclusions reached on age/live weight discussed previously, wherein on the whole younger, and therefore less mature, animals were shown to be less capable of exhibiting compensatory growth than older animals.

8.3.2. Changes in tissue proportions

Carcass tissues

First, what happens to bone growth when compensatory live-weight gains occur? It is valid to ask this question not solely because bone:muscle:fat ratios will be important as ultimate determinants of carcass quality, but also because the skeleton has vital functions in posture and movement and in containing and protecting soft tissues. Unless compensation in bone growth takes place within units of compensatory live-weight gain, then the ability to effect these functions and to match any enhanced soft tissue compensatory gains which occur, may be limited. There are two aspects in this consideration: bone strength, as reflected mostly in weight and therefore in mineralization, and bone size, that is length and diameter. In terms of mineralization there is evidence from the work of Wright and Russell (1991) of compensatory ash increases within units of compensatory live-weight gain at some, but not all, points in the recovery time period and these and other results from this carefully designed and executed experiment are discussed more fully below (Fig. 8.4, Tables 8.5 and 8.6). These results do not of course differentiate between ash deposited in bone and in soft tissues, and where studies of bone-weight gain have been made, compensatory

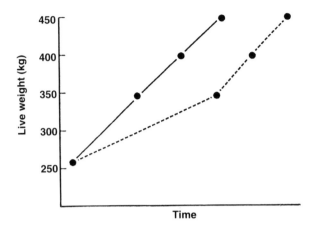

Fig. 8.4. Experimental design used by Wright and Russell (1991) to measure composition of compensatory live-weight gains. HH steers, ———; LH steers, - - - - -; each point (●) indicates the slaughter of six steers (i.e. after the slaughter of the initial group 18 steers were allocated to each of the H and L planes of nutrition at about 250 kg live weight and the six steers from each of the HH and LH treatments slaughtered at the live weights shown). The feeding level above 350 kg was such that at any given live weight food intake was the same for both treatments (see Tables 8.5 and 8.6 for results obtained) (reproduced by kind permission of the copyright holder: The British Society of Animal Production, Edinburgh, UK).

increases have not been found (e.g. Murray *et al.* (1974) in cattle and Drew and Reid (1975) in sheep). Increases in the size of skeletal parts, if not determined on bones from dissected animals, can be ascertained from radiographic studies and from measurements of height, length, width and girth taken on live animals as they grow (see also chapter 10). In the former, case studies on the horse have shown how this animal is apparently able to keep open epiphyseal plates in long bones during periods of food restriction, when it is ageing chronologically, to allow for the possibility of compensatory increases when food again becomes plentiful (Ellis and Lawrence, 1978a). External body measurements on both cattle (Table 8.7) and the horses referred to in the previous sentence (Table 8.8) also provide evidence of enhanced increases in skeletal size within compensatory live-weight gains. However a word of caution is necessary in both cases in terms of the way in which the results have been presented and the conclusions which can be drawn from them.

First, if the results presented in Tables 8.7 and 8.8 are considered, the following points emerge:

1. In spite of being held at near constant weight for 168 days, in all but one body measurement, there were increases in size, thereby implying a distortion in the external body conformation, the effects being less pronounced in measurements affected by long bone growth (e.g. height) than in those defining width, length

Table 8.5. Live weights (kg) and growth rates (g day⁻¹) of cattle fed to grow on a high plane of nutrition throughout (H) or a low plane of nutrition followed by a high plane of nutrition (LH) (see Fig. 8.4. for experimental design and Table 8.6 for composition of weight gains) (Wright and Russell, 1991).

	Plane of nutrition	
	LH	HH
Start to nominal weight range		
No.	18	18
Initial live weight	259	258
Final live weight	357	356
Live-weight gain	450	780
350–400 kg nominal weight range		
No.	12	12
Initial live weight	354	355
Final live weight	403	410
Live-weight gain	1350	980
400–450 kg nominal weight range		
No.	6	6
Initial live weight	410	407
Final live weight	447	458
Live-weight gain	1380	1200

and girth changes, which are in some cases more dependent on soft tissue deposition (these animals therefore becomes relatively tall, thin, narrow and shallow relative to their live weight) (Fig. 8.5).

2. The single exception to increase, that of circumference at the navel, is interesting in that a decrease occurred in the winter period at near constant live weight whilst the recovery was greatest in this measurement during the summer period (as this measurement is influenced considerably by gut fill it is a factor to bear in mind in relation to food intake considerations).

3. The effect on this latter measurement in a non-ruminant herbivore, the horse, is clearly less pronounced (Table 8.8) during a winter period of near constant live weight but the change in the summer again shows the most pronounced compensatory effect of all the measurements considered.

Therefore first impressions are of enhanced increases in skeletal measurements, and therefore probably in bone growth in size if not in density, within compensatory live-weight increases. However if the increases in skeletal measurements are now considered in relation to the actual live-weight changes recorded (Table 8.9), a somewhat different interpretation of the results emerges.

Table 8.6. Composition (g kg⁻¹) and estimated energy content (MJ kg⁻¹) of weight gains of steers in the experiment of Wright and Russell (1991) (for details of experimental design see Fig. 8.4 and for details of live weights and live-weight gains see Table 8.5).

	Live-weight period and treatment					
	Start to 350 kg		350–400 kg		400–450 kg	
	LH	HH	LH	HH	LH	HH
Empty body weight gain						
Water	630	553	663	422	491	744
Fat	92	229	108	311	291	67
Ash	57	35	12	94	62	−14
Protein	220	183	216	173	156	203
Energy	9.06	13.54	9.56	16.48	15.34	7.58
Carcass weight gain						
Water	596	542	693	444	437	739
Fat	108	236	106	256	369	94
Ash	62	34	24	73	57	3
Protein	235	189	176	226	136	164
Energy	10.00	13.95	11.56	15.56	15.27	7.74
Non-carcass weight gain						
Water	690	571	569	381	568	767
Fat	66	221	98	418	167	0
Ash	52	37	29	136	71	−67
Protein	190	169	303	63	194	300
Energy	7.39	12.89	11.27	18.00	11.16	7.25

The ratios of Table 8.9 show that in the case of both cattle and horses the increases in the body measurements were in nearly all cases smaller per unit of live-weight increase in the previously retarded animals when grazing summer pasture compared with the ratios obtained from the high plane animals in the winter periods. The important point here is that the experimental design allows the comparison to be based on similar initial weights but on different chronological ages. The possibility of a loss in synchrony between chronological and physiological age therefore emerges. The high plane animals in the winter period appear to have had a higher potential for bone growth in relation to live-weight increases compared with the previous low plane animals during the latter's subsequent compensatory periods, even though their live-weight increases were smaller. Also the better fed animals in the winter had similar overall advantages over those less well fed when comparative ratios for both winter and summer periods are compared. This is interesting because the ability of the less well fed animals in the winter period to exhibit growth in body measurements in that period, when live weight was more or less constant, was

Table 8.7. Body measurements (cm) taken on beef cattle at about 1 year of age prior to subjection in a winter period of 168 days' duration to different nutritional planes to give three different daily live-weight changes followed by a grazing period similar for all animals of 140 days. Body measurements and live weights at the end of the summer and winter periods are expressed relative to the measurements at the beginning of these periods where these are taken as 100 (Lawrence and Pearce, 1964a).

	Plane of nutrition in winter period		
	High	Medium	Low
No.	12	12	12
Start of winter period			
Live weight (kg)	224	224	223
Hooks width	35.6	35.3	34.9
Pins to hooks	39.5	38.9	39.0
Chest depth	52.9	52.6	52.4
Shoulder height	108.4	106.9	107.8
Elbow height	67.2	66.2	67.4
Navel circumference	174.6	180.8	177.4
Cannon bone circumference (fore)	15.3	14.8	14.9
End of winter period			
Daily growth in winter period (kg)	0.73	0.33	0.01
Live weight	155	125	101
Hooks width	117	110	104
Pins to hooks	114	108	106
Chest depth	115	110	105
Shoulder height	110	109	105
Elbow height	107	107	104
Navel circumference	113	101	93
Cannon bone circumference (fore)	110	109	103
End of grazing period			
Daily growth in grazing period (kg)	0.56	0.98	1.20
Live weight	123	149	175
Hooks width	107	113	117
Pins to hooks	107	111	111
Chest depth	107	110	113
Shoulder height	105	107	109
Elbow height	106	106	107
Navel circumference	103	113	122
Cannon bone circumference (fore)	107	110	116

clearly insufficiently great to give equalization in the ratios overall within the total (winter and summer) live-weight gains made. Therefore, whilst the results point to enhanced increases in skeletal size within compensatory weight increases in previously retarded animals, clearly considerable care is needed in

Table 8.8. Body measurements (cm) taken on filly foals at about 1 year of age prior to subjection in a winter period of 180 days' duration to different nutritional planes to give two different daily live-weight changes, at the end of the winter period and at the end of a subsequent grazing period of 180 days similar for all animals (Ellis and Lawrence, 1978a).

	Plane of nutrition in winter	
	High	Low
No.	18	18
Start of winter period		
Live weight (kg)	103.8	106.5
Hooks width	28.9	28.9
Pins to hooks	30.0	31.3
Chest depth	41.7	41.2
Withers height	104.2	105.6
Elbow height	69.2	69.7
Navel circumference	125.7	122.1
Cannon bone circumference (fore)	13.1	13.2
End of winter period		
Daily weight change in winter (kg)	+0.38	-0.01
Live weight (kg)	171.7	106.4
Hooks width	6.5	1.1
Pins to hooks	6.2	1.3
Chest depth	6.5	2.0
Withers height	12.0	3.5
Elbow height	5.3	2.3
Navel circumference	13.6	0.7
Cannon bone circumference (fore)	1.7	0.4
End of summer period		
Daily growth in summer (kg)	+0.53	+0.76
Live weight (kg)	266.5	244.1
Hooks width	6.9	10.1
Pins to hooks	5.3	7.9
Chest depth	6.1	8.9
Withers height	8.1	12.9
Elbow height	3.3	5.9
Navel circumference	29.7	42.0
Cannon bone circumference (fore)	1.9	3.3

the way in which data are presented and are interpreted. As mentioned previously, some further thought will be given later to interpretation of compensatory growth data.

If part of compensatory live-weight gain is likely to be due to an enhanced growth of bone in the carcass, to what extent are the soft tissues and their component parts involved? This is a very difficult question to answer and will

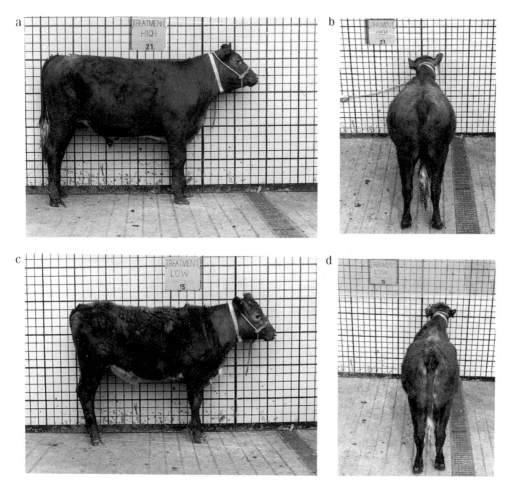

Fig. 8.5. Profile views (a, c) and rear-end views (b, d) of cattle after a winter period of 168 days' duration growing at 0.73 kg day^{-1} (a, b) or 0.01 kg day^{-1} (c, d) from initial live weights of 224 kg (a, b) and 223 kg (c, d) and similar initial ages of about 1 year. (Animals are representative of those used in the experiments of Lawrence and Pearce (1964a); see Tables 8.7 and 8.9 for details of body measurement changes.)

depend on a very large number of factors, such as the severity of the growth suppression in relation to the composition of the body initially: for example, adipose tissues could be depleted of lipid to varying extents and some organs and tissues containing protein may be depleted too. It follows that the proportions of soft tissues deposited in the carcasses of animals subsequently exhibiting compensatory live-weight gains might vary considerably and because of this and the complicating effects of other variables it is impossible to draw unequivocal conclusions.

One of the major complications is that of being able to compare animals at the same physiological age and to eliminate differences in food intake in relation

Table 8.9. Ratios (×10³) of increases in body measurements (cm) to live-weight increases (kg) in cattle initially about 1 year of age (Lawrence and Pearce, 1964a) and filly foals initially about 1 year of age (Ellis and Lawrence, 1978a) during winter periods for high plane nutrition animals and for summer grazing and for winter and summer grazing periods together for animals on a low plane of nutrition in the winter period (see Tables 8.7 and 8.8 for the other data from these experiments).

| | Animals from high plane of nutrition in winter period | | Animals from low plane of nutrition in winter period | | | |
| | | | Summer grazing period | | Winter and summer grazing periods combined | |
	Cattle	Horses	Cattle	Horses	Cattle	Horses
Duration of period (days)	168	180	140	180	308	360
Initial live weights (kg)	224.0	103.8	223.0	106.5	223.0	106.5
Live-weight gain (kg)	123.0	67.9	168.0	137.7	170.0	137.6
Ratios for:						
Hooks width	50.0	96.0	37.0	73.0	45.0	81.0
Pins to hooks	45.0	91.0	28.0	57.0	41.0	67.0
Chest depth	67.0	96.0	43.0	65.0	59.0	79.0
Shoulder/withers height	89.0	177.0	62.0	94.0	93.0	119.0
Elbow height	36.0	78.0	29.0	43.0	45.0	59.0
Navel circumference	184.0	200.0	222.0	305.0	145.0	310.0
Cannon bone circumference (fore)	13.0	25.0	14.0	24.0	17.0	27.0

to this. It is therefore opportune to be able to draw the reader's attention to an experiment in which the design allowed comparisons of the body composition of cattle exhibiting compensatory live-weight gains to be made when they were consuming the same amount of food over a similar live-weight range as the control animals (that is, there was no increase in food intake as a result of previous restriction). This experiment, conducted by Wright and Russell (1991), showed that following a period of food restriction the empty body-weight gains and carcass gains were relatively similar and that initially they were composed of increasing proportions of protein and water and a reduced proportion of fat compared with unrestricted cattle when both were given the same amount of food and were compared at the same live weight. A second phase followed, in which the proportion of fat increased and the proportions of water and protein decreased. Some details of this experiment are given in Fig. 8.4 and in Tables 8.5 and 8.6, and while no firm conclusions can be drawn to cover all eventualities the data do point to cattle attempting to regain a 'normal' body composition by altering the tissues which they deposit within units of compensatory live-weight gain. This, and other data, allows the tentative conclusion that in certain situations, and arguably from about 6 months of age onwards, and when growth has been reasonably good up to that point, periods of growth restriction, even to the point of maintaining live weight constant for about 6 months, will have little effect on carcass composition in cattle when live weights which are proportionately between about 0.4 and 0.5 of their mature size, are reached.

Non-carcass tissues

If enhanced growth in carcass tissues contributes to compensatory live-weight gains, do non-carcass parts and tissues exhibit a similar response and contribute as well? In the first place there is good reason to presume that some of the visceral organs might play an important role because of:

1. Their high contribution to total body weight (proportionately from about 0.07 in the horse to 0.17 in the shrew, with values for the liver and empty gastrointestinal tract of 0.07 for sheep and 0.10 for cattle).
2. The disproportionate contribution which they make to whole-body metabolism in relation to their size (proportional heat production, oxygen consumption and protein synthesis in relation to the total body is between 0.40 and 0.50 (Webster, 1989)).

Coupled with this there is good evidence of the visceral organs exhibiting quick responses to undernutrition by reducing their size (e.g. Drouillard *et al.*, 1991) and in reducing their metabolic activity (e.g. Lomax and Baird, 1983). The liver has been shown to lose considerable proportions of its weight in the early stages of restriction (Seebeck, 1973) and with the heart and the hide can be considered as a source of labile protein in times of nutritional stress (Winter *et al.*, 1976). If the responses in this direction are strong and positive it seems feasible that they might be equally strong and positive to enhanced nutrition. What is the evidence?

Again much will depend on the factors which have been considered already as complicating interpretations of carcass component increases, but it is generally accepted that at least for a short period of time some organs will exhibit a considerably enhanced growth rate after having been retarded to a greater extent than, for example, carcass components during undernutrition. In some cases this can be regarded as a direct response to increased metabolic activity. Thus organs not directly associated with digestive metabolism such as the heart and lungs may exhibit initial spurts of growth in response to increased metabolic activity while blood volumes may increase appreciably too. Organs more closely associated with digestion, such as the gastrointestinal tract and the liver, would perhaps be expected to show larger and more sustained enhancements of growth. In the case of the liver it is understandable that the initial response might be large when it is considered that increased stores of glycogen will be associated with four times their weight of water. Indeed in some cases there is evidence that enhanced growth in the liver might continue for a long period of time, giving some degree of 'over compensation' (Lawrence and Pearce, 1964b). In this work, referred to previously (see Tables 8.7 and 8.9), the liver weights at a common slaughter weight of 474 kg, at the end of the grazing period, were on average proportionately 0.08 heavier in the medium and low plane animals compared with those wintered on the high plane of nutrition. Also in this work the freely drained blood at slaughter increased progressively as the winter plane of nutrition decreased, proportionately 0.03 from high to medium planes of nutrition and 0.10 from low to high planes of nutrition. On the other hand, the contribution to the total live-weight increases over the 140-day recovery period at grass would have been quite small, about 0.45 kg on average in the case of the liver and about 12.5 kg in the case of the freely drained blood. The contributions to the initial compensatory live-weight gains could, however, have been considerably greater.

From the non-ruminant animal there is some supportive evidence of compensatory growth being associated mostly with non-carcass components (Tullis *et al.* 1986). These workers studied nitrogen retention in pigs given extravagant nitrogen intakes following low intakes and suggested from their results that the phenomenon of compensatory growth is associated principally with the replenishment of labile nitrogen stores in the skin, in the viscera and in the blood. This suggestion was based on the assumption that compensatory responses gradually disappeared, presumably as nitrogen stores became replete, and that skeletal muscle is likely to be relatively well protected during nitrogen deprivation. Other work with pigs tends to support these views (Ratcliffe and Fowler, 1980). In these studies pigs were restricted in growth in early life and the following compensatory live-weight gains which were exhibited on changing to a high plane of nutrition were, apart from increased food intake as pointed out previously, due mostly to increases in the internal organs and in the gastrointestinal tract. Because the latter organ is large in relation to live weight it is reasonable to assume that it could play a more important role than other organs in contributing to compensatory live-weight increases and that, accordingly, a separate and more detailed consideration is justified.

Lipid deposited in adipose tissue in and around the gastrointestinal tract is possibly used during periods of nutritional deprivation to meet immediate needs and then redeposited when food again becomes plentiful (Iason and Mantecon, 1993). Therefore a contribution to compensatory live-weight increases will be made. In terms of the gastrointestinal tract itself, however, it is difficult to draw conclusions about the part it might play in affecting compensatory live-weight changes. Several hypotheses have have been proposed, including that of atrophy leading to a reduced metabolic requirement for maintenance during periods of food shortage leading to rapid growth subsequently, in particular to cope with increased food intake, and these possibilities have been alluded to previously. Enticing as this hypothesis might be, unequivocal confirmatory evidence is not easy to find and Iason and Mantecon (1993) could not confirm it in growing lambs. Nevertheless, and particularly in the light of the work with pigs discussed previously, it seems feasible that the proposals of Hovell *et al.* (1987) are reasonably valid: that an enhanced rate of protein synthesis and deposition in the tract occurs initially when animals are switched from a low to a high plane of nutrition. Acceptance of this implies that there would be a contribution to live-weight gain at this stage. Certainly it is correct to regard this organ as highly demanding in energy (Webster, 1989; Ryan, 1990) and if, amongst those factors contributing to compensatory live-weight gains, an enhanced food intake is partly responsible, then an organ of appropriate size to cope with this is necessary.

8.4. Compensatory Growth and Overall Efficiency

The reviews of compensatory growth referred to at the beginning of this chapter do not allow conclusions to be reached on the comparative biological efficiency of animals grown without interruption at a rate at or near to their genetic potential compared with those retarded but then fed in such a way as to exhibit compensatory growth. Reasons for an inability to draw conclusions are not hard to find if the complications of the foregoing discussions are borne in mind. It is difficult to envisage that growing animals retarded at some point or points in their lives could be more efficient than animals fed continuously on a high plane of nutrition. The daily maintenance requirements of the former would at certain times be less than those of the latter, but the latter would usually carry their total maintenance costs over a shorter period of time and therefore if the composition of the growing parts were reasonably similar would be more efficient. This would be the likely outcome because the usual enhanced growth rates in the initial part of the recovery period would decline so that to reach a predetermined live weight at which comparisons are made of, for example, body composition and organ weight, a longer period of chronological time would be needed (see Fig. 8.6). However in species where a much greater degree of control over food intake is possible and where shorter periods of retardation in relation to overall growth are imposed, results show that restricted growth followed by compensatory growth can give a greater efficiency than continuous growth (Table 8.10).

A further complicating factor will be the way in which the animal is partitioning nutrients at any point in time in the recovery period. This needs no further elaboration here as the efficiencies of deposition of the various tissues is discussed fully in chapter 7, but the point to make is that differential tissue deposition within units of live-weight compensatory growth complicate any assessment of relative efficiency unless body compositional changes are in some way monitored regularly over the period under consideration. Note for example the shifting composition of gains made by cattle in the work of Wright and Russell (see Table 8.6) and referred to already.

Although there are difficulties in reaching conclusions on biological efficiency, there can be no doubt that to the animal owner a 'sensible utilization' of the phenomenon can yield him a greater economic efficiency in many

Table 8.10. Efficiency of utilization of food in turkeys given different amounts of dietary protein within the overall time interval from hatching to 14 weeks of age (Auckland et al. 1969).

	Diet*[†]									
	110		97.5		85		72.5		60.0	
	H	M	H	M	H	M	H	M	H	M
Live weights (kg)										
At 6 weeks	1.289	1.293	1.280	1.270	1.187	1.189	0.970	0.966	0.666	0.673
At 14 weeks	5.004	4.851	4.953	4.954	4.883	4.911	4.774	4.865	4.377	4.409
Percentage relative daily growth (6–14 weeks‡)	2.423	2.362	2.419	2.431	2.525	2.533	2.846	2.885	3.360	3.358
Food intake (g day^{-1})										
0–6 weeks	57.9		55.8		53.9		48.2		38.2	
6–14 weeks	218.2	216.6	221.4	221.2	213.6	208.7	208.0	215.8	191.7	196.0
g food g growth^{-1}										
0–6 weeks	1.96		1.92		1.99		2.20		2.60	
6–14 weeks	3.29	3.41	3.38	3.36	3.24	3.14	3.06	3.10	2.89	2.94
0–14 weeks	2.94	3.01	2.97	2.97	2.91	2.84	2.85	2.89	2.81	2.82
Overall efficiency: g dietary protein g growth^{-1} (0–14 weeks)	0.81	0.68	0.80	0.65	0.77	0.61	0.74	0.60	0.73	0.58

*0–6 weeks – calculated amino acid content of diet as percentage of requirement: 110, 97.5, 85, 72.5, 60.0.

[†]6–14 weeks – protein concentration in diet: H, 140%, and L, 110% of lysine required for maximum growth from 6 to 12 weeks of age.

‡((log$_e$ (weight at 14 weeks) – log$_e$ (weight at 6 weeks)/56 days) × 100).

situations for a production enterprise overall. The phrase 'sensible utilization' needs stressing because on welfare grounds it would be very wrong to subject animals to undue food restriction. The word 'sensible' implies that the phenomenon is well understood and that food restriction to retard growth when natural grazing is short or non-existent is such that animals do not suffer but are kept in a lean, fit and hard condition without too much supplementary feeding so that growth from grass grazed *in situ*, as a relative cheap nutrient source, is maximized. To go beyond such a generalization is difficult but some pointers to the responses that can be anticipated from cattle have been referred to already.

8.5. Compensatory Growth: Problems of Interpretation

In chapter 4 (section 4.2.1) the concept of the overall control of growth by homeostasis and homeorhesis was presented. From the foregoing discussion obviously both are likely to be involved in influencing compensatory growth: homeostasis in the short term and homeorhesis in the long term. Homeostasis would seem very likely to be important in affecting the postulated changes in metabolic rate, in directing nutrients to replenish depleted labile reserves and in stimulating directly or indirectly an increased food intake. The overall effect and route of influence of homeorhesis are not so easily identified, but the general thesis would be that disruption of the biological clock of the animal caused by removal of it from its genetically programmed and controlled growth pathway to maturity, needs to be redressed and that, accordingly, homeorhesis gradually affects this redress. Within this framework of both quick and longer-termed adjustment, from the previous discussion it is clear that interpretation of exactly what is happening when is very difficult and baselines for comparisons need to be chosen very carefully if interpretations are not to be misleading or at worst totally wrong. A few comments to elaborate on problems of interpretation therefore seem very appropriate and worthwhile. To illustrate problem areas of interpretation, and to correlate as closely as possible with the main thrust of the discussion presented so far, the model chosen will use cattle which are retarded for periods of about 6 months from an initial age no younger than about 6 months and then given about 6 months to recover on more or less an *ad libitum* intake of grass at pasture. Consideration of problems of interpretation will therefore be confined to a chosen segment of the sigmoid-shaped curve dictated largely by practical and/or experimental considerations.

In Fig. 8.6 four different hypothetical animals are considered in a winter period of food scarcity: animal 1 loses live weight, animal 2 maintains live weight constant, animal 3 increases in live weight but at a rate below animal 4 which is fed *ad libitum* to allow full expression of its genetic potential. The starting point S is common to all animals and the finishing point F is a predetermined live weight dictated by practical or experimental considerations but is proportionately somewhere between 0.40 and 0.50 of mature size. From the framework described in the previous paragraph, the entire time period

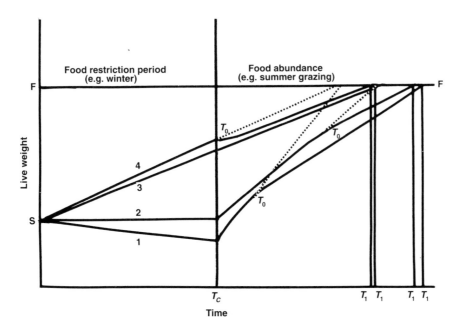

Fig. 8.6. Hypothetical growth retardations and subsequent live-weight gains in cattle. See text for explanation.

considered will be likely to coincide with the steepest part of the natural sigmoid-shaped growth curve. For convenience, progressive increases in live weight have been assumed to be smooth and it has been assumed that although animal 3 could initially lose some weight on being turned out to grass this would be small, it would be quickly recovered and would give overall a smooth progression to point F at time T_1 with the growth rates in the winter and summer periods being approximately equal.

For animals 1, 2 and 4 at, and for a short period after, time T_c, initially the live-weight changes would be likely to comprise different components. For example, animal 4 could have lost lipid from adipose tissue surrounding the gastrointestinal tract, from deposits surrounding the kidneys and from elsewhere within the body cavity in adapting from a high level of supplementary feeding to *ad libitum* grazing. This could also reduce gut fill. In extreme cases, if adaptation were very difficult, the lipid could be withdrawn from carcass tissues. This would be likely to be entirely different to the situation encountered in animals 1 and 2. Animal 1 would be likely to have its labile reserves quickly repleted and its gut fill increased dramatically, with also probably some repletion of body tissues. Animal 2 would experience the same happenings to a lesser degree. In contrast to these three animals, animal 3 would be likely to show less variation and to have a greater proportion of its live-weight gain in carcass components. Therefore, initially, each unit of live-weight change could hide a number of differences and compensatory live-weight gains in animals 1 and 2 would have a distinctly limited meaning relative to the live-weight changes in animals 3 and 4.

The times T_0 have been given fairly, but not totally, arbitrarily to animals 1, 2 and 4 to show where the live-weight gains could start to decrease and the preceding growth pathways to deviate from those which would have been present had live-weight gains been maintained at the same rates (as indicated by the dotted lines). In some cases, particularly for animals 1 and 2 where the deviations are later in the grazing time period than for animal 4, these points could coincide with a natural reduction in available herbage for grazing animals after the initial flush of grass but if this were the case then animal 3 would show a decrease in live-weight gain as well. Notwithstanding this possible complication, T_0 possibly represents a point in time about which gradually homeorhetic mechanisms assume a greater importance than homeostatic mechanisms. If this is so then it follows that better comparisons of growth rates would be after the T_0 points up to the terminal point of consideration (F). Such comparisons would avoid complications due to gut fill changes and repletion of labile reserves in non-carcass components and would reflect more accurately changes in tissues of the carcass. Even then, however, comparisons would not be straightforward unless detailed information were available on the actual tissues deposited within units of live-weight increases, because the starting physiological ages would probably be different.

Extrapolation at constant rates of the live-weight gains prior to T_0 points (the dotted lines) show how the differences in times taken to reach the T_1 points would be altered markedly if the initial growth rates were maintained. It is perhaps ironic that in the case of animal 2, comparisons of the projected live-weight gain with the actual live-weight gains in the winter and summer periods from T_0 to F for animal 4, are the only comparisons that would be very nearly totally valid with regard to, in particular, carcass component growth, amongst all those that can be made. This is so because on all occasions a common denominator would be an *ad libitum* supply of food and therefore full expression of genetic potential would be possible. The drawback would be that although at point T_0 the nominal live weight of animal 2 could be the same as that of animal 4 at point S, the former animal would be of quite a different body composition because of changes in composition at constant live weight over the winter period of restriction. Once again a valid comparison is foiled unless information on the composition of the live weights is known. However, it does highlight the fact that animals should be compared at similar physiological, rather than chronological, ages if reasonably valid comparisons of growth after retardation periods are to be made.

The gradual decline in the faster growth rates of animals 1 and 2, compared with animals 3 and 4, lead to different T_1 times to reach the common finishing weight F. Although this again brings into question the comparative compositions of units of live-weight gains at different points in the grazing period, and questions the ultimate composition of the weight F, it does give a better baseline for comparisons of the effects of growth retardation and recovery to be made.

The differences in the T_1 times to reach the predetermined F weights imply a prolongation of the growth period to reach this weight. If F had been higher or lower compared with S, then the differences between the T_1 times would also have differed to a greater or lesser extent. If conversely a common finishing time, rather than a common finishing weight, is chosen, then even greater problems of interpretation of compensatory live-weight gains arise because of differences in physiological, rather than chronological, age. The overall picture presented by Fig. 8.6 is one in which the homeorhetic mechanisms have not quite redressed the balance and equated physiological and chronological age in animals 1 and 2 compared with animals 3 and 4. But the point taken for comparison is arbitrary and at later points in time the redress could be more, if not totally, complete.

The points discussed above have been with sole reference to a carefully defined set of hypothetical circumstances in cattle. Most have wider implications both within and between species. They should highlight the great care needed in interpreting growth data in animals after periods of growth restriction. To compare like with like is rarely easy, quite often impossible. Therefore compensatory live-weight gains need the most careful consideration and component parts need to be analysed before any conclusions can be reached on precisely what has happened in the animal.

References

Allden, W.G. (1970) *Nutrition Abstracts and Reviews* 40, 1167–1184.

Auckland, J.N., Morris, T.R. and Jennings, R.C. (1969) *British Poultry Science* 10, 293–302.

Berge, P. (1991) *Livestock Production Science* 28, 179–201.

Bond, J., Hooven, N.W., Warwick, E.J., Hiner, R.L. and Richardson, G.V. (1972) *Journal of Animal Science* 34, 1046–1053.

Brookes, A.J. and Hodges, J. (1959) *Journal of Agricultural Science, Cambridge* 53, 78–101.

Carstens, G.E., Johnson, D.E. and Ellenberger, M.A. (1987) *Journal of Animal Science* 65, (Supplement 1), 263–264.

Carstens, G.E., Johnson, D.E., Ellenberger, M.A. and Tatum, J.D. (1988) *Journal of Animal Science* 66, (Supplement 1), 491–492.

Drennan, M.J., Conway, A. and O'Donovan, R. (1982) *Irish Journal of Agricultural Research* 21, 1–11.

Drew, K.R. and Reid, J.T. (1975) *Journal of Agricultural Science, Cambridge* 85, 535–547.

Drouillard, J.S., Klopfenstein, T.J., Britton, R.A., Bower, M.L., Gromlich, S.M., Webster, T.J. and Ferrell, C.L. (1991) *Journal of Animal Science* 69, 3357–3375.

Ellis, R.N.W. and Lawrence, T.L.J. (1978a) *British Veterinary Journal* 134, 322–332.

Ellis, R.N.W. and Lawrence, T.L.J. (1978b) *British Veterinary Journal* 134, 333–341.

Ferrer Cazcarra, R. and Petit, M. (1995) *Animal Science* 61, 511–518.

Hogan, A.G. (1929) *Research Bulletin Missouri Agricultural Experimental Station* No. 123, pp. 52.

Hogg, B.W. (1991) Compensatory growth in ruminants. In: Pearson, A.M. and Datson,

T.R. (eds) *Growth Regulation in Farm Animals. Series: Advances in Meat Research*, vol. 7. Elsevier Applied Science Publishers, Amsterdam, pp. 103–134.

Hovell, F.D. de B., Orskov, E. R., Kyle, D. J. and Macleod, N. A. (1987) *British Journal of Nutrition* 57, 77–88.

Huffman, C.F. and Duncan, C.W. (1944) *Annual Reviews of of Biochemistry* 13, 467.

Iason, G.R. and Mantecon, A.R. (1993) *Animal Production* 56, 93–100.

Joubert, D.M. (1954) *Journal of Agricultural Science, Cambridge* 44, 5–65.

Lawrence, T.L.J. and Pearce, J. (1964a) *Journal of Agricultural Science, Cambridge* 63, 5–21.

Lawrence, T.L.J. and Pearce, J. (1964b) *Journal of Agricultural Science, Cambridge* 63, 23–34.

Lush, J., Jones, J.M., Dameron, W.H. and Carpenter, O.L. (1930) *Bulletin Texas Agricultural Experimental Station* No. 409, 34pp.

Lomax, M.A. and Baird, D.G. (1983) *British Journal of Nutrition* 49, 481–496.

Meadowcroft, S.C. and Yule, A.H. (1976) *Experimental Husbandry* 31, 24–32.

Murray, D.M., Tulloh, N.M. and Winter, W.H. (1974) *Journal of Agricultural Science, Cambridge* 82, 535–547.

O'Donovan, P.B. (1984) *Nutrition Abstracts and Reviews. Series B. Livestock Feeds and Feeding* 54, 389–410.

Ratcliffe, B. and Fowler, V.R. (1980) *Animal Production* 30, 470 (abstract).

Ryan, W.J. (1990) *Nutrition Abstracts and Reviews. Series B. Livestock Feeds and Feeding* 60, 653–664.

Seebeck, R.M. (1973) *Journal of Agricultural Science, Cambridge* 80, 201–210.

Steen, R.W.J. (1994) *Animal Production* 58, 209–220.

Steensberg, V. and Ostergaard, P.S. (1945a) *Beretning fra Forsogslaboatoriet* 216, 150.

Steensberg, V. and Ostergaard, P.S. (1945b) *Forsogslaboratoriet Kobenhaven Beretning* 216, 149.

Thomson, E.F., Gingins, M., Blum, J.W., Bickel, H. and Schurch, A. (1979) In: *Energy Metabolism, Studies in Agricultural and Food Science*. (European Association for Animal Production No. 26.) Butterworths, London, pp. 427–430.

Tullis, J.B., Whiltemore, C.T. and Phillips, P. (1986) *British Journal of Nutrition* 56, 259–267.

Webster, A.J.F. (1989) *Animal Production* 48, 249–269.

Wilkinson, J.M. and Prescott, J.H.D. (1970) *Animal Production* 12, 443–450.

Wilson, P.N. and Osbourn, D.F. (1960) *Biological Reviews* 35, 324–363.

Winter, W.H., Tulloh, N.M. and Murray, D.M. (1976) *Journal of Agricultural Science, Cambridge* 87, 433–441.

Wright, I.A. and Russell, A.J.F. (1991) *Animal Production* 52, 105–113.

Wright, I.A., Russell, A.J.F. and Hunter, E.A. (1989) *Animal Production* 48, 43–50.

Growth and Puberty in Breeding Animals

9.1. Introduction

In this chapter the effect of growth rate on puberty in animals retained for breeding purposes will be considered. The *Shorter Oxford English Dictionary* defines puberty as 'the state or condition of having become functionally capable of procreating offspring'. This implies that puberty represents the time at which reproduction first becomes possible and will, therefore, in the female animal be characterized by ovulation and in the male animal by the production of semen with sufficient numbers of morphologically mature spermatozoa to fertilize ova. It is important at the outset to differentiate this state from that of sexual maturity, which is the state reached when the animal is able to express its full reproductive power. Thus the attainment of physiological puberty in the female animal may not be concomitant with the ability to conceive readily and to carry easily a viable fetus to full term, whilst in the male animal only a limited use may be possible if semen quality is to remain sufficiently high to give consistently high conception rates. In practice, and for reasons that will be discussed later, there is more often than not a time lag between puberty and the occasion on which the female animal is first used for breeding. Similarly the male animal will reach a stage where it can first be used successfully but it may be a considerable time after this before it can be used regularly with any expectation of achieving good conception rates.

In food-producing animals it is, for economic reasons, desirable to use as soon as possible those individuals that are retained for breeding purposes because the period from birth to the point where the female animal first conceives, and the male animal is used for the first time, is essentially non-productive and costly. The need is therefore one of attempting to induce puberty at as early an age as possible in order that the non-productive state can be transformed to the productive state as quickly as possible after this. Relative to this need the question of how far the manipulation of growth rate can affect age at puberty really has to be answered, for clearly the much-used adage of 'time is money' could nowhere be more appropriate. On the other hand, the

247

inducement of a precocious puberty must not be at the expense of a reduced performance subsequently in terms of, for example in the female, a poorer life-time milk yield or reproductive capability. A gain in one direction must not be the cause of a detrimental effect in another direction: a balance has to be struck so that any beneficial effects of early puberty are effects which are manifest overall in the animal's productive life span. This is true for both the female and the male but the fact that in breeding populations the female is numerically more important than the male dictates that considerations of growth and reproductive potential should be biased towards her rather than towards her male counter-part.

9.2. The Endocrinology of Puberty

In all animals the effects of the interplay between several hormones and the interaction of the hormones with target tissues and glands are the factors which ultimately determine puberty. Many other factors can play a part (see section 9.3), but their effects will be secondary in that they will exert their effects on this hormonal axis in the first place.

In the female puberty represents the ultimate defeat of the previous suppressive effects of oestradiol on the hypothalamic–hypophyseal axis. As a result of this defeat the first surge in gonadotrophin release is induced and there is a consequential stimulus to ovulation. The suppressive effects are often referred to as the 'negative feedback effects' and indicate an immaturity of response of the hypothalamus to oestradiol. As the negative feedback effects of oestradiol on the hypothalamus progressively weaken in the prepubertal period, the hypothalamus becomes increasingly more strongly orientated towards secreting gonadotrophin-releasing hormone with the result that the anterior pituitary releases gonadotrophins. Therefore the ascendency of the hypothalamus over the previous suppressive effects of oestradiol leads to the position where 'positive feedback effects' dominate.

If the ewe lamb is taken as an example, the suppressive effects are present to a limited extent *in utero* but by about the 5th week after birth they have reached a maximum. The balance then changes progressively with the negative feedback gradually weakening and the positive feedback gradually becoming stronger. During this protracted interplay between the two feedback mechanisms, the secretion of follicle stimulating hormone remains more or less constant whilst the pulsatile secretion of luteinizing hormone is characterized by the frequency and amplitude of the pulses increasing dramatically during the peripubertal period as the responses of the hypothalamus–hypophyseal axis heighten. When a circhoral rhythm is achieved, representing an hourly pulse frequency, a threshold appears to be reached which permits continuity of the maturation process in the follicles and which leads finally to the first ovulation. Endocrinologically, puberty therefore represents the climax of a long transition period from sexual quiescence to sexual function controlled by a complex but changing interplay between hormones secreted from the gonads, the hypothalamus and the anterior pituitary, with the hypothalamus finally conditioning the anterior pituitary to release sufficient gonadotrophin to induce ovulation. The main features of the process are represented schematically in Fig. 9.1, but for full details reviews such as those of Quirke *et al.* (1983) and Adam and Robinson (1994) should be consulted.

The events leading to puberty in the male are basically very similar to those which occur in the female, the pituitary–hypothalamus–testis axis moderating the change from sexual quiescence to sexual function. Luteinizing hormone is released from the pituitary under the influence of gonadotrophin-releasing hormone from the hypothalamus and acts upon Leydig cells in the testis to facilitate testosterone production. In turn, testosterone has a major part to play in influencing both spermatogenesis and sexual behaviour. Contrarily, follicle-stimulating hormone has a less well defined mode of action but is probably involved not only in the initiation of spermatogenesis but also in the

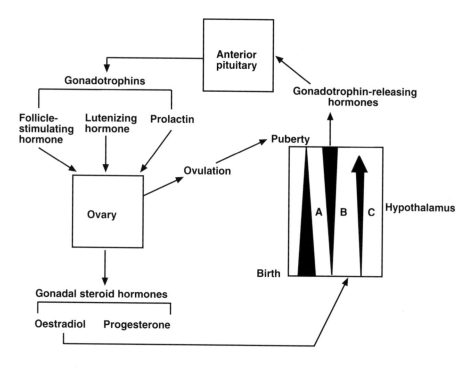

Fig. 9.1. Schematic representation of the main features involved in the hypothalamus-hypophyseal axis controlling the onset of puberty in female animals. The decreasing depressing negative feedback effect (A) and the increasing response to the positive feedback effect (B) between birth and the onset of puberty are reflections of the increased maturity (C) of the hypothalamus in responding to oestradiol secretion by the ovaries. The resulting secretion of gonadotrophin-releasing hormones stimulates the anterior pituitary to release gonadotrophin hormones, which in turn signifies the onset of puberty by stimulating the first ovulation.

maintenance of the process as well. In some species there is a synergistic effect between prolactin and lutenizing hormone in stimulating steroidal hormone production. For further detail the reader is referred to reviews such as those of Haynes and Schanbacher (1983) and Adam and Robinson (1994).

Perhaps the best descriptive summary of the endocrinological control of puberty is that of Ryan and Foster (1980): 'Puberty is the eventual persistence of gonadotrophin release in the presence of a functional gonad'.

9.3. Factors Affecting Puberty

Apart from growth rate a number of factors can influence the hypothalamus–hypophyseal axis (described above in section 9.2) and therefore affect puberty (Fig. 9.2). Some of these are more likely to be direct in their effect than are others. Several are more species specific than are others, for example photoperiod. All tend to confound studies on the relationships between growth rate and reproductive capability and because of this the effects of growth rate on puberty will be considered here mostly under species headings, in order that the relative effects of the non-growth factors may be elucidated as far as the present state of knowledge allows.

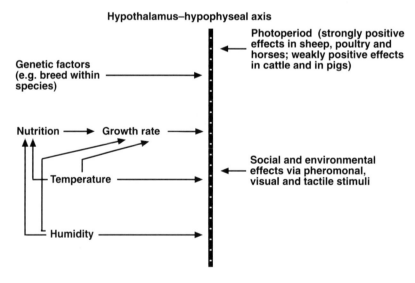

Fig. 9.2. Factors affecting the hypothalamus–hypophyseal axis and conditioning the onset of puberty in animals. Factors which tend to be mostly direct in effect are on the right of the axis, those which can be both direct and indirect are on the left of the axis.

9.4. Effects of Growth Rate on Puberty

9.4.1. General

In most species any suggestion of a consistent relationship between age and live weight at puberty is conspicuous by its absence. The relationships between growth rate, age and live weight at puberty are very complex and it is virtually impossible to separate the effects of growth rate *per se* from those of live weight and/or age. Selection for high growth rates under domestication has probably led to puberty being attained at younger ages and sometimes at lighter live weights, and therefore at lower proportions of mature live weight, compared with the situation in unselected populations from which such animals have been derived (Fig. 9.3; Foxcroft, 1980). Therefore animals selected for high growth rates may reach puberty whilst still relatively immature in terms of live weight and may distort to some extent the hypothesis that the inflection point of the growth curve usually follows closely the attainment of puberty and that the secretion of steroid hormones at the time of sexual maturation is itself responsible for the changing pattern of growth (Brody, 1945). The possibilities of this type of distortion are explored in Fig. 9.3.

Relative to this there must be a physiological mechanism which controls puberty by changing the sensitivity of the hypothalamus to oestradiol, but how this mechanism is triggered by growth rate, live weight and/or age remains largely unresolved. As might be expected, the possibilities of other triggers have been contemplated and several which are correlates of growth rate have been proposed (Fig. 9.4). The complexity of the situation is all too evident and must be borne in mind in the next sections which deal with the effects of growth rate on puberty in the different common farm animal species. Nevertheless, it is clear that growth rate could be an important trigger in affecting the hypothalamus, either in its own right, or via the other five triggers postulated, each of which in turn could act as a direct trigger independently of growth rate.

9.4.2. Cattle

For all cattle, of all breeds and of both sexes, the effects of growth rate on puberty are important. However, it is in the dairy heifer that the attainment of early puberty may have most significance because of the large numbers of animals involved in milk production in most countries and because of the need to reduce the length of the rearing period as much as possible in order that the cost of rearing dairy herd replacements may in turn be reduced. In consequence this particular section is biased heavily towards the dairy heifer.

Although cattle breed throughout the year, research work has indicated that some aspects of reproduction, including puberty, are affected by environmental factors. There is evidence that age at puberty is modified by season of birth (see review of Hansen, 1985). For example, Roy *et al.* (1980) reared Friesian heifers on a high plane of nutrition and found that those born during periods of

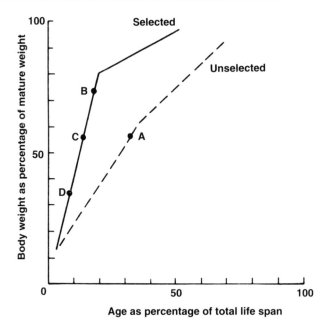

Fig. 9.3. Possible relationships between age and body weight at puberty based on theoretical growth curves. The figure represents theoretical growth curves from selected and unselected populations. Point A represents the attainment of puberty in relation to growth before selection. Points B, C and D represent the position of the growth curve at which puberty might occur following either selection between or within species (reproduced from Foxcroft (1980) by kind permission of the copyright holder, Dr G.R. Foxcroft).

increasing daylength reached puberty about 2 months earlier than those born during periods of decreasing daylength. Therefore spring-born heifers reached puberty earlier than winter-born heifers with those born in the summer and in the autumn months in an intermediate position. They found also that the phase of the moon had a marked influence on the frequency of occurrence of the first oestrus, where four distinct peaks at approximately 7-day intervals and positioned in time by the occurrence of the full moon, were detectable within the lunar cycle. Other work (Greer, 1984) has not confirmed these findings and work conducted in environmental chambers, and which has been programmed to simulate the diurnal fluctuations in temperature and photoperiod of the four seasons of the year, has shown that heifers born in the autumn may in fact be younger at puberty than heifers born in the spring (Schillo *et al.*, 1983). Nevertheless, in this work heifers exposed to temperatures and photoperiods of spring and summer after 6 months of age were younger at puberty than heifers exposed to autumn and winter conditions. From these studies it is difficult to know which of the environmental stimuli were responsible for the seasonal effects found, but other controlled work has shown that photoperiod can act alone in influencing puberty (Hansen *et al.*, 1983). In this work a day with 18 h

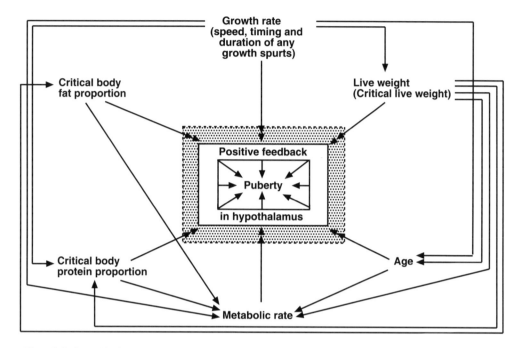

Fig. 9.4. Possible factors which may act as trigger mechanisms in the establishment of the positive feedback mechanism in the hypothalamus and through this to the onset of puberty. Double-headed arrows indicate those possible triggers which may act directly on the hypothalamus as well as being affected by growth rate to act in an indirect manner. The shaded area surrounding the area designating the positive feedback in the hypothalamus, bounded by the broken line and containing the double-headed arrows, represents the unknown physiological mechanism which responds to the trigger mechanisms and influences the positive feedback in the first place. The possibilities of the existence of: (i) a critical live weight and (ii) the speed, timing and duration of growth spurts, affecting the hypothalamus, are indicated in parentheses to differentiate them from ranges of live weights or overall growth rates which may have an effect on the hypothalamus (see text for further discussion).

of light beginning at 5 months of age induced puberty at younger ages than in animals exposed to natural autumn and winter photoperiods. As has been mentioned earlier (see chapter 4), the effects of photoperiod on the pineal gland relative to the secretion of melatonin and to growth rate stimulation are imperfectly understood and because of this and the possible influence of other seasonally controlled hormone secretions, for example prolactin, the relationships between growth rate, photoperiod, temperature and puberty remain a matter for conjecture.

Whilst it may be difficult to separate the effects on puberty of growth rate from those of other environmental factors, there is a large body of evidence which points to a curvilinear relationship between nutritional level, and thereby indirectly to growth rate, and puberty with an optimal level of feeding lying somewhere between overfeeding and underfeeding (Foxcroft, 1980). There is little doubt that heifers which grow faster reach puberty at younger ages, but not necessarily always at lighter live weights, compared with slower growing animals. It is very difficult to place this relationship in a quantitative framework, for there is considerable variation between individual animals and between different breeds and crosses. In the latter instance differences found indicate that in later maturing breeds such as the Holstein/Friesian, a high plane of nutrition giving a fast growth rate may induce puberty at lighter live weights than would otherwise occur, whilst in earlier maturing breeds the earlier puberty might be associated with heavier live weights. The differences in body composition between early and late maturing breeds at similar live weights lead back to the possibility of a critical body fat or protein proportion being the ultimate trigger relative to the induced faster growth rate, as represented previously in Fig. 9.4. In cross-bred animals the breed of the sire has been shown to affect significantly both the age and the weight of the heifer at puberty (e.g. see Morris *et al.*, 1993), but overall it would appear reasonable to predict that heifers that are older at puberty will in most cases also be heavier; positive, but not high, correlation coefficients of between 0.3 and 0.4 have been found between age and weight at puberty.

The point at which growth is accelerated may also influence age and weight at puberty. Early enhancements of growth rate, as for example in the preweaning stages of suckled animals, have shown consistently a more positive influence on the early attainment of puberty than have later enhancements of growth rate in the postweaning period. Foxcroft (1980) points out that as this trend is normally accompanied by heavier live weights at puberty, a relatively higher proportion of mature live weight will have been reached (see Fig. 9.3).

How early can a fast growth rate induce puberty in dairy heifers? The answer seems to be very early from the work conducted at various centres, but particularly in Israel. A number of studies by Amir and his colleagues, and referred to by Amir and Halevi (1984), have shown that puberty can be induced as early as 4.5–5.5 months of age from very fast growth rates, with an age of 8 months at puberty being an easily achieved target. Under experimental conditions heifers bred from at about 8–9 months of age performed satisfactorily. Under UK conditions the guidelines used often for weights at first service and at first calving are those given in Table 9.1. Assuming for the Friesian/ Holstein a birth weight of about 40 kg and that first calving should not be later than about 2 years of age at 530 kg live weight (but perhaps 30–50 kg more for the pure Holstein or for the cross-bred animal with a high proportion of Holstein blood), the implied overall growth rate from birth to 1.25 years of age at first service at about 325 kg live weight would be no greater than about 625 g daily. This is a moderate growth rate for this type of animal and, on the assumption that the growth rate within the 1.25 years did not fluctuate greatly, would probably imply, from the deductions that are possible from published work, that puberty would be induced at about 8–9 months of age. Therefore the time lag between puberty and first breeding would be about 6–7 months.

To achieve a first calving at about 2 years of age in the Friesian/Holstein heifer

Table 9.1. Guidelines used in the UK for weights (kg) of different breeds at first service and first calving.

Breed	Weight at first service	Weight at first calving
Jersey	230	335
Guernsey	260	390
Ayrshire	280	420
Friesian/Holstein*	325	530

*Possibly 30–50 kg more for the pure Holstein or the cross-bred animal with a high proportion of Holstein blood.

is in most cases not difficult. However, in this breed, as in other breeds, the growth rates necessary to achieve this goal must not be ultimately to the detriment of various other factors in the productive life span of the animal. The advantages from early puberty and early breeding of reduced rearing costs, increased returns to the owner and reduced generation intervals, and therefore quicker genetic evaluations of bulls, must be set against the milk yields obtained after early calving, the ultimate growth and size of the heifer which are realized and the productive life of the animal in the herd. Puberty induced at about 8–9 months of age, with service at about 15 months of age and calving at about 2 years of age, should have advantages overall, and not disadvantages, for the Friesian/Holstein heifer (Table 9.2). One temporary disadvantage not indicated by the data in this table is that first lactation yields may be less in animals calving at 2 years of age than at later ages. Therefore if a moderate growth rate overall of about 625 g per day over the first 9 months of life will induce puberty at about 200 kg live weight, it follows that to achieve puberty at about 5 months of age at the same live weight, a

Table 9.2. Comparisons of calving Friesian heifers at different ages (based on Kilkenny and Herbert, 1976).

	Age at calving (years)		
	2	2.5	3
Herd life (years)	4.0	3.8	3.8
Lifetime yield (litres × 10³)	18.576	17.802	17.496
Yield per day in herd (litres)	13.07	13.07	13.21
Yield per day of life (kg)	8.67	7.85	7.25
Number of heifers being reared per 100 cows at different replacement rates and ages at first calving			
15% replacement rate	30	38	45
20% replacement rate	40	50	60

daily growth rate of about 1150g would be necessary. Again this is within the capability of the Friesian/Holstein genotype, but it is extremely debatable as to the weight or age at which such an animal should be served. For example, if the growth rate of about 1150g daily were to be sustained, then the target weight at first service of 325 kg (Table 9.1) would be reached in a further 4 months, that is at about 9 months of age. Again this would be feasible in the context of the Friesian/Holstein genotype and the previously mentioned Israeli work confirms this. However, it is extremely doubtful if milk yield in the first lactation, at the least, would be as high as in heifers calving at later ages and there is experimental evidence to support this supposition. The possible reasons behind this, of fat infiltrating the developing udder tissue and reducing the milk-secreting capacity, are of interest in the context of the trigger mechanisms and the onset of puberty mentioned earlier. Undoubtedly, a growth rate of 1150g daily to induce puberty at this age would involve a considerably greater rate of fatty tissue growth than would lower growth rates giving puberty at older ages but perhaps at similar live weights. What must be avoided is to serve heifers at a constant age irrespective of size, for here animals may be too small. This will avoid problems of dystocia at parturition and also problems arising from competition for available nutrients between the growing fetus and the growing heifer which may have detrimental effects on the subsequent productivity of the heifer. Overall it would appear that the present state of knowledge does not allow confidence in proposing other than that overall growth rates from birth of 625–650g daily, to give puberty at about 8–9 months of age at about 200 kg live weight, with service at about 320–330 kg live weight and calving at about 530 kg live weight, are near to the ideal for the Friesian/Holstein heifer when all factors are considered. Puberty can be stimulated at an earlier age by inducing a faster growth rate but the advantages of this are questionable.

9.4.3. Pigs

As in the case of cattle, it is very difficult to separate the effects of growth rate from those of live weight and age. There are also other complicating effects of season of year, genotype and, very importantly, social environment. Because of the complex way in which all of these factors interact and because very few experiments have been designed to accommodate all variables present, it is extremely difficult to reach a conclusion on the effect of growth rate on the attainment of puberty in the gilt.

The European wild pig is a seasonal breeder, with farrowing distributions resulting from seasonal ovarian activity. The sow exhibits anoestrus in the summer and autumn months with the result that farrowing is usually once yearly in the April to May period. Sometimes there is a bimodal, rather than a unimodal, farrowing distribution. In such cases the two peaks of farrowing are in January to February and in August to September. Gestation lengths of between 112 and 126 days are reasonably similar to those in the domesticated pig but ovulation rates averaging 5.25, and litter sizes of about 4.5, are very different (Mauget, 1982). Gilts become pregnant at between 18 and 24 months of age at live weights which are greater than 25 kg and the sex organs regress noticeably in the anoestrous periods (Aumaitre *et al.*, 1982). Although social factors play a part in influencing breeding activity and puberty, it appears likely that seasonal effects, possibly mediated both directly and indirectly through photoperiod changes, dictate the occurrence of puberty very strongly in that the time of birth, together with the adequacy of food supplies subsequently, controls the point in time at which a threshold live weight is reached and at which puberty is attained. Food supplies, and other associated climatic factors, such as temperature, may play largely an indirect part, in that a gilt born in the late spring–early summer does not reach the threshold live weight necessary to precipitate puberty about 9 months later because the food available to her has been inadequate to promote the growth rate necessary to achieve that threshold live weight. She therefore reaches puberty later and becomes pregnant for the first time within the age range and above the minimum live weight mentioned above. Compared with the gilt born in August–September, however, she will mature sexually more quickly, by responding to the increasing day length in the early part of her life.

To what extent is this seasonality reflected in the domesticated pig? There is in the broad sense considerable evidence of 'reproductive inefficiency' in the autumn and summer months in the domesticated pig in many countries of the world (see references cited by Mauget, 1982). Specifically, for example, farrowing intervals have been found to be longer (Hurtgen and Leman, 1981). However, there is limited evidence only that a seasonal breeding pattern exists and there is no firm evidence that spring-born gilts reach puberty at earlier ages than gilts born at other times of the year. Experiments that have separated the effects of photoperiod on the one hand from temperature on the other, have indicated that increasing daylength does enhance puberty but that high temperatures of summer months may have the opposite effect. Attainment of puberty in the spring-born gilt may therefore be enhanced by increasing daylength but retarded by high temperatures, whilst in the autumn-born gilt the decreasing daylengths may be to the detriment of sexual development but the cooler temperatures to its advantage. The effects of the two environmental factors may cancel each other out and give the apparent similarity noted in attainment of puberty between gilts born in the spring and in the autumn.

If date of birth has very little, if any, effect on puberty, is there any effect of the different genotypes which have been developed under domestication? In this case there is only limited evidence of differences between breeds and the general consensus is that differences between individuals within breeds are likely to be at least as great as differences between breeds. On the other hand, the effects of heterosis have been demonstrated consistently with the cross-bred gilt exhibiting an earlier onset of puberty than the pure-bred gilt, and gilts from line-breeding programmes exhibiting puberty earlier than gilts from inbreeding programmes.

If photoperiod and genotype have limited effects only on puberty, the effects of social environment are very much more marked and their influence can be profound. Contact with a boar, particularly if mature, in the later prepubertal stages is generally agreed as likely to be the most powerful single factor that can be used in stimulating puberty. Introduction of the boar at about 3–4 months of age is likely to have a very small effect but a boar introduced at about 160–165 days of age will induce puberty after a similar period compared with introducing a boar in the very late prepubertal stages at about 6 months of age. However, because at 6 months of age the gilt is near to the age where puberty occurs, the onset is likely to be only fractionally sooner compared with that in the unstimulated gilt. Notwithstanding the fact that the actual moving and relocation of gilts can act as powerful stimuli to puberty, the previous sentence gives a strong hint that puberty is age related. This is the equivocal view held generally and it appears that within wide limits growth rate and/or live weight are of secondary importance in determining puberty. This is not to say that wide variation in age at puberty cannot be found. It can, and Hughes (1982) cites variation from 102 to 350 days and above with variation in live weight from about 55 to about 120 kg. It is obviously difficult, and to some extent meaningless, to attempt to give a mean age at which puberty might occur, but a consensus view is that most gilts under 'normal' conditions attain puberty at between 170 and 200 days of age. The live weights may vary considerably, with poorly grown gilts reaching puberty at lighter live weights than gilts which have grown more quickly, but there is probably a lower live-weight threshold below which gilts will not attain puberty. Relative to the inability of growth rate to affect puberty, the evidence available does not indicate that any particular nutrient (e.g. protein, vitamins or minerals) involved in promoting growth is of any significance. In terms of energy intake, which is to a large extent synonymous with plane of nutrition, the evidence available indicates that intakes below half of those possible under *ad libitum* feeding may give the lower live-weight threshold referred to above.

If the consensus view is accepted that chronological age, rather than physiological age or growth rate, determines puberty, it is ironic that the experimental techniques which have been used in most cases to detect puberty must give it a doubtful validity! This is so because attainment of puberty has been measured by the response of the gilt to the boar, an animal which, as previously stated, can have the greatest single effect on puberty. Therefore, and as Hughes (1982) points out, unless other parameters indicative of puberty are measured, the consensus view must remain open to doubt.

Gilts reaching puberty at between 170 and 200 days may weigh between 65 and 100 kg live weight. It is well known that the ovulation rate at this first oestrus will be low but that it will increase up to about the third oestrus with the result that litter sizes may be higher if mating is delayed. As a consequence most gilts are not mated until the 3rd oestrus when their live weights have increased to 110–120 kg and when they are about 8 months of age. Depending on how the ultimate productivity of the gilt is viewed, it is questionable whether this delay between puberty and first service will represent a real improvement in productivity of either the individual gilt or of the herd as a whole. The crux of the problem revolves around the benefits to be derived from the increased litter size at delayed matings after puberty compared with the time lost in keeping the gilt out of production for an extra 21 or 42 days. If the gilt is mated at first oestrus, she will be both younger and smaller than if mating had been delayed, and in this case it is obvious that a satisfactory growth rate should have been obtained in the prepubertal period to give a live weight at puberty of no less than about 90 kg. This approach is unlikely to retard subsequent growth but raises the question of the 'once-bred' gilt as an inherent part of a production system.

In the first place the 'once-bred' gilt must obviously grow well to reach about 90 kg at puberty. If made pregnant at this point, she will use the food given to her to grow her own body tissues, and to nourish her growing fetal mass and the other products of conception, and will yield a carcass which will be proportionately about 0.75 of her live weight. This carcass may differ but little from that obtained from the 'manufacturing' or 'heavy hog' type pig slaughtered at between 110 and 120 kg live weight, although some trimming of mammary tissue may be slightly to its detriment. Mating at a chronologically controlled puberty, but ensuring that the growth rate of the animal gives a certain live weight at that puberty, therefore opens up the prospects of an alternative production system in which the costs of keeping female breeding stock over a number of expensive, from the point of view of feeding costs, pregnancies is eliminated. Growth rate in the gilt is clearly important, not from the point of view of inducing puberty, but from the point of view of obtaining an animal of an appropriate size which is capable of producing one litter and a carcass which, in size and in composition, is acceptable for the meat trade.

9.4.4. Sheep

Changing photoperiod is a dominant factor in controlling the reproductive physiology of sheep (see the review of Adam and Robinson, 1994), and because of this it is even more difficult to unravel the effects of the various factors which affect puberty compared with the situation in both cattle and pigs. Although sheep have probably received more attention, as measured by the volume of published work, than have either cattle or pigs, the effects on puberty of growth rate are not understood well. It is necessary with sheep to consider within the overall framework of changing photoperiod the effects of growth rate relative to the other factors, such as age and social behaviour, which were considered also for cattle and pigs. Furthermore, it has to be appreciated that not only are the behavioural signs of the first oestrus in the ewe lamb weak and less conspicuous than are those in the adult ewe, with a high incidence of silent heats, but also that the duration of overt oestrus is usually shorter and later in the breeding season. Lastly, the strong positive effects of the ram noticeable with adult ewes is very weak, if present at all.

The first complication to face up to is that breed may have an effect on puberty (Dyrmundsson, 1973). Also heterosis can have a significant effect. This is perhaps not surprising when the onsets and durations of breeding seasons in various breeds are considered (Hafez, 1952; Fig. 9.5). Although there is much variation between individuals within breeds, the possibility of a breed effect existing becomes most striking when a broad comparison is made of British breeds on the one hand compared with those of the Merino type on the other. In this comparison breeds of British origin are usually observed to attain puberty at an earlier age than do breeds of the Merino type. Generally speaking, origin of breed relative to distance north or south of the equator appears to exert an effect, in that whilst ewe lambs of most breeds originating in both the northern and southern hemispheres will normally attain puberty, other circumstances permitting, in the autumn and summer months irrespective of their times of birth, in breeds of tropical origin puberty is less well defined seasonally. Even though at higher latitudes changes in photoperiod appear to exert a major effect on puberty, there is an inherent underlying rhythm which overrides light treatments if they are imposed artificially and which dominates events ultimately. In this particular connection live weight appears to be important and the changing photoperiod appears to be able to exert its influence only if the ewe

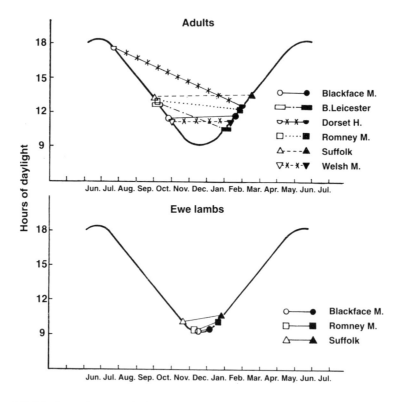

Fig. 9.5. The date of onset and of cessation of the breeding season in relation to the curve of daylight hours in adults and ewe lambs of the Blackface Mountain, Border Leicester, Dorset Horn, Romney Marsh, Suffolk and Welsh Mountain breeds (reproduced from Hafez (1952) by kind permission of the copyright holder, Cambridge University Press).

lamb has achieved a certain minimal or threshold live weight (Foxcroft, 1980). Strain within breed, however, may be as influential as breed, but again live weight appears to play an important role, although in certain breeds (e.g. Merino; Tierney, 1979) strains selected for high fertility may have a potential for attaining puberty at younger ages than would have been expected from their live weights at those ages. It might be anticipated that environmental temperature could play a part in influencing puberty in the context of breed origin and photoperiod, as discussed above, but the evidence of a direct effect is very weak (Drymundsson, 1983).

If, as hinted at above, live weight is an important factor influencing puberty, what is the magnitude of this influence in relation to age? Once again it is very difficult to separate the effects of age from those of live weight. In terms of age, the existing literature suggests that most ewes attain puberty in the age range 6–18 months. Obviously this is a very wide range and there are various factors which influence the age at which any one ewe achieves this point in her sexual development. Time of birth can play an important role in that early spring-born lambs tend to reach puberty earlier in the season than do lambs born later in that season, but at both greater ages and live weights. However, if the ewe lamb born in the spring is to reach puberty in the first autumn of her life, then it is important that she receives good nutrition and grows quickly. Compared with her contemporaries born at the same time but which receive poorer nutrition and, as a consequence, grow more slowly, she will be both younger and heavier at puberty. If the poorer feeding and consequential depressed growth continue between mid-summer and mid-winter before better feeding (and consequentially improved growth) is introduced, then although the threshold size for puberty may be reached in the following spring, puberty will be delayed until daylengths decrease in the following autumn (Foster *et al.*, 1985). Relative to this response, ewe lambs born in the autumn and which grow quickly usually fail to reach puberty in their first winter of life and also have to wait for the decreasing daylengths of the next autumn to precipitate this event. Exceptions to this general rule have been recorded, some ewe lambs born in the late summer–early autumn have been known to attain puberty at 3–4 months of age. In some cases conception has been reported at this age. Therefore if age at puberty varies widely then so too does live weight, most ewe lambs achieving this goal when proportionately between 0.50 and 0.70 of mature weight has been realized. Heritability estimates of live weight at puberty tend to be reasonably high at 0.50–0.70, but the standard error of estimate is usually large. Therefore puberty does not occur at either a constant age, at a constant live weight or at a particular time of the year.

The contention is held in some quarters that total body protein may be a more precise determinant of sexual development than total live weight. Whether or not this has any foundation, the attainment of a certain body protein mass could be related closely to the growth rate achieved in the chronological time scale and therefore, ultimately, to nutrition. In terms of nutrition it is possible only to contemplate differences in intake of a reasonably well-balanced diet for the size and age of the animal in question – that is the overall plane of nutrition.

As previously discussed, the evidence points quite strongly to a high plane of nutrition advancing puberty, with a low plane of nutrition having the opposite effect, particularly if imposed when the ewe is very young. Puberty induced by a high plane of nutrition, and therein inducing fast growth rates, will also give larger reproductive tracts and more multiple ovulations. 'High' and 'low' are relative terms and it is impossible to quantify with any precision their intended meaning but an indication of the differences inferred may be obtained by studying the differences in growth rates between twin and single ewe lambs reared on their mothers but with equal access to herbage and other food. In this type of comparison there is a great deal of evidence to support the hypothesis that the shared milk supply from the mother of the twin ewe lambs retards growth sufficiently to give a greater age and a lower mean live weight at first oestrus than in the single ewe lamb. This implies that the stage at which the growth differential is greatest is important, as it is likely that the greatest difference in growth rate would be in the early stages when the lambs are totally dependent on their mother's milk for all nutrients. In the later stages of the growth cycle the mother's milk would supply a decreasing proportion of total nutrients as the lambs consume progressively more herbage, or other food is made available.

If growth rate can affect puberty, how, if at all, can this be manipulated to allow an earlier use of the ewe lamb in the flock? Are there disadvantages? If so, do they outweigh the advantages gained from earlier breeding? The norm is that in both the UK, and in many other countries with similar systems of sheep production, ewes are managed to lamb-down for the first time at about 2 years of age. As a consequence their lifetime production could be enhanced if they were to be mated to produce a lamb crop in their first year of life. Generally speaking ewe lambs that are born in the spring and that fail to attain puberty before 7–8 months of age are not considered for inclusion in the breeding ewe population in their first year of life.

It has already been mentioned that ewe lambs exhibit weaker and shorter oestrous periods than do adult or yearling ewes. Additionally, ewe lambs exhibit less regular oestrous cycles, but this and the other disadvantages are counter-balanced to a certain extent by the fact that those which attain puberty early exhibit their last oestrus late in the breeding season. Early birth dates and fast growth rates are therefore conducive to an extended breeding period in ewe lambs. If fast growth rates are important in this respect, they are also important in the context that the incidence of silent heats is likely to be lower than in ewes which have grown at slower rates. In spite of these advantages conferred by a fast growth rate, conception rates and lambing rates are invariably lower than the rates found in adult ewes. The concept of a threshold live weight enters any consideration of this relationship, because although in ewe lambs the greater the live weights and the greater the ages at mating, the better the lambing performance is likely to be in terms of the numbers of ewes lambing and the numbers of lambs born, such a relationship is probably dependent on live weight and may be virtually non-existent once a certain threshold live weight is reached.

Lambs born to ewes mated at less than 1 year of age tend to be less viable than those born to older ewes. Higher mortality rates are often evident, particularly in the neonatal period, and a higher proportion of ewes are likely to exhibit a poor mothering ability, particularly in the periparturant period. Size of lamb, both large and small, can be important. If large, problems with dystocia may ensue; if small, for example in the case of twins, there may be problems in dealing with the new environment into which they have been thrust. On the other hand, those ewes that are bred from early are often better mothers as yearlings and, additionally, are more reliable breeders. The fleece weights and milk yields in the first lactations of ewes lambing at about 1 year of age are likely to be less than those from older ewes, but the growth rates of their lambs, if singles, are usually very similar to those of twins, but less than those of single lambs reared by yearlings and by older ewes. Subsequently, the reproductive ability of both the ewe lamb mother and her female offspring do not appear to be deleteriously affected, providing nutrition and consequential growth are good. If not, the pregnancy and lactation may considerably retard temporarily growth and development, with similar live weights to those of ewes bred first as yearlings achieved at between 2 and 2.5 years of age.

The obvious advantage from mating ewes early is that in breeding programmes the generation interval will be shortened and the rate of genetic improvement increased. In terms of productivity there is increasing evidence that life-time performance is not affected deleteriously. In fact, to the contrary, the evidence overall is that productivity may be enhanced.

In the ram lamb there are overt signs of sexual activity at a very young age but a huge variation from 12 to 45 weeks at which puberty is actually attained. Compared with the ewe, photoperiod appears to be without effect on puberty (Dyrmundsson, 1987), with spermatogenisis commencing in spring-born ram lambs during mid-summer and about 140 days before the attainment of puberty in ewes born at the same time and before natural daylength declines. However, this apart, other factors that affect puberty in the ewe also affect puberty in the ram. Thus high planes of nutrition inducing fast growth rates give puberty at younger ages and at heavier live weights than do low planes of nutrition and corresponding slower growth rates. Nevertheless, it is difficult to reach firm conclusions on the importance of growth rate in determining puberty, and assessments of puberty attainment on the basis of testicular morphology, and of spermatozoa numbers and their morphology in seminferous tubules and in ejaculates, indicate a very wide range of live weights at which the central hypothalamic–hypophyseal axis is sensitized.

9.4.5. Horses

As with sheep, changing photoperiod is the most important factor in controlling the reproductive physiology of the mare. However, in contrast to the ewe, the mare comes into oestrus in response to increasing daylength. Whilst changing photoperiod forms the overall framework for ovarian activity, there is a dearth of information on the effect on puberty of factors which may operate within it. Evidence on age at puberty is very scant and usually expressed in the broadest generalizations imaginable. An age of 16–18 months is often cited (Joubert, 1963; Ginther, 1979), but there can be little doubt that breed, temperature and live weight all play a part (Arthur, 1969). In the thoroughbred that has been well fed, and that has in consequence grown quickly, the onset can be as early as 12 months of age (Arthur, 1969).

The indirect implication that growth rate may affect puberty, as mentioned above, has been substantiated in an experiment conducted by Ellis and Lawrence (1978) using weanling New Forest filly foals. In this work the filly foals in the autumn, when approximately 6 months of age, were given a common diet for 180 days so that the overall plane of nutrition either allowed a reasonably good growth rate or held weight constant over a winter period in which daylength first decreased and then increased (see also chapter 8 for

Table 9.3. Live weights and dates at first oestrus in New Forest filly foals grown at different rates between 6 and 12 months of age (Ellis and Lawrence, 1978).

	High plane nutrition		Low plane nutrition	
	Date at first oestrus	Weight	Date at first oestrus	Weight
Weight at 6 months of age (kg)		104		106
Weight change between 6 and 12 months of age (kg)		68		0
Weight change between about 12 and 18 months of age at pasture (kg)		95		138
Individual dates and weights (kg) at onset of oestrus	3 April	202	13 May	149
	3 April	189	21 May	145
	3 April	161	29 May	166
	8 April	154	12 June	158
	8 April	164	10 July	178
	14 April	151	31 July	137
	16 April	158	31 July	169
	24 April	193	21 Sept.	206
	28 April	145	21 Sept.	250
	28 April	157	12 Nov.	222
	30 April	162		
	2 May	170		
	5 June	183		
Mean live weights*		168.5		178.1

*1 SD ($P < 0.05$) = 9.6.

reference to this work). Subsequent to this period, from April through to November, when daylength first increased and then decreased, all the fillies were run on good pasture and each filly was teased on alternate days with a 4-year-old Welsh Mountain stallion for the detection of first oestrus as indicated by the willingness of each filly to 'stand' to the stallion with her tail raised, and/ or to urinate, or to 'wink' her clitoris. The results of this experiment, given in Table 9.3, indicate a delayed first oestrus, at an increased live weight, in the previously retarded animals, although their growth rates were higher in the summer period than were those of the animals that had grown well previously. In addition, the total number of heat periods recorded was greatest for those fillies that had grown quickly originally between about 6 and 12 months of age, although the shortest heat period was the most common in both groups (Table 9.4). The average length of the heat periods for the fillies which grew quickly originally was 5.4 days and the average length for those fillies that were held in a weight-constant condition over the same period was 3.5 days. Cycle lengths are presented in Fig. 9.6, and indicate that in those fillies which were not retarded in the winter months the most common length of cycle was around

Table 9.4. Numbers of heat periods according to length of heat (days) subsequent to filly foals growing at different rates between 6 and 12 months of age (Ellis and Lawrence, 1978).

Length of heat (days)	Heat periods (no.)	
	High plane nutrition*	Low plane nutrition[+]
1	39	20
2	5	4
3	7	6
4	6	3
5	2	2
6	3	2
7	9	1
8	5	2
9	5	–
10	3	1
11	3	–
14	1	–
17	1	3
20	1	–
29	1	–
30	1	–
31	1	–
40	1	–
42	1	–
Totals	95	44

Total growth between 6 and 12 months of age: *68 kg; [+]0 kg.

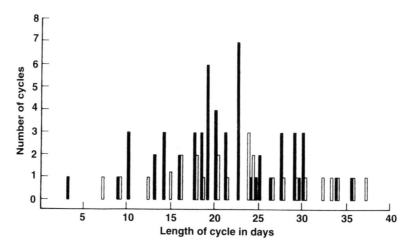

Fig. 9.6. Numbers of oestrous cycles according to length of cycle (days) in New Forest ponies. Solid columns represent those fillies reared on a high plane of nutrition from about 6 to 12 months of age and hollow columns those fillies reared on a low plane of nutrition in the same period (see Table 9.3. for details) (reproduced from Ellis and Lawrence (1978) by kind permission of the copyright holder, Bailliere Tindall).

20–21 days. In this particular context the data for the previously retarded fillies is of limited value, because in those that exhibited one heat period only, the length of the cycle was obviously not ascertainable. The evidence from this experiment therefore indicates that growth rate between about 6 and 12 months of age may be important in affecting puberty. Nevertheless, it is important to note that in the subsequent period of adequate nutrition, when photoperiod was changing, only 13 of the original 18 animals which grew quickly attained puberty, all before daylength began to decrease, whilst only 10 of the original 18 animals which had previously been retarded attained the same sexual goal, four of these before the onset of decreasing daylength.

9.4.6. Poultry

In the case of poultry puberty has, in most cases, different implications from those considered in the case of the common farm mammals. With pullets intended for laying flocks, the criteria that are important are the age at which the first egg is laid and the age at which a specified level of production, usually 50%, is attained. These criteria are adopted widely as measures of sexual maturity, but in terms of the definition of puberty given at the beginning of this chapter, strictly speaking it is the age at which the first egg is laid that is meaningful. However, sexual maturity combining both criteria is important in practice relative to the overall level and efficiency of production achieved ultimately.

The reproductive physiology of birds is strongly tuned to changing photoperiod and in modern systems of production the photoperiod of pullets intended for egg-laying flocks is carefully controlled up to about 20–21 weeks of age. During this period various lighting treatments are used, but one which is fairly commonly used in practice is to change the initial 20h of light and 4h of dark, imposed daily in the first week after hatching, to a day of 6h of light and 18h of dark by the time 20–21 weeks is reached, and then to reverse this subsequently in the laying phase after 20–21 weeks. This lighting regime is likely to give near to the best egg production, all other factors being equal. Within this framework of imposed photoperiod changes growth rate can have a marked effect.

Sexual maturity of growing pullets can be retarded either by restricting overall food intake or by restricting energy intake. There is also evidence that restricted intakes of protein and/or amino acids can exert a similar effect. In practice the intakes of overall balanced diets are in most cases restricted to retard growth rate and to delay sexual maturity. The restriction is usually applied from an age of 6–8 weeks until the pullets have reached about 21 weeks of age. Earlier restriction of growth rate before 8 weeks of age gives a greater reduction in live weight at sexual maturity and a greater delay in sexual maturity itself, but the effects are usually not very pronounced. If the restriction of food intake is

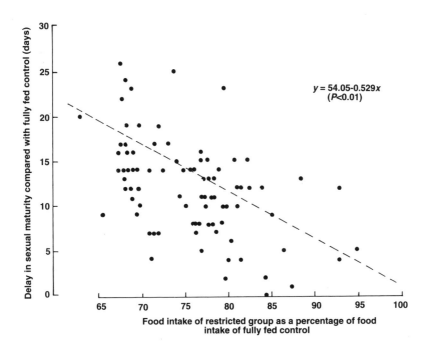

Fig. 9.7. Relationship between food intake during the rearing period and delay in sexual maturity in the domestic hen (Lee *et al.*, 1971; for original data see references in the same paper) (reproduced by kind permission of the copyright holder, British Poultry Science).

continued beyond about 24 weeks of age, the age at which the first egg is laid may not be altered but the age at which 50% egg production is achieved may be increased and subsequent egg production may be decreased.

The advantages of restricting growth rate in the rearing period lie in the resulting reduced overall food consumption for the rearing and laying periods taken together and, partially as a consequence of this, in the smaller food intakes for each egg produced. If overall benefits are obtainable from this approach, it is pertinent to ask if they can be quantified. The answer to this question is 'yes' and Figs 9.7 and 9.8. quantify the effects of food, and therefore indirectly of growth, restrictions in delaying sexual maturity and in affecting the live weight at which sexual maturity is reached. In addition to these benefits, pullets restricted in growth during the rearing period reach higher peaks of production than do those which grow more quickly. Also for any given period of time after sexual maturity has been reached, a higher average rate of lay is usually achieved, even though egg production overall may differ but little when the terminal point of egg production is a fixed age.

If the picture presented so far is bright, the reader may well be thinking that somewhere there are some disadvantages inherent in this approach. Possibly the chief disadvantage rests in the higher mortality rates and the greater susceptibility to disease from restricting growth rate in the rearing period. Many, but not

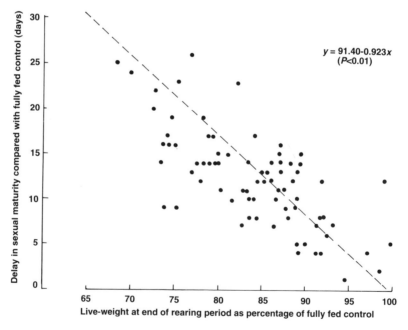

Fig. 9.8. Relationship between live weight at end of rearing period and delay in sexual maturity in the domestic hen (Lee *et al.*, 1971; for original data see references in the same paper) (reproduced by kind permission of the copyright holder, British Poultry Science).

all, studies have shown these drawbacks, but to some extent such disadvantages tend to be counterbalanced in the laying period, when mortality rates are lower than in those pullets which grew quickly through more generous feeding in the rearing period. In terms of egg size, chronological age rather than physiological state or rate of lay is the determining factor. Thus egg weight at any given age appears to be unaffected by restricted growth during the rearing period, but because maturity is delayed and the whole production curve is displaced, there is often a decrease of about 1 g in the average weight of all eggs laid.

References

Adam, C.L. and Robinson, J.J. (1994) *Proceedings of the Nutrition Society* 53, 89–102.

Amir, S. and Halevi, A. (1984) Early breeding of dairy heifers under farm conditions. In: *The Reproductive Potential of Cattle and Sheep. Joint Israeli–French Symposium, Rehovot (Israel), 21–23 February, 1984. Institut National de la Recherche Argronomique, Paris*, pp. 77–103.

Arthur, G.H. (1969) *Equine Veterinary Journal* 1, 153.

Aumaitre, A., Morvan, C., Quere, J.P., Peiniau, J. and Vallet, G. (1982) *14th French Swine Research Days*, pp. 109–124.

Brody, S. (1945) *Bioenergetics and Growth*. Hafner, New York.

Dyrmundsson, O.R. (1973) *Animal Breeding Abstracts* 41, 273–289.

Dyrmundsson, O.R. (1983) The influence of environmental factors on the attainment of puberty in ewe lambs. In: Haresign, W. (ed.) *Sheep Production*. Butterworths, London, pp. 393–408.

Dyrmundsson, O.R. (1987) Advancement of puberty in male and female sheep. In: Fayez, I., Marai, M. and Owen, J.B. (eds) *New Techniques in Sheep Production*. Butterworths, London, pp. 65–76.

Ellis, R.N.W. and Lawrence, T.L.J. (1978) *British Veterinary Journal* 134, 205–211.

Foster, D.L., Yellon, S.M. and Olster, D.H. (1985) *Journal of Reproduction and Fertility* 75, 327–344.

Foxcroft, G.R. (1980) Growth and breeding performance in animals and birds. In: Lawrence, T.L.J. (ed.) *Growth in Animals*. Butterworths, London, pp. 229–247.

Ginther, O.J. (1979) *Reproduction Biology of the Mare. Basic and applied aspects*. Ginther, Wisconsin.

Greer, R.C. (1984) *Animal Production* 39, 59–63.

Hafez, E.S.E. (1952) *Journal of Agricultural Science* 42, 189–265.

Hansen, P.J. (1985) *Livestock Production Science* 12, 309–327.

Hansen, P.J., Kamwanja, L.A. and Hauser, E.R. (1983) *Journal of Animal Science* 57, 985–992.

Haynes, N.B. and Schanbacher, B.D. (1983) The control of reproductive activity in the ram. In: Haresign, W. (ed.) *Sheep Production*. Butterworths, London, pp. 431–451.

Hughes, P.E. (1982) Factors affecting the attainment of puberty in the gilt. In: Cole, D.J.A. and Foxcroft, G.R. (eds) *Control of Pig Reproduction*. Butterworths, London, pp. 117–138.

Hurtgen, J.P. and Leman, A.D. (1981) *Veterinary Record* 108, 21.

Joubert, D.M. (1963) *Animal Breeding Abstracts* 31, 295.

Kilkenny, J.B. and Herbert, W.A. (1976) *Rearing Replacements for Beef and Dairy Herds*. Milk Marketing Board, Thames Ditton, Surrey.

Lee, P.J.W., Gulliver, A.L. and Morris, T.R. (1971) *British Poultry Science* 12, 413–417.

Mauget, R. (1982) Seasonality of reproduction in the wild boar. In: Cole, D.J.A. and Foxcroft, G.R. (eds) *Control of Pig Production*. Butterworths, London, pp. 509–526.

Morris, C.A., Baker, R.L., Hickey, S.M., Johson, D.L., Cullen, N.G. and Wilson, J.A. (1993) *Animal Production* 56, 69–84.

Quirke, J.F., Adams, T.E. and Hanrahan, J.P. (1983) Artificial induction of puberty in ewe lambs. In: Haresign, W. (ed.) *Sheep Production*. Butterworths, London, pp. 409–429.

Roy, J.H.B., Gillies, C.M., Perfitt, M.W. and Stobo, I.J.F. (1980) *Animal Production* 31, 13–26.

Ryan, K.D. and Foster, D.L. (1980) *Federation Proceedings* 39, 2372–2377.

Schillo, K.K., Hansen, P.J., Kamwanja, L.A., Dierschke, D.J. and Hauser, E.R. (1983) *Biology of Reproduction* 28, 329–341.

Tierney, M.L. (1979) Genetic aspects of puberty in Merino ewes. In: Tomes, G.J., Robertson, D.E. and Lightfoot, R.J. (eds) *Sheep Breeding*, 2nd edn. (Revised by William Haresign.) Butterworths, London, pp. 379–386.

Measuring Growth

10.1. Introduction

The previous chapters of this book have described the ways in which farm animals grow, change in shape and in composition and ultimately reach a stage either when they can be used for breeding purposes or at which they are slaughtered to yield products, notably carcass meat, for direct human consumption. This discourse on growth processes has included considerations of changes in total live weight as well as changes in mass in the various tissues within an overall live-weight change. In both practice and in experimental work a number of different techniques are used for measuring these different aspects of the growth process in its entirety. They include methods which are both objective and subjective, which are applied to the live animal and to the carcass of the slaughtered animal and which predict in the live animal the expected tissue deposition in the carcass of that animal when it is slaughtered. A list of techniques used currently and those which may find increasing use in the future is given in Table 10.1

10.2. Measurements on the Live Animal

10.2.1. Live weight

The recording of live weight is probably the most widely used technique of all, both in practice and in experimental work, in determining the growth rates of animals and in predicting their likely body compositions and, therefore, the rates at which various tissues have grown. In practice the recording of live weight is important if animals of suitable body composition to meet various grading and other requirements are to be marketed at the correct time and profits are to be maximized. For example, in the UK the recording of the live weights of lambs is a very important procedure in indicating the proximity of animals of different breeds and crosses to optimal live weights commensurate

271

Table 10.1. Techniques used in measuring growth in farm animals and in other animals (O = objective technique; S = subjective technique).

Live animal only	
Radioactive isotopes	O
Dilution techniques	O
Balance studies	O
Slaughtered animals only	
Dissection procedures	O
Specific gravity determination	O
Live and slaughtered animals	
Weight	O
Linear and circumference measurements	O
Ultrasonic, X-ray and other imaging techniques	O
Probes	O
Visual assessments based on handling and scoring	S

with meeting grading criteria. Similarly, in cattle and pig production programmes progress towards planned terminal points is monitored by recording live weights which form not only a basis for making decisions on food adjustments but which indicate also when the terminal point of the production cycle has arrived. In experimental work the recorded live weights of animals are used very widely in not only measuring responses to imposed treatments but also in providing criteria against which other variables may be set and correlated with. Indeed the consensus view from a great deal of research work is that live weight is the most important single predictor of many carcass traits and is the variable that is most important to hold constant in prediction equations which include other variables.

The recording of live weights is, in relation to other procedures, a technique which is easy to effect and which is not costly. Nevertheless, it is a technique which can give very misleading results unless careful consideration is given to a number of factors. The true growth of any animal, whether under practical or experimental conditions and estimated from recorded live weights, will depend on the validity of the actual live weights recorded and it is here that there is considerable scope for error to creep in and to invalidate to varying degrees the calculated growth rates. The validity of recorded live weights depends on three factors: the precision of the weighing machine, human error and the extent to which apparent changes in live weight represent true changes in the weight of the carcass and organ tissues in relation to fluctuations in the quantities of food present in the gastrointestinal tract. Whilst the first two factors are important, careful attention to all detail by the operator should ensure that the recorded live weights are accurate. However, the third factor presents a much more difficult challenge, particularly with the ruminant animal. In the growing pig the proportion of live weight attributable to the contents of the gastrointestinal tract may vary between 0.02 for the 90-kg pig and 0.05 for the 20-kg pig. In the ruminant animal the contents of the rumen and reticulum proportionately account for at least 0.10–0.15, and frequently up to 0.23, of the total live weight, with the contents of the rest of the tract accounting for a further 0.02–0.03 of the total live weight. At all times, in all species, there will be considerable variation and the technique of recording live weight must attempt to accommodate such variation.

In the pig used in experimental work a variation of plus or minus 1-kg around recorded live weights may have a profound influence on calculated growth rate differences resulting from the imposition of, for example, different dietary regimes. A simple example will illustrate this point. Suppose that the growth responses to two different dietary treatments in pigs between 20 and 90 kg live weight is to be investigated and that 12 well-matched pigs, from the point of view of sex, previous background and live weight, are to be individually given each of the two diets. With an experimental design such as this it is not unreasonable to expect that the 20–90-kg live-weight interval could be traversed, on average, in 90 days to give an average daily live-weight gain of 778 g, that a standard error of difference between the two treatment means could be about 11 g in daily live-weight gain and that, accordingly, a least significant difference for a 0.05 probability level of 22 g daily would exist. Therefore, in this case, in the period of 90 days, the least significant difference of 22 g in daily growth rate would account for only about 2-kg difference in actual live weight, a difference that could easily be suspect and to varying degrees inaccurate if care were not taken in standardizing the weighing technique used and/or perhaps taking a mean of 3 successive days' weighings as the starting and finishing points of the experiment. The presence of about 1 kg of food in the gastrointestinal tract of the 20-kg pig and of about 2 kg of food in the 90-kg pig is clearly of potential importance in giving misleading results unless techniques are standardized.

Variation in live weight of ruminant animals attributable to digesta present in the gastrointestinal tract has been alluded to briefly above. This variation is certainly the largest single cause of error in estimating the live weights of ruminant animals and is influenced by many factors. In the first place the amount of digesta present depends on the quantity and quality of the food ingested and, if the animal is at pasture, the grazing system to which it has been exposed. Also, and particularly if the animal has not had constant access to pasture, the time that has elapsed since the last intake of food will have a marked effect. Diet type influences live weight through, in particular, the retention time of the digesta in the gastrointestinal tract. Quality is here very important as, generally speaking, with equal intakes, the digesta from high-quality (digestible) diets tend to pass through the tract more quickly, and therefore to form lower proportions of total live weight, than do digesta from poor quality diets. In relation to changes in live weight attributable to true tissue growth, changes in live weight due to variations in digesta present are extremely rapid in the short term. In this particular context, Hughes (1976) concluded that an active period of ingestion in the ruminant animal can result in a proportional hourly increase in live weight of 0.019 and that, conversely, in cattle that are fasted the proportional hourly loss of live weight may be 0.006–0.007 over a 3- to 4-h period. He points out, however, that the rate of loss will depend on the initial level of fill and that whilst, for both cattle and for sheep, a 12-h loss may be proportionately 0.05–0.06 of initial live weight, the 36-h loss will be about twice this figure. Such dramatic changes in live weight are clearly of considerable importance and necessitate the standardization of weighing procedures. How may this be done?

Fasting animals before weighing, in an attempt to equalize gastrointestinal fill, is one method which may be used. Fasting will initially give a rapid loss in live weight, being greater in heavier than in lighter animals. According to Hughes (1976), the ideal approach is to standardize the fasting interval. The interval does not have to be too long, as there is a strong correlation between live-weight loss in cattle in the first 6 h and live-weight loss by the end of 24 h (Hughes, 1976). However the withdrawal of water as well is unlikely to have a consistent effect on live-weight changes.

Another way of minimizing live-weight fluctuations is to select for standardization a weighing time when live-weight variations due to fill are minimal. In grazing animals the digesta present in the gastrointestinal tract are likely to be minimal in quantity and least variable at the beginning of the day and highest and most variable at the end of the day. In relation to this, weighing as near as possible to the start of the day is obviously the preferred time to adopt as a standard. In non-grazing animals the comparable time is before feeding. By sticking to these times in routine procedures, accuracy should be improved. Further attempted refinement using multiple weighings may actually increase the error due to stress and disturbance decreasing growth rates (Hughes, 1976).

10.2.2. Body measurements

Measurements taken on the live animal have been used extensively for a variety of reasons in both experimental work and in practice. Some are linear and are taken either with various types of measuring rods or sticks or with calipers. Others record circumferences and are taken with flexible tapes. Most of the linear measurements reflect primarily the lengths of the long bones of the animal. Overall they indicate, when taken sequentially over a period of time, the way in which the animal body is changing shape and have been used as predictors of both animal live weight and carcass composition. Mostly they have been used on cattle and horses, to a much smaller extent on sheep and to a very limited extent on pigs and poultry.

Examples of measurements which may be taken are given in Fig. 10.1. and Table 10.2. The accuracy of such measurements depends on a number of factors.

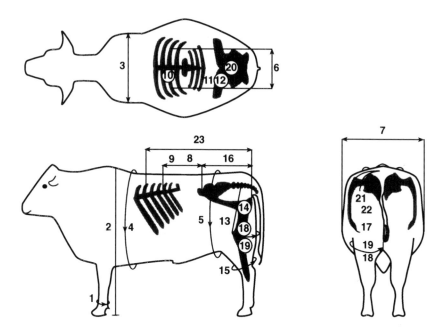

Fig. 10.1. Examples of body measurements (adapted from Fisher, 1975a). See Table 10.1 for details.

Fisher (1975a) proposes that there are three sources of error in taking any one of these measurements:

1. Correct identification and location of end reference points in linear measurements.
2. Anatomical distortion produced by the animal changing either position or posture or by by changing muscle tone.
3. Error involved in actually taking the measurement at any one position, which will be minimal for caliper measurements but greater for measurements using a flexible tape over surfaces, particularly if they are concave.

The accuracy of any one of the measurements described in Table 10.2. depends on whether it reflects size of skeletal units only, is a reflection of the development of both soft tissues and skeletal units or is a reflection of the development of soft tissues only. The order of decreasing accuracy is from skeletal measurements to skeletal plus flesh measurements to soft tissue measurements (Table 10.3). In cattle and in sheep the prediction of tissue deposition in carcasses, particularly soft tissue, and the distribution of tissues within regions of the body, from live body measurements, is poor. Also, because animals vary in both shape and in size, generally they cannot be used as accurate predictors of body weight. If this was not the case, that is the animal retained

Table 10.2. Details of body measurements shown in Fig. 10.1 (adapted from Fisher, 1975a).

Number	Measurement	Description	Equipment
1	Circumference of cannon	Circumference at narrowest point along metacarpus or cannon bones	Tape
2	Height at withers	Highest point over the scapulae vertically to the ground	Measuring stick
3	Width of shoulders	Widest horizontal width across the shoulder region	Large calipers
4	Heart girth	Smallest circumference	Tape
5	Rear flank girth	Smallest circumference just anterior to the hind legs in the vertical plane	Tape
6	Width of ribs	Distance between points on either side of the animal between the 12th and 13th ribs on a line passing through the vertical tip of the tuber coxae and parallel with the ground	Pincer calipers
7	Width of paunch	Widest horizontal width of the paunch at right angles to the body axis	Large calipers
8+9	Length of loin	Length from the point between the 12th and 13th ribs, lying on the line drawn horizontally through the ventral tip of the tuber coxae, to the ventral tip of the tuber coxae	Small calipers (8) and tape (9)
10	Depth of rib point	Vertical depth of the point between the 12th and 13th ribs, at the level of the ventral tip of the tuber coxae, from the dorsal mid-line	Tape
11-12	Depth of hooks	Vertical depth of the ventral tip of the tuber coxae from the dorsal mid-line	Small calipers (11) and tape (12)

No.	Measurement	Description	Instrument
13+14	Depth of patella from base of tail	Oblique vertical distance of the anterior point of the patella from the depression between the first coccygeal vertebra and sacrum in the mid-dorsal line	Large calipers (13) and tape (14)
15	Circumference of hind leg	Measured at the level of the junction of the gastrocnemius muscle and its tendon, at right angles to the tibia	Tape
16	Length of pelvic girdle	Distance between the ventral point of the tuber coxae and the ventral tuberosity of the tuber ischii	Pincer calipers
17	Depth of patella from dorsal mid-line	Vertical depth of the anterior point of the patella from the dorsal mid-line	Tape
18+19	Length from patella to posterior mid-line	Horizontal distance from the anterior point of the patella to the posterior junction of the buttocks	Pincer caliper (18) and tape (19)
20	Width of rump	Distance between the points on either side of the animal located at one-half of the distance measured from the ventral point of the tuber coxae to the ventral tuberosity of the tuber ischii by means of a tape	Pincer calipers
21+22	Depth of rump	Vertical distance from the dorsal mid-line to the point located halfway between the tuber ischii and tuber coxae measured by means of a tape	Small calipers (21) and tape (22)
23	Length of hindquarter	Distance between the point located on the 10th rib, on the horizontal line passing through the ventral tip of the tuber coxae, and the ventral tuberosity of the tuber ischii	Tape
24	Skinfold thickness at flank	Taken at the thin fold of the skin stretching from the lateral abdominal region to the anterior face of the hind leg and at a point halfway along this fold	Small calipers
25	Skinfold thickness	Taken at the most ventral point of the brisket fold, usually anterior to the forelegs	Small calipers

Table 10.3. Class of measurement and associated accuracy (Fisher, 1975b).*

Measurement class	Measurement description	Mean (mm)	Residual SD (mm)
Skeletal tissue only	Circumference of cannon[+]	210.5	2.06
	Tuber coxae to tuber ischii[‡]	469.2	5.60
	Sacrococcygeal joint to patella[‡]	531.1	6.53
Soft tissues only	Circumference of hind leg[+]	505.6	18.55
	Patella to posterior mid-line[‡]	503.5	19.86
Skeletal + soft tissues	Heart girth[+]	1799.5	15.24
	Width shoulders[‡]	494.7	12.96
	Sacrococcygeal joint to patella[‡]	631.5	10.36

*Based on duplicate measurements by one operator on 15 Hereford steers.
[+]Taken with tape.
[‡]Taken with calipers.

a constant shape and varied in size only, then as long as the basic geometry of the animal is understood, dimensions or tissue volumes could be predicted from other dimensions or tissue volumes. Fisher (1975b) refers to this problem and points out that in addition to a range of different sizes in animal bodies, there are also ranges in tissue shapes and proportions. To account for some of this allometric variation all measurements may be expressed in relation to a standard skeletal dimension which is assumed to be a constant proportion of total skeletal size. Therefore a measure of basic body size must be used as an independent variable if predictions are to be made of absolute tissue weights from one or more body measurements. Nevertheless, and as Fisher points out, it is important that dimensional parity should be maintained for all variables. For example, all linear measurements should be cubed if related directly to tissue volume, or in practice tissue mass, as these are already in the cubic dimension.

The body measurement which exhibits the highest correlation with body weight is heart girth, but even here the relationship is anything but straightforward. The error involved in estimating the live weight of cattle from heart girth measurements is likely to between three and six times greater than in direct weight determinations (Johansson and Hildeman, 1954). The size of the error may be diminished by making allowances for variations in breed, age, size and body condition. For animals growing over a wide weight range, the relationship between live weight and heart girth is curvilinear. Linearity may only be assumed in groups of similar aged animals. In adult animals the regression coefficient of live weight on heart girth increases with increasing body size and increasing fatness. A degree of subjective appraisal of body condition is therefore necessary in the first instance. The data of Table 10.4 quantify variations in the relationship between live weight and heart girth in cattle. A further complication, related to the rate of passage of digesta and resulting

Table 10.4. Interpolated values for average live weight and for regression coefficients for cattle over 1 year old classified on the basis of heart girth into classes of 10-cm range (5 cm on either side of the figure given in the first column) and divided into condition groups within each class (Johansson and Hildeman, 1954).

Heart girth (cm)	High condition		Normal condition		Low condition		SD from regression line (kg)
	Mean live weight (kg)	Regression coefficient (kg cm^{-1})	Mean live weight (kg)	Regression coefficient (kg cm^{-1})	Mean live weight (kg)	Regression coefficient (kg cm^{-1})	
120	135	3.5	120	3.2	105	3.0	12
130	190	4.1	180	3.8	175	3.5	15
140	240	4.7	230	4.4	220	4.1	20
150	300	5.3	285	4.9	270	4.6	25
160	365	5.9	345	5.5	325	5.1	30
170	440	6.5	410	6.0	385	5.6	30
180	515	7.1	485	6.6	450	6.1	30
190	600	7.7	555	7.1	515	6.6	30
200	695	8.3	645	7.7	675	7.1	35
210	805	8.9	740	8.2	695	7.6	40
220	930	9.5	850	8.8	770	8.1	45

gastrointestinal tract fill (see section 10.2.1) is whether the animal is grazing pasture in a rainy or a dry period. In this context Goodchild (1985) has shown, from work conducted in Tanzania, that the ratio of heart girth to empty body weight, and therefore indirectly to live weight, is greater in the dry season than in the wet season. Notwithstanding these limitations there is little to be gained by including one or more other body measurements as variables in prediction equations of live weight from heart girth measurements, the partial correlations of various body measurements other than heart girth with live weight tending to be very low when heart girth is held constant. The significance of heart girth as the best single predictor of live weight is exemplified by the findings of Jones *et al.* (1989). In this work, for horses varying between 230 and 707 kg live weight, $\log_e$ values of chest girth as a predictor of live weight gave adjusted R^2 values of 90.5% with the best prediction equation using a single factor assuming the form $\log_e$ live weight (kg) = $-7.60 + 2.66 \log_e$ chest girth (cm).

A further, but in some ways rather different, measurement which is sometimes taken is that of anal skinfold thickness of cattle. The skin and the subcutaneous fat lying between the ischium and the base of the tail constitute this measurement and if the animal is fat the fold bulges on either side of the anus. Calipers are used to record the measurement when the fold is pinched up by the fingers of the hand. Charles (1974), from work with Aberdeen Angus, Hereford, Friesian and Charolais cross-bred steers totalling 37 in number, proposed that a change of +0.5 cm in anal fold thickness reflects proportional changes of –0.02 muscle, +0.03 fat and –0.01 bone plus fascia in the carcass and a change in killing-out proportion of +0.01. However, the data were obtained from relatively few animals and unless further data are obtained from both larger and more diverse (in terms of breeds) populations the accuracy of the technique must remain open to doubt. For example, later work (Simm *et al.*, 1983) with Hereford bulls found a correlation of –0.92 between anal fold thickness and percentage of muscle in the carcass, whereas Charles found a correlation of –0.04 between the same two variables. The fact that the steers used by Charles ranged widely in age, live weight, breed type and body composition, compared with the bulls used by Simm *et al.*, perhaps indicates one of the major disadvantages of the technique, namely that prediction equations may be limited to defined types of populations of animals only.

10.2.3. Visual appraisal of live animal conformation

Man has for many years used his senses of touch and sight in assessing the degree of 'finish' of cattle, sheep and pigs to meet certain market requirements. The subjective appraisal of animals in this way reflects the deposition of soft tissues, in particular fatty tissue, in the animal. Many systems of scoring animals are based on this approach and all are subject to considerable error and are of limited value when small differences between animals exist. The repeatability of scoring is poor, both within and between individuals assessors, and as a tool to assist in the selection of animals for meat production programmes it has a limited potential because of its inability to predict muscle growth. Indeed, in the past it may have hindered progress because of the bias towards fatty tissue growth (the 'well-rounded' type of animal conformation) which has been inherent within it. Nevertheless, the scoring technique has been a very useful management tool in cattle, sheep and pig production, where the condition scoring of breeding females has assisted considerably in formulating appropriate feeding regimes. The usefulness of live animal scores in predicting growth and growth-related traits is indicated in the data of Table 10.5. The limitations of score assessments are evident but when consideration is given to the fact that the scores in the first place reflect a bias towards fatty tissue growth, then they are clearly of little use in predicting growth in animals where muscle accretion rates are of greatest importance.

The use of body scoring techniques as tools in the management of female breeding stock offers considerably greater advantages than in predicting growth

Table 10.5. Average phenotypic and genetic correlations of live animal scores with growth, and growth-related traits and heritability estimates for live animal scores, of Aberdeen Angus, Hereford and Shorthorn breeds of cattle in the USA (data adapted from the report of Petty and Cartwright, 1966).

Correlations	Weaning score		Yearling pasture score		Final feedlot score	
	$r_P^\dagger$	$r_G^\dagger$	r_P	r_G	r_P	r_G
Birth weight	0.15	0.36	—	—	0.15	0.07
Growth from birth to weaning	0.33	0.36	—	—	—	—
Weaning weight	0.39	0.39	0.20	0.02	0.22	0.19
Feedlot growth	0.00	0.08	−0.16	−0.03	0.39	0.31
Pasture growth	—	—	0.25	0.50	—	—
Final feedlot weight	0.31	0.43	—	—	0.41	0.38
Yearling pasture weight	0.21	−0.03	0.40	0.30	—	—
Yearling pasture score	0.36	0.58	—	—	—	—
Final feedlot score	0.40	0.68	—	—	—	—
Heritability						
By parent–offspring regression						
Mean	0.24		0.16		0.18	
Range	0.00–0.50		0.00–0.32		0.07–0.46	
By paternal–half-sib correlation						
Mean	0.36		0.23		0.46	
Range	0.05–0.75		0.13–0.85		0.12–0.92	

$^\dagger r_P$ = phenotypic correlation coefficient; r_G = genetic correlation coefficient.

in tissues in carcass-yielding animals, because fairly wide differences in body tissue deposition, with reasonably large errors, are acceptable as guides to feeding regimes which should be adopted in management procedures. The degree of accuracy required is much smaller and so the errors associated with scoring are relatively less important. This contrasts strongly with the position in animals which are to yield carcass meat, where small differences in tissue deposition are of great importance in a number of different contexts, including, on the one hand, the interpretation of responses to imposed experimental treatments and, on the other hand, the values of the carcass to the farmer. Many different scoring systems have been developed and an example of one such system for use in cattle is detailed in Table 10.6.

Using the system of condition scoring shown in Table 10.6 on various breeds and crosses of cow, Wright and Russell (1984a) correlated score with body composition determined directly by dissection after slaughter. They found in Hereford × Friesian, Blue-Grey, Galloway and Luing cows that a unit change in condition score was associated with a change of 2242 (SE 103) MJ of body tissue energy but that in British Friesians the comparable figure was 3478 (SE 392). The difference between genotypes reflects to a large extent the differences in the proportions of fat stored in the main depots of the body giving, in turn, differences in the relationships between condition score and body fat. The British Friesian cows had, compared with the others, a higher proportion of their total fat in the intra-abdominal depots and the lowest proportion in subcutaneous fat, therein resulting in them being fatter at any given condition score.

The regression analyses of body components on condition score give, in addition to the energy changes detailed above, estimates of the composition of

Table 10.6. System of body condition scoring of cattle (Lowman *et al.*, 1976).

Score 0	The animal is emaciated. No fatty tissue can be detected and the neural spines and transverse processes feel very sharp
Score 1	The individual transverse processes are sharp to the touch and easily distinguished
Score 2	The transverse processes can be identified individually when touched, but feel rounded rather than sharp
Score 3	The transverse processes can only be felt with firm pressure and areas on either side of the tail head have some fat cover
Score 4	Fat cover around the tail head is easily seen as slight mounds, soft to the touch. The transverse processes cannot be felt
Score 5	The bone structure of the animal is no longer noticeable and the tail head is almost completely buried in fatty tissue

Table 10.7. Change in composition of the empty body associated with one unit change in condition score (Wright and Russell, 1984a).

	Breed	
	Hereford × Friesian, Blue Grey, Galloway, Luing	British Friesian
Water (kg)	22.2	22.2
Fat (kg)	52.6	84.1
Ash (kg)	1.18	1.18
Protein (kg)	7.35	7.35

body tissue changes associated with changing condition score, and those calculated by Wright and Russell are given in Table 10.7. From these and other data the indications are that body condition scoring can be a very useful management aid in predicting body composition, in particular fat content, and therefore because the energetics can be quantified, in estimating nutrient input for a given set of conditions.

10.2.4. Dilution techniques

The inverse relationship between the proportions of water and fat in the animal has been pointed out in previous chapters. Dilution techniques measure water and therefore, indirectly by difference, give an estimate of fatty tissue. Each technique measures the dilution of body water by introducing a known amount of a tracer into the animal, by allowing the tracer to equilibrate with body fluids and finally by determining in a sample of body fluid, usually the blood, the concentration of the tracer. Inherent in this approach is the assumption that the tracer is distributed evenly amongst the various body tissues.

The tracers used can be either in liquid of gaseous forms. If gaseous tracers are used the animal has to be enclosed in a chamber and the amount of gas given which is absorbed into fatty tissue is measured. A number of different gases have been tried for this purpose and the one that appears to hold greatest promise is krypton. However, compared with liquid tracers, gaseous tracers present more problems in application and on the whole are preferred less. Of the liquid tracers tried, heavy water (D_2O) and tritiated water (TOH) are generally preferred to various other tracers including antipyrene and N-acetyl-4-aminoantipyrene.

Dilution techniques using tracers are difficult to apply and must be regarded as tools for the experimenter only. Even for the experimenter, however, there are some considerable problems. For example, although tritiated water is a preferred tracer it is radioactive and has a long half-life. As a consequence, the animal's carcass at slaughter is unfit for human consumption and must therefore be wasted. Standardization of technique is also very important if meaningful results are to be obtained because total body water, including that present in the bladder and in the digesta of the gastrointestinal tract, is determined. Also it is very important that differences in body water according to age, breed, sex and previous nutritional history, are accommodated. Overall, the dilution techniques are probably of less use in establishing absolute values than in work where comparative measurements are desired and some indications of how they in turn compare with other techniques is given later (see section 10.2.10)

Robelin and his coworkers in France, in producing new data and in reanalysing old data, have done much to give a surer footing to the prediction of body composition from the measurement of body water using dilution techniques. The allometric equations proposed by Robelin and Geay (1978) for predicting the protein and water content of the fat-free body, and the relationships established between lipid and water content of the empty body by Robelin and Theriez (1981), are given in Table 10.8. In relation to these equations, Robelin (1984) points out that the protein content of the fat-free body is remarkably similar and constant for a wide range of species and that, consequently, the lipid and the protein content of the empty body may be predicted with a reasonable degree of accuracy. From the practical point of view, the limitations have been mentioned above: only full body weight and total body water can be measured *in vivo*. Standardization of technique is therefore clearly of considerable importance if comparative values between individual animals or between groups of animals are to reflect true differences in body composition compared with differences in total body composition where the proportions of total water are divided between body tissue and other non-carcass tissues and media. Accepting these limitations, various estimates of body lipid and body protein from water space (Table 10.9) show that body lipid decreases by 0.9–1.3 kg when the water space decreases by 1 kg.

Table 10.8. Allometric equations relating protein and water content to fat-free body weight (FFB) in growing cattle (Robelin and Geay, 1978) and prediction equations of water content of empty body (EBW) from lipid content of empty body in growing cattle and sheep (Robelin and Theriez, 1981).

Protein (kg) = 0.1259 FFB$^{1.096}$ (residual coefficient variation = 2.8% proteins)

Water (kg) = 0.8477 FFB$^{0.974}$ (residual coefficient variation = 1.1% water)

Cattle: Water % EBW = 74.7 − 0.824 lipids % EBW

Sheep: Water % EBW = 77.4 − 0.866 lipids % EBW

Table 10.9. Equations of prediction of body lipids (L, kg) and proteins (P, kg) from body weight (BW, kg) and water space (WS, kg) in various types of animals (Robelin, 1984) (see original reference for origins of equations).

Animals	*n*	Equation	RCV (%)*
Growing pigs	81	L = 0.934 BW – 1.316 WS – 0.22	7.9
Growing cattle	42	L = 0.769 BW – 0.943 WS	13.8
Mature ewes	38	L = 0.904 BW – 0.913 WS – 6.0	12.8
Mature cows	20	L = 0.903 BW – 1.135 WS	8.7
Mature cows	18	L = 0.828 BW – 0.904 WS – 15.1	14.8
Growing cattle	42	P = 0.124 BW + 0.058 WS	4.7
Mature ewes	38	P = 0.048 BW + 0.076 WS + 1.055	2.3
Mature cows	20	P = 0.088 BW + 0.075 WS	2.5

*Residual coefficient of variation as percentage of the mean of dependent variable.

10.2.5. Neutron activation analysis

This technique has been proven in the human medical field and may, in spite of the high costs involved in using whole-body counting chambers, find an increasing use in agricultural studies in the future. The method allows the measurement of gamma radiation, or photon production, from neutron irradiation in hydrogen, nitrogen, oxygen, sodium, phosphorus, chlorine and calcium and in the naturally occurring radioactive isotope of potassium (^{40}K). Using ^{40}K, high repeatability between measurements has been demonstrated (Carr *et al.*, 1978), and accurate predictions of muscle content are possible. It is likely that in the future the technique may complement rather than replace other methods of assessing the distribution of tissues within the animal body (East *et al.*, 1984).

10.2.6. Probes

Probes have been used mostly in pigs to measure backfat thickness. The precision in predicting muscle backfat thickness is high but there are valid objections to the technique on welfare grounds unless local anaesthesia is induced in the areas where the skin incisions are made for the insertion of the probe. The probe is valuable in carcass assessment work (see section 10.3).

10.2.7. Balance studies

Balance studies can give short-term estimates of protein and fat deposition in the animal body. As the name implies, a balance sheet has to be drawn up of particular elements entering and leaving the body in order that retention may be calculated by difference. The work involves sophisticated equipment, is expensive and is only suitable for estimating the deposition of protein and fat under experimental conditions.

A combined carbon and nitrogen balance will allow an estimate to be made of the storage of both protein and fat in any particular period of time. The animal must be in positive balance, that is the excreted amounts of carbon and nitrogen must be less than the intakes. The amount of body protein stored is calculated by multiplying the amount of nitrogen retained by 6.25, therein assuming that all body proteins contain 160g nitrogen kg^{-1}. Body proteins also contain about 520g carbon kg^{-1} and therefore the amount of carbon stored as protein can be calculated. The amount of carbon in fat is taken as 746g kg^{-1} and thus the remaining carbon which is stored as fat is calculated by dividing the carbon balance, less that stored as protein, by 0.746. Details of this approach are given in Table 10.10.

Using this approach it is not possible to predict accurately either the total muscle deposited or its location within the animal. Similarly, the fat deposition may be either in carcass or non-carcass tissues or in both. Also the total protein stored in the body may be in carcass or non-carcass tissues, or in both, and may also be in tissues other than muscle. Furthermore, the variation in the chemical

Table 10.10. Calculations involved in determining protein and fat storage in animals from carbon and nitrogen balance data.*

	Carbon	Nitrogen
Intake	X	y
Excretion in faeces	X_f	y_f
Excretion in urine	X_u	y_u
Brushings	X_b	y_b
Excretion in methane (if ruminant animal)	X_m	—
Excretion as carbon dioxide	X_{co_2}	—
Balance (stored) (X' for carbon and Y' for nitrogen)[†]	$X' = X - (X_f + X_u + X_m + X_b + X_{co_2})$	$Y' = y - (y_f + y_u + y_b)$
Protein stored (P)	$= Y' \times 6.25$	
Carbon used for protein gain (C_P)	$= P \times 0.520$	
Carbon available for fat gain (C_f)	$= X' - C_P$	
Fat stored (F)	$= C_f \div 0.746$	

*A time scale is not indicated, but it is unlikely that balances such as this would last, at the most, for more than 2 or 3 days. Usually the data is collected for a 24-h period only.
[†]Assumes a positive balance, i.e. X and Y are both greater than the total losses of carbon and nitrogen.

composition of tissues, evident from chapter 3, clearly presents other problems. For example, because of variation in the proportional protein content of skeletal muscle the use of an average figure as a multiplier of the protein storage figure could be very misleading and would, in any event, not allow any differentiation between muscle tissue on the one hand and non-muscle tissue, such as blood, on the other hand, or between skeletal and non-skeletal muscle. In addition, the variation in the lipid proportion of skeletal muscle does not allow any estimate to be made of true adipose tissue growth involving lipid deposition in connective tissue in depots, compared with lipid deposition in muscle tissue.

10.2.8. X-ray and computed tomography

X-rays have long been widely used in the human and animal fields for diagnostic purposes. In these cases the clinician has been interested primarily in locating fractures in bones and in locating and visualizing abnormalities in other tissues, organs and regions of the human and animal body. To use X-rays for measuring growth as indicated by tissue mass or volume has therefore necessitated the development of new techniques which have progressively recorded the shape and size of tissue masses by serially scanning the body and by feeding the details of the pictures obtained into a computer for the purpose of integrating the series into a whole: hence the term computerized tomography (CT). Programmes of this type have been developed successfully for animal use and the principle of X-ray computerized tomography (CT) for the prediction of body composition has been reviewed by Skjervold (1982).

The following is a brief description of the method of CT. The animal or human is placed in the centre of a large wheel and an X-ray tube and special detectors for it rotate a full 360 degrees around the body and scan it. By linking the detectors to a computer this allows, by attenuation of the X-rays, the calculation of the density in each of about 65,000 small squares or cubes known as pixels which constitute the image of the object in cross-section. Therefore the image of the cross-section displayed on the monitor reflects the densities of the parts of the body under consideration. The densities are known as CT values and are set to vary from -1000 for air to $+1000$ for hard bone, with water being ascribed the mid-point of this range at zero. Obviously, with animals, temporary immobilization, not necessarily by administering drugs, is necessary if a scan is to be obtained.

To use the scan for the prediction of body or carcass composition, the density matrix has to be used in a quantitive manner and for this purpose a frequency distribution is produced of the CT values of each scan (Fig. 10.2) and the information of the frequency distributions is then used to predict body or carcass compositions.

Work with goats suggests that CT is a valuable *in vivo* method for predicting body fat and energy content but that it is less valuable for predicting the contents of water and protein in the body (Sørensen, 1992). Examples of X-ray tomographs of pigs and sheep and of the relationships of predicted body

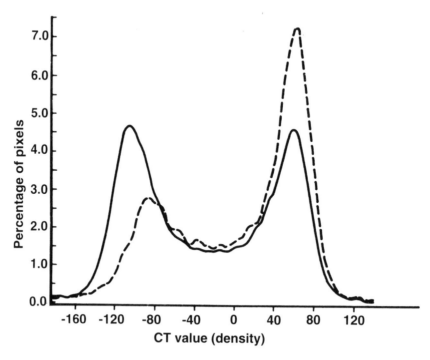

Fig. 10.2. Frequency distribution of CT values for the average of two scans of a fat line pig (——) and a normal Norwegian Landrace pig (– – –) of equal live weight. Both pigs were given food *ad libitum* and the CT values were recorded at the same live weight. The percentage of pixels in each class (class width 6 CT values) is given on the ordinate. The figure demonstrates that the fat line pig had a much higher percentage of its pixels in the 'fat area' of the CT range (from about −160 to −40) than the normal pig. The opposite is true for the 'muscle area' of the CT range (from about +30 to about +90) (reproduced from Standal (1984) by kind permission of the copyright holder, Elsevier Applied Science Publishers Ltd).

and carcass compositions from CT based on actual dissection data from the same animals are given by Allen and Vangen (1984) and Sehested (1984).

10.2.9. Nuclear magnetic resonance

Again this is a technique which currently is used widely in the human medical field but which increasingly is being considered for use with animals. The technique examines proton distribution and binding in the body. Nuclear magnetic resonance (NMR) CT relies on the induction of resonance in protons in the body by placing it in a strong magnetic field. The signals that are emitted are a reflection of the body's reaction to the high frequency disturbance that the magnetic field induces and are, therefore, products of the matter of the body itself. This contrasts to X-ray techniques as described above (section 10.2.8), which measure the absorption rate of ionizing radiation: for example, calcified bone tissue presents an inpenetrable barrier to X-rays whereas other tissues are to varying degrees penetrable. The technique is capable of observing only those protons which are relatively mobile and the major contributions to an NMR image are the hydrogen nuclei (containing a single proton) of water molecules and lipids. Sensitivity to other nuclei in addition to protons is also exhibited but because these are less abundant than hydrogen nuclei, high resolution NMR images cannot be obtained readily. An example is the nucleus of the phosphorus atom.

An NMR machine consists of a magnet with a space in the centre sufficiently large to accommodate either a human or an animal (Fig. 10.3). The body is exposed to a non-uniform magnetic field, the purpose of which is to label different parts of the body with different Zeeman field strengths. Consequently the nuclei in these parts respond with different, but recognizable, frequencies which enable the structure to be determined (Andrew, 1980). As a whole all biological systems provide strong NMR signals, but the strength of the signal varies from tissue to tissue within the organism as a whole. The strongest signals are obtained from fluids, lipids and soft tissues and the weakest signals from teeth and cortical bone. Signals of intermediate strength are obtained from muscle and tendon.

The image achieved is a result of the magnetic field aligning the nuclei of atoms which have an odd number of protons, or an odd number of neutrons, or both. Cross-sectional images of the body in any plane therefore illustrate the distribution density of hydrogen nuclei. The tissues are therefore visualized according to the proton density which they exhibit and to the relaxation time T_1. If the proton is part of a water molecule it can move fairly freely in the cell or tissue so that the relaxation time T_1 is fairly long. This will contract to the situation of the single proton attached to a hydrogen atom which is an integral part of a large molecule where the forces holding it give a greater rigidity to its position and, in consequence, a shorter relaxation time T_1. As Fuller *et al.* (1984) point out, the freedom of motion of a water molecule reflects the molecular and ionic composition of its environment because both ions and large organic molecules are able to bind layers of water molecules over their surfaces. Because of this the water molecules are more restricted in their movement and the T_1 values are shorter than in unstructured water. Free lipids in adipose tissue cells, but not the lipids of membranes, are held relatively loosely and because of this the relaxation times of the protons are sufficiently long to affect the NMR signal from the tissue. Nevertheless, the times are shorter than those for water protons. Consequently, adipose tissue is seen as a fast-relaxing tissue whilst, in contrast, muscle tissue with very little free lipid, has a much longer relaxation time T_1 (Fuller *et al.*, 1984). Ettinger *et al.* (1984) cite T_1 relaxation times of about 225 ms for muscle containing proportionately about 0.075 water and about 150 ms for adipose tissue containing proportionately 0.10–0.30 water.

Fuller *et al.* (1984) used the technique to image live and dead pigs. After obtaining images at nine sites along the body, the pigs were slaughtered and

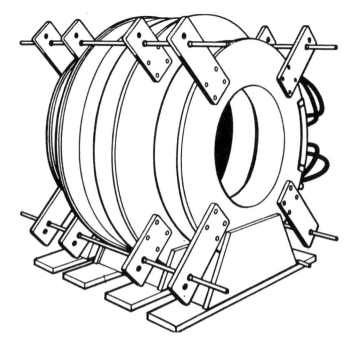

Fig. 10.3. A four-coil, air core, water-cooled, resistive electromagnetic for human whole-body nuclear magnetic resonance imaging (reproduced from Andrew (1980) by kind permission of the copyright holder, The Royal Society, London).

then, after freezing, sectioned at the nine locations for the purpose of comparing photographs of the cut surfaces with the NMR images. Figure 10.4 shows the NMR images based on proton density and T_1 relaxation times of sections through the thorax and rump. Comparative measurements of T_1 in living and dead adipose and muscle tissues are given in Table 10.11.

The technique of NMR holds exciting prospects for the future but programmes that will allow the estimation of whole-body tissue volumes using serial scans and computer integration (as with CT) will have to be developed. Andrew (1980) details the advantages over X-ray and other techniques:

1. The technique is non-invasive.
2. There is no ionizing radiation.
3. There are probably no hazards.
4. The electromagnetic radiation penetrates bone tissue without significant attenuation.
5. The density distribution of the most abundant chemical element in all biological systems, that is hydrogen, is measured with tissue discrimination.

As with CT, temporary immobilization of the animal will in most cases be necessary but the biggest obstacle to overcome is cost. NMR scanners are

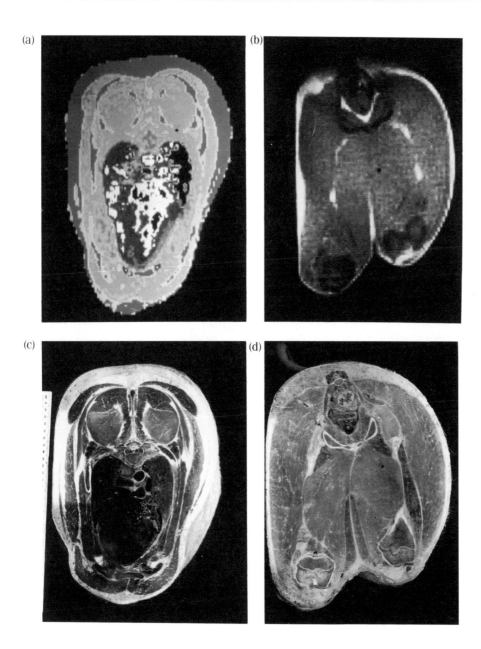

Fig. 10.4. Sections through the thorax (a) and rump (b) of a live pig as displayed by NMR imaging by proton density (p) and relaxation times (T_1) compared with carcass cross-sections of thorax (c) and rump (d) (by courtesy of Dr M.F. Fuller, Rowett Research Institute, Aberdeen).

Table 10.11. Measurements of T_1 relaxation times (ms) in living and newly dead pig adipose tissues and muscles (Fuller *et al.*, 1984).

	\multicolumn{4}{c}{T_1 relaxation times}			
	Alive		Dead	
	Mean	SE	Mean	SE
Adipose tissue				
Shoulder position 1	135	1.8		
Shoulder position 2	139	1.7	135	2.3
Thorax	138	1.6		
Mid-back 1	142	1.7		
Mid-back 2	138	2.7	137	1.9
Loin	148	3.3	132	3.3
Rump	155	2.5		
All sites	140		135	1.9*
Individual muscles				
m. triceps	245	3.5	241	3.5
m. superspinatus	288	5.0		
m. trapezius	260	5.0		
m. longissimus dorsi				
at thorax	236	2.2		
at last rib	233	2.5		
at loin	234	2.2	232	7.1
m. psoas	246	7.1		
m. rectus femoris	235	3.5		
m. biceps femoris	238	3.5	239	3.5
m. semimembranosus	239	4.1		

*SE of differences.

immensely expensive, even for use in the human field where cost is relatively unimportant, and until much cheaper versions are made available, the potential which the technique holds for determining body composition of animals may not be realized.

10.2.10. Ultrasonic techniques

Various ultrasonic machines are used widely in breeding schemes, particularly with pigs, and in various research programmes. For a general review, see Goddard (1995) and Nyland and Mattoon (1995). The type of machine available varies from the simple, which gives a reading of tissue depth, perhaps subcutaneous fat on the back, to the complex, which is capable of producing two-dimensional images of cross-sections through the animal body. Pulses of ultrasound pass through different tissues at different, but specific to the particular tissue, rates. An interface between two tissues partly reflects and partly transmits the ultrasound pulse. The sound pulses are converted to electrical pulses which are displayed on suitable oscilloscopes as spikes on a baseline. When used for animals the scanning has mostly imaged the echoes which have emanated from the tissue interfaces and which are relatively strong compared with the weaker echoes derived from within tissues (Fig. 10.5). In most cases the prediction of body composition has depended on ultrasound echoes bounced off tissues within the animal at one or two points only, mostly in the rib region. In pigs but not to the same extent in cattle, the depth of backfat measured in this way is closely correlated with carcass composition. When considering carcass quality, it is the proportion of muscle that is important above all other considerations in the case of pigs, cattle and sheep. How well do ultrasound techniques applied to the live animal body predict the muscle that has accumulated, and therefore the proportion that may be expected, in the carcass?

In cattle, the proportion of the chemically determined fat and crude protein, and derived estimates of energy for the carcass and whole empty body, have been shown to be between 0.67 and 0.87 accountable for by ultrasound measurements recorded at the hindquarters and shoulder regions (Ivings *et al.*, 1993). However to really answer the above question, it is important to appreciate that the accuracy of prediction will depend on the type of machine that is used. Several studies have been carried out in this area with both cattle, sheep and pigs. For example, Kempster *et al.* (1979) compared four ultrasonic machines of varying sophistication for predicting the body composition of live pigs, and in later work (Alliston *et al.*, 1982) three of these ultrasonic machines were used in similar studies. The residual standard deviations for the prediction of carcass lean proportion from the measurements taken with the machines are given in Table 10.12. From these results Kempster *et al.* (1982) concluded that there was little to justify the more costly Scanogram (Fig. 10.6), Ilis Observer and Danscanner machines compared with the cheaper Sonatest, which gave a reasonable prediction of carcass lean content. Kempster *et al.* (1982) also concluded that in cattle and in sheep, predictions were not likely to be as good as those in pigs but pointed out, importantly, that the precision achieved will depend on the particular population studied, the precise details of the scanning technique used and the skill and the experience of the operator.

Ultrasonic techniques have been compared with other techniques such as live weight (from heart girth measurements), rib skin fold thickness and anal fold thickness in predicting the lean content of beef sires (Simm *et al.*, 1983), and with live weight, skeletal size, total body water as estimated by deuterium oxide dilution and blood and red cell volumes, in predicting the body compositions of mature beef cows (Wright and Russell, 1984b). In the former case, where 97 Hereford bulls were assessed using Scanogram and Danscanner machines, higher correlations with carcass lean were obtained using these machines than with any other technique (Table 10.13). Multiple regression equations which included live weight and the best anatomical sites for ultrasonic measurements

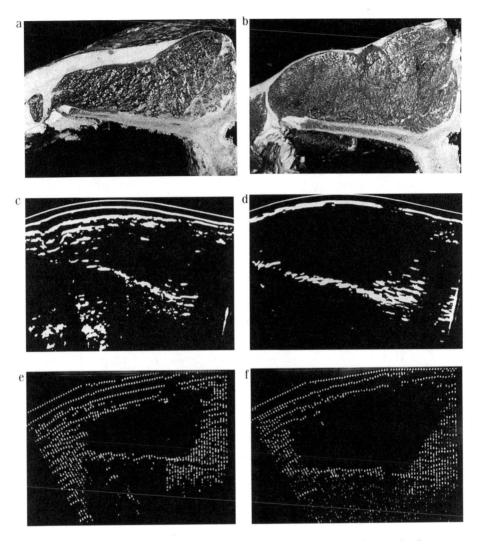

Fig. 10.5. Images produced by two ultrasonic scanning machines on live cattle of contrasting fatness compared with the cut surface of the same rib area and one of the machines (the Scanogram) being used to measure fat thickness and area on a bull. Carcass cut rib surface and images from fat animal (a, c, e) and from lean animal (b, d, f). a, b, Carcass cut rib surfaces. c, d, Images from Scanogram machine. e, f, Images from Danscanner machine. Images taken from the Commission of the European Communities Report 'Ultrasonic techniques for describing carcase characteristics in live cattle' (Bech Andersen *et al.*, 1981).

gave the best predictions of lean content in the carcass. In the work reported by Wright and Russell (1984b), the Scanogram machine was used to measure fat depth at three points over the eye muscle. From this work, in relation to the prediction of body composition, it is doubtful if the ultrasonic measurement of

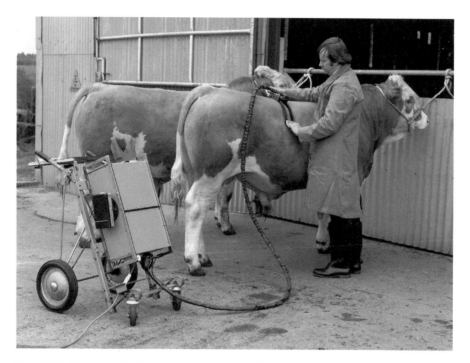

Fig. 10.6. Photograph of Scanogram in use on live bull (by courtesy of the Meat and Livestock Commission).

Table 10.12. Residual standard deviations for the prediction of carcass lean percentage of live pigs from measurements with different ultrasonic machines* (Kempster *et al.*, 1982).

Trial	1a[†] SC	1a[†] SN	1b[†] SC	1b[†] IL	2[†] SC	2[†] SN	2[†] DN
Residual sᴅ of lean percentage	3.94	3.94	4.12	4.12	1.63	1.63	1.63
Residual sᴅ for live weight (W)	3.73	3.73	3.67	3.67	1.62	1.62	1.62
W + best fat thickness	2.56	2.72	2.26	2.61	1.35	1.29	1.33
W + best fat area[‡]	2.78		2.61	2.72	1.44		1.45
W + m. longissimus area	3.37		3.38	3.40	1.62		1.62
W + best combination of measurements examined	2.29	2.69	2.25	2.36	1.35	1.29	1.33

*SC = Scanogram; SN = Sonatest; IL = Ilis observer; DN = Danscanner.
[†]1a = Kempster *et al.* (1979): 143 pigs from differnt companies in the Meat and Livestock Commission's commercial pig evaluation scheme. Data were pooled within sex and company. 1b = As above but 38 pigs only. 2 = Alliston *et al.* (1982).
[‡]Fat area over the m. longissimus at last rib.

Table 10.13. Correlations of carcass lean content with individual live animal measurements (Simm *et al.*, 1983).

Trait	Correlation coefficient*	Residual standard deviation (g kg^{-1})
Live weight	−0.10, −0.34, −0.22, −0.43	28.54, 28.76, 31.63, 13.59
Live weight from heart girth	−0.33	27.37
Skin fold thickness		
13th rib	−0.18	28.45
Anal	−0.04	28.93
Scanogram fat over m. longissimi thoracis and lumborum		
Danscanner machine	−0.60, −0.34	25.62, 13.21
Scanogram machine	−0.58, −0.66, −0.62, −0.66, −0.22	23.56, 26.46, 25.25, 26.45, 13.87
Age	−0.14	28.66

*With the exception of live weight, all data are adjusted to a constant live weight. Correlations and residual sd for ultrasonic data are means of values at two to four anatomical locations (10th rib, 13th rib, 1st and 3rd lumbar vertebrae). The range of correlation coefficients with their residual SD relate to results obtained from different groups of bulls, within the overall 97 examined, using in some cases different operators.

subcutaneous fat depth alone is of much value. Combined with measures of live weight it may be more useful, but not necessarily more so than predictions of body composition based on the other techniques used. For example, in predicting body fat content, live weight, followed closely by condition score, were better predictors than was the ultrasonic measure of fat depth at the 12th to 13th rib site, and adding ultrasonic readings into prediction equations with other variables added measurably, but not greatly, to the precision of prediction of both fat content and the content of other chemical components of the body.

In contrast to the above results obtained with mature beef cows, there is evidence that the amount of fat in the empty body of the sow at 6–8 weeks after weaning can be accurately predicted from the live weight of the animal and the depth of backfat measured ultrasonically at a distance of 45 mm from the mid-line at the level of the last rib (King *et al.*, 1986). The simple correlation coefficients of backfat depth determined *in vivo* by ultrasound with the proportions of water and fat in the empty body were −0.92 and +0.93 respectively and correlations of this magnitude indicate the usefulness of the technique in the context of not only experimental work but as a management tool in practice as well, although in the latter case it is doubtful if it has any advantages at all over condition scoring (see section 10.2.3).

10.3. Measurements on the Carcass

10.3.1. General

In practice many, but not all, pigs, sheep and cattle are sold on a deadweight basis. In such cases the relationship of the weight of the saleable carcass to the live weight of the animal is very important. Of equal or greater importance is the quality of the carcass. In this case both subjective and objective techniques are used in attempts to predict the proportions of the major tissues, bone, muscle and fat, and to relate these proportions to payments which reflect differences in quality according to the anticipated proportions of muscle and fatty tissue present. The recent major preoccupation of humans with the subject of diet and disease has lead inevitably to the concept of quality being synonymous with minimal levels of fatty tissue. Because of this, in the more recent past most techniques have been aimed at describing quality in the context of the measurable levels of certain fatty tissues on the basis that defined minimal levels are desirable in their own right and that they may be important predictors also of muscle mass. In the UK, grading schemes for pigs are based on carcass backfat measurements and the producer is heavily penalized if his/her pigs have backfat thicknesses which are outside the maxima for the top grades. Many other schemes for pigs use a similar approach but in the case of cattle and sheep, although some schemes use backfat thickness as one component of a grading index, many do not rely solely on this parameter as a basis for quality payment but rely on a combination of other variables assessed by both subjective and objective techniques.

The weight of the carcass and its composition is of no less importance to the research worker than it is to the practical farmer. In breeding programmes, and in nutritional and other experiments, the live-weight growth responses in most cases give only partial answers to the questions that have been posed. For there to be both practical reality to the research work and a scientific explanation of the live growth responses obtained, the rate of tissue deposition and the ultimate quantities deposited in the carcass must be determined. However, even this does not take matters far enough in some cases, where the final interpretation of experimental findings may depend on chemical analyses of the deposited tissues. To accommodate such a wide range of needs it is hardly surprising that the techniques used in research work are extremely varied, on the whole relatively sophisticated and that, in the end, they have given to the practical field the techniques that are now used commonly to assess carcass quality. Compared with the subjective techniques of visual appraisal and the objective techniques of measurement by rule and by probe of varying type, the techniques of complete and partial dissection, of chemical analysis and of density or specific gravity are those which are often used to supplement simple measurements taken on the whole or split carcass. Some are partially or wholly destructive in the sense that the carcass is spoiled and has little or no residual value. When this occurs the cost is very high and many of the techniques used have been developed in an attempt

Table 10.14. Techniques used in carcass evaluation studies.

1		Carcass weights; killing-out (or dressing-out) proportion
2		Visual appraisal (scoring) of the carcass
3		Density or specific gravity of the carcass
4		Measurements on the intact, centre split or quarter split carcass
	i	Linear
	ii	Fat thickness by probe, caliper or rule
	iii	Muscle thickness by probe, caliper or rule
	iv	Cross-sectional areas of muscle by caliper, rule or planimeter
	v	Ultrasonic measurements to give information on measurements ii and iii above
	vi	Weights of perinephric and retroperitoneal fat (KKCF)
5		Cutting of carcass into anatomical or commercial joints
6		Dissection of commercial or anatomical joints
7		Dissection of complete (half) carcass
8		Chemical analysis of carcass

to predict the results that would have been obtained had such costly techniques been used.

Those techniques which are used most commonly are summarized in Table 10.14. The choice of any one technique depends on many factors and in most cases attempts at estimating tissue proportions in carcasses are in some way a compromise between the ideal and the possible that has been dictated by economic and other factors. Kempster *et al.* (1982) suggest that the ideal evaluation would be applied to all animals if cost were no obstacle, and any departure from this must be that in which techniques are used which get as near as possible to the results that would have been obtained had the ideal been achieved. To realize the ideal and to give a maximum amount of information, they suggest that the carcass should be divided into standardized commercial joints and that tissue separation within joints should then be carried out. This provides information on joint proportions and on the distribution of tissues in the joints. Additionally, it provides information which can be readily correlated with chemical analyses and if information on the anatomical grouping of muscles is required, this can be realized from this approach by summating the weight of muscle portions which occur in the different commercial joints. The following sections consider the techniques that can give ideal evaluations, together with those which attempt to predict the results that the ideal should reveal in the context of how tissues have been deposited in the animal during its growing phases as reflected in the weight, shape and composition of the carcass which remains after slaughter.

10.3.2. Carcass weight and killing-out (or dressing-out) proportion

Carcass weight is one of the most important variables, if not the most important variable, to be included in any equation which predicts carcass composition. As with live weight, the relative ease with which it may be recorded hides the difficulties that are inherent in the actual act of recording it if interpretable results are to be obtained. The main problem in recording carcass weight is that of standardizing the time after slaughter at which the weight is recorded. Immediately after slaughter the carcass is hot and the subsequent cooling and shrinking which takes place for about 24h in a chiller room, amounting to variable losses in weight but up to 20g kg^{-1} in many cases, must be taken into account if baselines are to be valid for the expression of carcass composition and killing-out proportions. Furthermore, the carcass weight will depend in the first instance on the dressing procedure adopted, on exactly what is or is not regarded as saleable carcass, particularly in respect of the degree of fat trimming practised. If an appropriate period of time for the carcass to cool and to reach a constant 'cold' weight cannot be allowed, then it is possible to make appropriate deduction from a conversion scale to calculate the cold weight. Cold weights, that is weights realized after 24h in a chiller room, are usually used in all practical and research situations and so the numerator of the killing-out or dressing-out proportion is given a standardized base. The denominator, that is live weight, must also be recorded under standardized conditions if valid comparisons are to be made. The many factors which may influence the validity of recorded live weight have been discussed in section 10.2.1, but in the particular context of the use of live weight as the denominator in this ratio, the length of time between the recording of the final live weight of the animal, in particular in relation to the method of feeding in the ruminant animal (pasture or indoor feeding) and its time of slaughter, is clearly of considerable importance.

In other chapters the effects of breed within species, of sex within breed and of nutrition in relation to speed of growth and adipose tissue deposition, on killing-out proportion, were discussed and the positive correlation of the latter with adipose tissue proportion was noted. In ruminant animals in particular there may be a further dimension to this relationship. Fast growth rates usually give not only higher proportions of adipose tissue in the carcass but also higher proportions of internal adipose depots as well. These could restrict digestive capacity and therefore the higher killing-out proportion could mirror a compound effect with, in turn, the point made above concerning method of feeding being tied in with this effect. For example, compared with an animal reared inside to grow very quickly on a high proportion of concentrates, an animal of similar live weight at pasture may have a higher proportion of its live weight as gut fill and may lose a higher proportion of this fill in any given period between recording live weight prior to slaughter and the time of slaughter itself. Its killing-out proportion may therefore be lower, not only because of this but also because its smaller deposits of internal adipose tissue will not have restricted digestive capacity to the same extent. Admittedly this is a generalization and there will be cases where exceptionally fast growth rates at pasture will have given large deposits of internal adipose tissue. Overall, however, it is a generalization that holds good in most circumstances.

10.3.3. Specific gravity or density

This is a technique which can be used with the carcasses of any species but which in reality has been used almost exclusively with pig carcasses. The carcass is weighed in both air and in water and the volume of the carcass is obtained from its displacement of water on the basis that the difference in weight in air and in water is equivalent to the volume of water displaced. The density or specific gravity of the carcass is then calculated as the ratio of the weight in air to the volume.

The basic assumption of this approach is that different tissues have different densities and that, because of this, the overall density of the carcass will indicate the relative proportions of the major tissues, muscle and adipose, present. In this particular context, the density of muscle is usually taken as about 1.10 and that of adipose tissue as about 0.90. It follows that a change in carcass adipose tissue content of about $10\,g\,kg^{-1}$ carcass will be reflected in a change in carcass density of about 0.002.

Although specific gravity is relatively easy to determine and there are no deleterious effects on the carcass, the precision is low unless great care is taken to standardize the procedures involved. If this is done it can be as good a predictor of body composition as any of the other measurements discussed below, for it is actually measuring the total adipose tissue in the carcass.

10.3.4. Measurements taken by ruler and by probe

Various measurements which may be taken on the carcass by ruler and by probe have been detailed in Table 10.14. Some of these are better than others as predictors of carcass composition. Many are used as bases in carcass classification schemes to define quality and to determine monetary value. This section attempts to describe what are probably the more important of these measurements and to show how the results that are obtained from them correlate with carcass composition obtained from complete dissection techniques. Linear measurements, such as carcass length, width and depth, are alone of little or no value as predictors of carcass composition. Carcass length has been used extensively in some countries in the past to select pig carcasses suitable for curing for bacon and ham. At best, however, it may be moderately useful in differentiating between broad types of carcass which are either short and blocky or long and thin. In these cases the differences in meat (muscle + adipose tissue) to bone ratios could be identified in the broadest possible terms.

Measurements of subcutaneous fat thickness over muscle, particularly where the m.longissimus provides a suitable base in the rib and in the loin areas, have proved to be valuable techniques for predicting carcass composition and for giving standards on which quality payments are based. In experimental work the sectioning of the carcass provides a cut face on which measurements may be taken directly by rule and by caliper. Under practical conditions the depth of subcutaneous fat can be measured by various instruments, ranging in complexity from the simple sharpened calibrated probe which is forced into the fatty tissue until it reaches the muscle tissue underneath, through various types of relatively simple optical probes such as the intrascope, to more complex probes which either rely on conductivity of light reflectance or in electrical conductivity between fatty tissue and muscle tissue. The cost of the latter types of probe is very much higher than is the cost of the simple optical probe but they offer the possibility of improving data handling and saving labour costs when used commercially in grading schemes.

Probes have been used most successfully on pig carcasses where, because of a more homogeneous distribution of backfat and a firm consistency, the prediction of lean in the carcass has been better than in sheep and in cattle carcasses. The precision of different probes in measuring both fat and muscle thicknesses will be reflected in the relationship between repeated measurements and may depend on the point(s) of the carcass which is (are) actually probed. For example, Fortin *et al.* (1984) compared two instruments which both relied on the differences in reflectance between muscle and fat to record automatically fat and muscle thickness in pig carcasses, in relation to measurements recorded with a simple ruler. Fat and muscle thicknesses were recorded at the last rib, measured caudally, and between the 3rd and 4th ribs, measured from the last rib. The fat and muscle thicknesses recorded, compared with those recorded using a simple ruler, are given in Table 10.15. The results show that one instrument gave higher readings than the other for fat thickness at both locations but that for muscle thickness one gave the highest reading at one location and the other at the second of the two locations. Compared with the measurements recorded by ruler, one instrument gave a negative mean bias (underestimation for fat thickness), the other did not. Error variances of repeated measurements show that the precision of fat measurements was significant for one instrument, but not for the other, while for muscle thickness, the larger error variance indicates a lower precision at the last rib. However, it will be observed that all the differences, although statistically significant, were in fact very small.

Other work (Kempster *et al.*, 1985) compared the same two probes with an optical probe and highlighted differences between them according to a number of factors, including location of probe. In this work pigs were deliberately selected to be of widely different live weights to give big differences in carcass muscle (lean) content (449–532g kg^{-1}) and the carcasses were subsequently dissected and the three probes compared as predictors of carcass lean content. The residual standard deviations for the predicted lean content of the carcasses (Table 10.16), when variation in carcass weight was held constant, show that the results for the two automatic probes were better at the 3rd/4th last rib position than at the last rib position. Comparing these two probes at the latter position, one instrument gave precision which was similar to that achieved with the optical probe but the other was significantly inferior. Although, again, the differences were small there was clearly evidence at this time that some caution needed to be exercised in using probes of different types to predict the lean content of carcasses under a wide range of conditions.

Notwithstanding these earlier findings, which indicated the limitations of

Table 10.15. Comparison of two probe instruments with a simple ruler in measuring fat and muscle thickness (m. longissimus) (mm) in pig carcasses (Fortin *et al.,* 1984).

Instrument	HG[+] Mean	HG[+] SD	FOM[+] Mean	FOM[+] SD	Ruler Mean	Ruler SD	HG	FOM	Effect of probe
	Means and standard deviations						Error variance (mm^2) of repeated measurements		
Fat thickness									
last rib	20.4	5.4[‡]	21.4	4.6[‡]	21.9	5.5[‡]	1.75	1.19	*
3rd to 4th last rib	21.8	5.5	23.2	4.9	23.1	6.2	3.54	0.93	**
Effect of location							**	NS	
Muscle thickness									
last rib	53.5	10.9[‡]	51.1	8.0[‡]	–	–	49.43	57.78	NS
3rd to 4th last rib	45.8	7.3	47.3	6.6	–	–	24.06	13.86	**
Effect of location							**	**	

[+]HG = Hennessy Grading Probe; FOM = Fat-O-Meater.
NS = $P > 0.05$; * = $P < 0.05$; ** = $P < 0.01$.
[‡]Measurements significantly different for the two locations ($P < 0.01$).

certain techniques in measuring backfat thickness in pigs and in predicting lean content of the carcass from these measurements, there is now much evidence that probe measurements of backfat thickness are better than any other measurements for this purpose. Indeed in some of the earlier work (Kempster and Evans, 1979) it was found, in carcasses of widely different weights, that although visual conformation score, carcass length and m.longissimus depth were relatively valueless as predictors of carcass lean content (residual standard deviations of 3.84, 3.77 and 3.80 respectively), nevertheless fat thickness measurements taken on the exposed surface of the split carcass provided a less precise prediction (residual SD, 2.89) than did optical probe measurements (residual SD, 2.20) taken at a distance of 65 mm from the dorsal mid-line at various positions along the carcass. The highest precision was achieved when the probe was used at the last rib (now commonly referred to as the P2 measurement and taken at a distance of 60 mm from the dorsal mid-line) and at the 13th rib. Overall these two measurements were more precise than any other pair of measurements in predicting the total lean in the carcass and the lean content of various joints. This was borne out in later studies (Diestre and Kempster, 1985) and regression equations of carcass lean content on carcass weight and P_2 fat thickness, and values estimated from the equations, are given in Tables 10.17–10.19.

At the beginning of 1989 a pig carcass grading scheme was introduced

Table 10.16. Residual standard deviations (SD) for the prediction of carcass lean content (g kg^{-1}) from an optical probe compared with two probes relying on light reflections between muscle and fatty tissue, after removal of variation in carcass weight (Kempster *et al.* 1985).

	SD of carcass lean content	Optical probe	Prediction from			
			Fat-O-Meater Probe		Hennessy Grading Probe	
			Fat thickness	Fat and muscle thickness	Fat thickness	Fat and muscle thickness
Last rib[+]	61.9	30.8	37.1	36.8	40.6	39.9*
3rd/4th last rib[+]	61.9	30.8	31.4	29.1***	36.6	35.9*

* = $P < 0.05$; *** = $P < 0.001$.
[+]Overall mean from six different abattoirs.

Table 10.17. Regression equations of carcass lean content (Y, g kg^{-1}) on carcass weight (W, kg) and fat thickness (P$_2$, mm); values estimated from the equations (Diestre and Kempster, 1985).

Sample	Carcass weight (kg)	Prediction equations	Residual SD
1	66	$Y = 591.7 + 0.384\,W - 7.44\,P_2$	25.2
2	47*	$Y = 598.8 - 7.33\,P_2$	25.4
	72†	$Y = 595.2 - 6.33\,P_2$	26.4
	93‡	$Y = 574.0 - 5.43\,P_2$	27.8
3	61	$Y = 557.4 + 0.840\,W - 7.78\,P_2$	29.4
4	78	$Y = 636.9 - 0.330\,W - 6.47\,P_2$	27.3
5	70	$Y = 610.6 + 0.510\,W - 8.32\,P_2$	30.3

*Pork-weight pigs, 61 kg live weight.
†Bacon-weight pigs, 91 kg live weight.
‡Heavy-weight pigs, 118 kg live weight.

Table 10.18. Carcass lean contents estimated from equations* in Table 10.17 (Diestre and Kempster, 1985).

	Carcass weight (kg)						
	50		70			90	
P$_2$ fat thickness sample	10	15	10	15	20	15	20
1	537	499	544	507	470	515	478
2	548	509	556	514	471	537	491
3	552	515	550	515	479		
4			539	497	456	510	478
5	521	486	525	492	460	498	466

*Values for carcass weight and P$_2$ were applied to the equation.

within the European Community. This covers all abattoirs slaughtering more than 200 pigs weekly on a yearly average basis and dictates that lean proportions have to be declared on carcasses by predictive methods based on objective measurements. Any predictive method has to have a coefficient of determination (R^2) greater than 0.64 and a residual SD of less than 25 g kg^{-1} lean carcass weight. In anticipation of this legislation and the need to have accepted by the European Community predictive methods based on sound objective measure-

Table 10.19. Means and standard deviation values for carcass composition determined by dissection (Diestre and Kempster, 1985).

Sample	Lean in carcass (g kg^{-1})		Subcutaneous fat (g kg^{-1})		Separable fat* (g kg^{-1})	
	Mean	SD	Mean	SD	Mean	SD
1	499	44.2	184	44.2	258	58.8
3	475	53.3	194	54.3	272	71.5
4	469	65.9	221	59.9	306	78.2
5	502	68.5	182	65.2	256	84.4

*Subcutaneous + intermuscular + flare fat computed for each set of data.

ments for use in Great Britain, Cook *et al.* (1989) furthered the studies referred to above by examining the worth of m. longissimus (rib muscle) depth and the depths of subcutaneous fat at the last rib (P_2) and at the 3rd/4th from last rib as predictors of carcass lean content using four different machines: the optical probe and three automatic probes, the Fat-O-Meater (from Denmark), the Hennessey Grading Probe II (from New Zealand) and the Destron PG-100 probe (from Canada). The four instruments gave similar levels of accuracy of predicting lean content in the carcass but the Canadian machine just failed to reach the required statistical criteria for approval for the scheme. Approval was sought, and given, for adoption for Great Britain of some of the prediction equations obtained from the three other machines (Table 10.20) and the residual SD values (g kg^{-1}) for the best combination of measurements for predicting lean were 23.1, 21.8, 23.7 and 25.5 for the optical probe, the Fat-O-Meater, the Hennessey Grading Probe II and the Destron PG-100 probe, respectively,

In the case of beef and sheep carcasses, fat thickness measurements, as either bases for grading schemes or as predictors of carcass composition, have not been used widely in the past but increasingly these and/or other predictors are needed to satisfy markets where concepts of quality have become all important. As with pig carcasses, it would appear likely from available evidence that in beef carcasses the depth of subcutaneous fat over the m.longissimus at certain points is likely to be of principal importance in any prediction equation, although it is unlikely that its importance will be as great as in pig carcasses. Kempster *et al.* (1982) suggest that the best fat thickness measurements taken by probe can provide as precise a prediction of carcass lean proportion as visual scores given by experienced operators.

Table 10.20. Regression equations for predicting lean proportions (g kg^{-1}) from carcass and probe measurements accepted by the Pigmeat Management Committee of the European Community for use in Great Britain* in the Community Grading Scheme (Cook *et al.*, 1989)[†].

Machine	Constant	Carcass weight (kg)	Fat at P$_2$ (mm)	Rib fat (mm)	Rib muscle (mm)
Optical Probe	652	0.76	-11.5	—	—
	645	0.95	-6.9	-4.2	—
Fat-O-Meater	587	0.78	-5.8	-3.2	1.8
Hennessey Grading Probe II	622	0.55	-6.2	-4.6	1.6

*To obtain European Community reference lean proportion add 3 to constant in each equation.
[†]Prediction equations based on following: 162 carcasses (about 20 from each of eight abattoirs). Mean carcass weights (kg ± SD) (and lean in carcass (g kg^{-1} ± SD)) by dissection were 51.1 ± 4.40 (558 ± 44.0), 67.7 ± 4.14 (559 ± 39.2) and 93.8 ± 3.41 (472 ± 37.2) for 63 (pork), 80 (bacon) and 19 (heavy hog) carcasses in each group respectively. Within each abattoir carcasses were sampled on a stratified basis which, weight apart, included fat depth and sex. About twice as many lean and fat carcasses as average carcasses were selected to provide more stable planes for the regressions, and to be representative of commercial slaughter practice a sex distribution of about one castrated male to two gilts to one entire male was chosen for the bacon weight range, with more entire males than castrated males in the pork range and no entire males in the heavy hog range.

10.3.5. Visual appraisal (scoring)

Subjective appraisal of subcutaneous fat cover on carcasses is used mostly in predicting the fat content of beef and sheep carcasses. Scores based on such appraisals are used in some countries as the bases of grading schemes on which quality payment rests. The precision depends heavily on the experience of the assessor and a scale varying from 1 to 10 is usually considered to be ideal.

In beef carcasses it would appear that visual judgements of skilled assessors in predicting fat content is approximately as precise as taking simple measurements on the cut surface of the carcass (Harries *et al.*, 1974). The correlation coefficients which these workers found between scores for fat thickness and other subjective criteria, using five expert judges, and objectively derived data from dissection procedures, are given in Table 10.21. The highest correlations were clearly for subcutaneous and kidney knob and channel fat (KKCF) and the data indicate further that although there were small differences between judges, the most striking feature was the consistency in the relationship of their scores for the proportion of subcutaneous fat to the actual proportions derived from dissection procedures. In Table 10.22 regression equations are given using the average visual scores of the five judges as predictors of adipose tissue in the carcass side. These equations show that when side weight is included as a further

Table 10.21. Correlation coefficients between judges' scores and objective data (Harries *et al.*, 1974).

Score	Objective variable	Judge's score				
		1	2	3	4	5
Muscle to bone ratio	Muscle to bone ratio	0.47	0.45	0.41	0.33	0.37
Proportion of lean	Percentage of lean	-0.22	-0.09	-0.14	0.60	-0.16
Proportion of subcutaneous fat*	Total subcutaneous fat	0.89	0.86	0.84	0.83	0.81
Proportion of subcutaneous fat*	Percentage of subcutaneous fat	0.91	0.94	0.89	0.91	0.92
KKCF[+]	Percentage of KKCF[+]	0.87	0.88	0.88	0.86	0.92
Overall conformation	Total muscle	0.25	0.23	0.22	0.07	0.02
Overall conformation	Percentage of lean	-0.35	-0.38	-0.33	-0.40	-0.51
Overall conformation	Muscle to bone ratio	0.52	0.46	0.36	0.35	0.36
Overall conformation	Percentage of subcutaneous fat	0.52	0.54	0.46	0.49	0.60
Overall conformation	Proportion high- to low-priced parts	0.17	0.18	0.17	0.17	0.16

*Using photographic standards.
[+]KKCF = kidney knob and channel fat.

variable to a visual score of subcutaneous fat, the prediction of the total quantity of subcutaneous fat is improved (equations 1 and 2). The results from equation 5 indicate that the judges succeeded to a very large extent in balancing their assessments for differences in total side weight. However, the use of scores as predictors of total fat is obviously not as good as in the prediction of subcutaneous fat (equation 6), although once again carcass side weight can clearly improve the accuracy of prediction (equation 7).

10.3.6. Jointing and dissection techniques

As mentioned earlier this must be regarded as the ideal were it not for the time and cost involved. The techniques described above have all attempted to predict the results achievable by jointing and/or by dissection procedures and the limits to the levels of precision likely to result have been exposed. Joints can be derived by using various anatomical points for reference or by using commercial cutting techniques, and the compositions in different joints give different precisions in predicting carcass compositions in different species. But first, as an example, some detail of how measurements of composition may be obtained from an anatomical dissection of beef carcasses is given.

Brown and Williams (1981) described a technique for measuring the

Table 10.22. Details of multiple regressions to predict fatty tissues from the average expert visual assessments, alone and in combination with other measurements (Harries *et al*, 1974).

	Dependent variables	Independent variables	Regression coefficients	Standard errors	Proportion of variance	Residual standard deviation	Intercept
1	Subcutaneous fat	Average visual score (subcutaneous fat)	3.365	0.207	79.0	1.923	+1.601
2	Subcutaneous fat	Side weight	0.089	0.006	94.3	0.997	-8.292
		Average visual score (subcutaneous fat)	3.055	0.110	–	–	–
3	Subcutaneous fat	Side weight	0.085	0.007	94.4	0.983	-7.840
		Visual score (subcutaneous fat)	2.872	0.152	–	–	–
4	Subcutaneous fat	Side weight	0.088	0.007	94.5	0.980	-8.124
		Visual score (subcutaneous fat)	2.737	0.188	–	–	–
5	Percentage of subcutaneous fat	Average visual score (subcutaneous fat)	2.585	0.087	92.5	0.811	+1.800
6	Total fat	Average visual score (subcutaneous fat)	5.098	0.550	54.5	5.107	+16.284
7	Total fat	Side weight	0.250	0.014	91.9	2.146	-11.562
		Average visual score (subcutaneous fat)	4.224	0.236	–	–	–
8	Total fat	Side weight	0.246	0.015	91.9	2.151	-11.095
		Average visual score (subcutaneous fat)	4.035	0.332	–	–	–
9	Total fat	Side weight	0.221	0.016	92.9	2.05	-10.174
		Average visual score (subcutaneous fat)	3.789	0.259	–	–	–
		Kidney knob, channel fat	0.662	0.205	–	–	–
10	Percentages of total fat	Average visual score (subcutaneous fat)	3.594	0.187	83.8	1.739	+15.094
11	Percentages of total fat	Average visual score (subcutaneous fat)	3.056	0.193	88.0	1.497	+9.349

composition of beef carcasses using an anatomical procedure, and whilst it is not intended to convey the impression that this is the only procedure available, it is one that probably gives the best of all worlds. In this procedure, prior to jointing, the kidney knob and channel adipose tissue (KKCF or perirenal fat), the adipose tissue lying in the channel between the symphysis pubis and the sacrum (the retroperitoneal or pelvic fat) and the adipose tissue (if present) which loosely adheres to the dorsal surface of the sternum, are all removed. The carcass is first cut into forequarters and hindquarters by using a saw to cut through the vertebral column at the middle of the 13th thoracic vertebra immediately behind the last rib. A knife is then used to cut through the m.longissimus in the same plane and the cut is continued along the caudal edge of the last rib and through the flank. Subcutaneous fat is then removed and apportioned to defined areas to give an indication of distribution and the forequarter and hindquarter are then divided as shown in Fig. 10.7 to give two joints in the former (thoracic limb and neck plus thorax) and two in the latter (lumber and abdominal, and pelvic). It is interesting to note at this juncture that this is a small number of joints compared with the number which would result from a commercial cutting exercise where the hindquarter might have five joints (loin, rump, round, leg, thin and thick flanks) and the forequarter might have as many as seven joints (wing ribs, foreribs, chuck, clod and sticking, brisket, plate and shin). In the jointing exercise to give two joints in each quarter, careful attention is paid to muscles and tendons which have insertions in one joint but their bulk in the other. For example, the thoracic limb includes the scapula, the humerus, the radius/ulna and carpals all with their associated muscles, and its removal needs careful attention to detail if the placement of specific muscles, such as the m. latissimus dorsi which is included in the neck and in the thorax joint from the point of view of dissection, but whose tendon is inserted into the humerus, is to be correct. Subsequent to separation into the four anatomical joints each is dissected into its components. For the determination of total muscle weight, all adherent intermuscular fat is removed, the tendons are removed at right angles to the limit of the red muscle and the tendinous tissue sheaths are removed from the abdominal muscles. The dorsal edge of the scapula and the cartilages of the costal bones are cleared of all traces of muscle, tendon and fat but are not scraped. The periosteum therefore remains in position and as a consequence the total bone includes all cartilage. The adipose tissue is in fact intermuscular fat which has been dissected from the bones and muscles after other tissues such as tendons and the major ligaments, blood vessels and lymphatic nodes, have been removed.

Using this, or other techniques similar to it, enables both a composite and a sectional picture to be built of the proportions of the major tissues in the carcass. It does not give any information on, for example, either the protein and water contents of the muscle or the lipid content of the adipose tissue and its fatty acid make-up. Chemical analyses will be needed to give information in these areas.

The precision with which joint composition will predict total carcass composition depends in the first place on the precision of the process of halving,

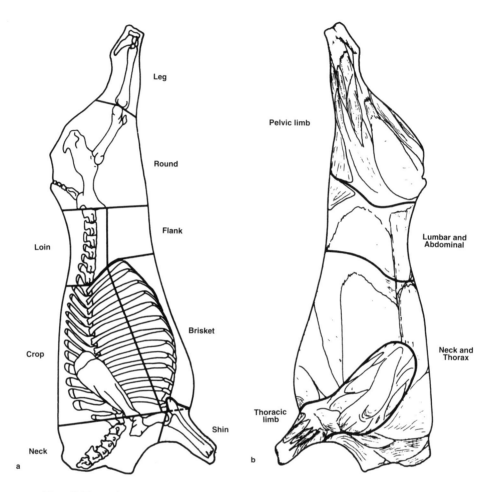

Fig. 10.7. Beef carcass jointing and dissection. a, Subdivision of subcutaneous fat depot. b, Demarcation of four anatomical joints (reproduced from Brown and Williams (1981) by kind permission of the copyright holder, School of Veterinary Science, University of Bristol).

quartering and jointing. High repeatability is essential if precision is in turn to be high. In beef carcasses the weight of the actual joint, as well as the weight of the carcass, has been shown to be an important variable to include in any prediction equation for some joints, but not for others (Kempster and Jones, 1977; Table 10.23). For example, reference to Table 10.23 shows a big advantage for the wing rib and sirloin joints, little advantage for the leg joint and advantage for the shin joint. Kempster *et al.* (1982) in reviewing this area concluded that when all factors were considered, including cost, the fore and wing ribs offered the best compromise between cost and precision in predicting total lean or muscle content of the carcass. In pig carcasses, they concluded that the ham

Table 10.23. Residual SD of carcass lean weight or percentage from different sample joints with the prediction equation constructed in different ways (Kempster and Jones, 1977)*.

	Type of equation		
	A1 (kg)	A2 (kg)	B (%)
SD	7.81	7.81	2.79
Residual SD			
Shin	2.66	2.66	2.56
Coast	2.57	1.34	1.13
Clod + sticking	2.76	1.82	1.61
Forerib	2.86	1.58	1.34
Pony	2.71	1.39	1.17
Leg	2.85	2.74	2.42
Thin flank	3.19	1.84	1.53
Rump	2.83	1.82	1.50
Sirloin	2.79	1.75	1.46
Wing rib	2.94	1.81	1.50
Top piece	2.07	1.37	1.21

*Results are pooled within 17 breed type × feeding system groups.
A1: $Y_W = a + b_1 x_1 + b_2 x_2$
A2: $Y_W = a + b_1 x_1 + b_2 x_2 + b_3 x_3$
B: $Y_P = a + b(100 x_2 / x_3)$
where Y_W = weight of lean in side; Y_P = weight of lean in side as a percentage of side weight; x_1 = side weight; x_2 = weight of lean in joint; x_3 = weight of joint.

joint, followed next by the hand joint, offered the best compromise between precision and money with, in the case of the latter joint, a fat thickness measurement taken over the m.longissimus at the last rib (P_2) giving a noticeable improvement in precision (residual SD reduced from 1.97 to 1.53). In sheep carcasses the picture is less clear but the conclusion reached was that the breast joint offered relatively high precision in relation to cost. The best predictor of lean content was the leg joint, but it was concluded that this did not offer the best value for money.

References

Allen, P. and Vangen, O. (1984) X-ray tomography of pigs, some preliminary results. In: Lister, D. (ed.) *In Vivo Measurement of Body Composition in Meat Animals* Elsevier, London, pp. 52–66.

Alliston, J.C., Kempster, A.J., Owen, M.G. and Ellis, M. (1982) *Animal Production* 35, 165–170.

Andrew, E.R. (1980) *Philosophical Transactions of The Royal Society, London. Series B* 289, 471–481.

Bech Andersen, B., Busk, H., Chadwick, J.P., Cuthbertson, A., Fursey, G.A.J., Jones, D.W., Lewin, P., Miles, C.A. and Owen, M.G. (1981) *Ultra-sonic Techniques for Describing Carcase Characteristics in Live Cattle* CEC, Luxemburg (EUR 7640).

Brown, A.J. and Williams, D.R. (1981) *Beef Carcass Evaluation – measurement of composition using anatomical dissection. Meat Research Institute Memorandum No. 47.* Meat Research Institute, Langford, Bristol.

Carr, T.R., Walters, L.E. and Whiteman, J.V. (1978) *Journal of Animal Science* 46, 651–657.

Charles, D.D. (1974) *Research in Veterinary Science* 16, 89–94.

Cook, G.L., Chadwick, J.P. and Kempster, A.J. (1989) *Animal Production* 48, 427–434.

Diestre, A. and Kempster, A.J. (1985) *Animal Production* 41, 383–391.

East, B.W., Preston, T. and Robertson, I. (1984) The potential of *in vivo* neutron activation analysis for body composition measurements in the agricultural sciences. In: Lister, D. (ed.) *In Vivo Measurement of Body Composition in Meat Animals.* Elsevier, London, pp. 134–143.

Ettinger, K.V., Foster, M.A. and Miola, V.J. (1984) Future developments in the *in vivo* measurements of body composition in pigs. In: Lister, D. (ed.) *In Vivo Measurement of Body Composition in Meat Animals.* Elsevier, London, pp. 207–233.

Fisher, A.V. (1975a) *Livestock Production Science* 2, 357–366.

Fisher, A.V. (1975b) *EEC Seminar: Criteria and Methods for Assessment of Carcass and Meat Characteristics in Beef Production Experiments, Zeist*, pp. 43–55.

Fortin, A., Jones, S.D.M. and Haworth, C.R. (1984) *Animal Production* 38, 507–510.

Fuller, M.F., Foster, M.A. and Hutchison, J.M.S. (1984) Nuclear magnetic resonance imaging of pigs. In: Lister, D. (ed.) *In Vivo Measurements of Body Composition in Meat Animals.* Elsevier, London, pp. 123–133.

Goddard, P.J. (ed) (1995) *Veterinary Ultrasonography.* CAB International, Wallingford.

Goodchild, A.V. (1985) *Animal Production* 40, 455–463.

Harries, J.M., Pomeroy, R.W. and Williams, D.R. (1974) *Journal of Agricultural Science* 83, 203–211.

Hughes, J.G. (1976) *Animal Breeding Abstracts* 44, 111–118.

Ivings, W.E., Gibb, M.J., Dhanoa, M.S. Fisher, A.V. (1993) *Animal Production* 56, 9–16.

Johansson, I. and Hildeman, S.E. (1954) *Animal Breeding Abstracts* 22, 1–17.

Jones, R.S., Lawrence, T.L.J., Veevers, A., Cleave, N. and Hall, J. (1989) *Veterinary Record* 125, 549–553.

Kempster, A.J. and Evans, D.G. (1979) *Animal Production* 28, 87–96.

Kempster, A.J. and Jones, D.W. (1977) *Journal of Agricultural Science* 88, 193–201.

Kempster, A.J., Cuthbertson, A., Owen, M.G. and Allison, J.C. (1979) *Animal Production* 29, 485–491.

Kempster, A.J., Cuthbertson, A. and Harrington, G. (1982) *Carcase Evaluation in Livestock Breeding, Production and Marketing.* Granada, London.

Kempster, A.J. Chadwick, J.P. and Jones, D.W. (1985) *Animal Production* 40, 323–329.

King, R.H., Speirs, E. and Eckerman, P. (1986) *Animal Production* 43, 167–170.

Lowman, B.G., Scott, N. and Somerville, S. (1976) Condition scoring of cattle. Revised ed. *Bulletin of the East of Scotland College of Agriculture* No. 6.

Nyland, T.G. and Mattoon, J.S. (1995) *Veterinary Diagnostic Ultrasound.* W.B. Saunders, Philadelphia.

Petty, R.R. and Cartwright, T.C. (1966) *A Summary of Genetic and Environmental Statistics for Growth and Conformation Traits of Young Beef Cattle. Department Technical Report of the Texas Agricultural Experimental Station*, No. 5.

Robelin, J. (1984) Prediction of body composition *in vivo* by dilution technique. In:

Lister, D. (ed.) *In Vivo Measurement of Body Composition in Meat Animals.* Elsevier, London, pp. 106–112.

Robelin, J. and Geay, Y. (1978) *Annales de Zootechnie* 27, 159–167.

Robelin, J. and Theriez, M. (1981) *Reproduction Nutrition Development* 21, 335–353.

Sehested, E. (1984) Computerized tomography of sheep. In: Lister, D. (ed.) *In Vivo Measurement of Body Composition in Meat Animals.* Elsevier, London, pp. 67–74.

Skjervold, H. (1982) *CEC Workshop, Copenhagen.* 15–16 December, 1981.

Simm, G., Alliston, J.C. and Sutherland, R.A. (1983) *Animal Production* 37, 211–219.

Sørensen, M.T. (1992) *Animal Production* 54, 67–74.

Standal, N. (1984) Establishment of CT facility for farm animals. In: Lister, D. (ed.) *In Vivo Measurements of Body Composition in Meat Animals.* Elsevier, London, pp. 43–51.

Wright, I.A. and Russell, A.J.F. (1984a) *Animal Production* 38, 23–32.

Wright, I.A. and Russell, A.J.F. (1984b) *Animal Production* 38, 33–44.

11 The Future

What can we expect in the way of technological developments over the next 10 years? Will the consuming public still wish to eat meat when it is clear that many groups appear to live happily on non-meat diets? Can systems of production overcome fears about welfare? Will the back-to-nature school of thought gain sufficient political momentum to reverse many of the current scientific developments? Finally, will medical opinion give meat consumption a clean bill of health. These are questions which challenge many in animal science at the present time.

Some trends are clear already. For example, those countries which have developed a high level of technology have become models for other countries which have lower levels of technical resources. Carbon-copy production systems of developed countries have already been installed in countries where people barely have adequate housing and nutrition themselves. This represents an enormous moral dilemma because in some cases technical elaboration has become mixed up with status and political symbology.

There is a danger that this book could contribute to the view that only meat production at the very frontier of technology can be justified. The truth is that although it is very important to understand the basic principles and underlying science, the final assembly of the ideas into a functioning system depends on the specific environment, which includes such diverse features as tradition, politics, finance, technical ability and market possibilities.

11.1. Future Demand for Meat and Meat Products

11.1.1. Change in ethical views

In Europe and North America, producers of meat from pigs, poultry and cattle are in fierce competition with each other. The response of the industry has been to intensify and use every possible device to secure a share of the market by supplying the cheapest possible product at an acceptable quality. The pursuit of low cost has, however, damaged the image of intensive production in the eyes of the public.

Some systems of production are so intensive that they are perceived as unduly repressive, and beyond the bounds of reasonable human behaviour towards animals. Even within those countries where welfare of livestock is an important issue, some producers have been reluctant to move towards those systems which are widely regarded as offering improved welfare, sometimes on the grounds of cost, but sadly in a few instances because the producers are insensitive to the arguments and do not see the future consequences of their failure to react.

Politicians in different countries may eventually consider it necessary to introduce legislation which greatly limits what a farmer is allowed to do on a livestock unit. To ensure that farmers conform will then require some form of inspection and licensing. Unresponsive farmers may find themselves unable to continue in stock farming and much as they might regard such measures as undue interference in their chosen way of life, they will perhaps find very little sympathy from the general public.

Considerations such as these throw a shadow over the way in which technology may be used to alter the efficiency of growth and remove the basic tenet of a decade ago, when increased efficiency was regarded as being the only way forward.

11.1.2. Changes in the perception of meat as a healthy food

Human nutritionists have not produced consistent nutritional objectives for the consuming public. Sometimes this has worked for the benefit of meat producers, but more recently it has tended to incline away from their interests. In the UK, a number of government reports on diet and human health have been published recently. Virtually without exception, they have recommended a reduction in the intake of fat with the emphasis upon fat in animal products. The hazards which those reporting thought they saw were an association between fat intake and the incidence of a number of potentially killer diseases such as coronary heart disease, certain vascular disorders including an increased risk of impaired blood supply to the brain which could culminate in a stroke, enhanced risk of cancer of the large intestine, and a greater tendency towards obesity and its associated disorders. Although some members of the medical profession regarded the recommendations as based on debatable evidence and as not taking sufficient account of other correlated factors in a sophisticated life style, these reports and similar ones in other countries have had a profound effect on the attitude of the public to the quantity and quality of the meat which they consume.

Fat consumption

The switch from accepting that fat contributes substantially to the flavour and eating quality of meat and to the satisfaction derived from the meal as a whole, to the view that almost any visible fat is verging on the immoral, has been nothing short of a revolution. Its effects on production and processing methods have been profound, and it will continue to dominate the argument about the role of meat in human health. It is of little import whether the arguments are right or wrong or exaggerated, if the balance of demand has swung to meat products which have virtually no fat. There is no sign that this trend is going to be significantly reversed, although there are some counter arguments. For example, it has been claimed that meat from very lean pigs lacks succulence because it lacks intramuscular fat, that is fat actually within the muscle.

In the USA hamburgers and streaky bacon have had a special role in the 'great American breakfast', but such is the weight of reaction against animal fat that even this traditional market may also diminish unless the product is changed. It is clear that the demand from the consumer is for lean meat and for joints which have been very attentively trimmed of fat, so that it is either hardly visible or appears as an even, very thin, layer over the outside of the joint or piece.

Growth of vegetarianism and the consumption of non-meat diets

Whilst in many countries the populace is striving to increase its proportion of meat in the average diet, in others vegetarianism is being held up as a desirable nutritional objective. It must be conceded that with modern nutritional understanding, and the use – if necessary – of vitamin and mineral supplements, a perfectly satisfactory diet can be followed which involves the consumption of little or no meat. This of course disposes of a favourite view which used to be widely held, that meat was essential for a healthy diet and was essential for a sense of well-being.

It is an absolute right of individuals under most circumstances to eat what they choose. For some people there are genuine religious reasons why they do not wish to eat meat. Unfortunately, a few of the proponents of the vegetarian ethic who have no basic religious conviction on the subject still manage to elevate their cherished opinions to that of an almost religious crusade, harnessing any possible argument to bring the consumption of meat into disrepute.

It is difficult to be absolute about any of these issues, but it seems that the basic scientific position is that meat eaten in moderation and without too much attendant fat can make a valuable contribution to a nutritious and interesting diet. However, opponents of meat eating often gain leverage for their position by publicizing any dubious aspect of the production chain. They are often helped in this by the casual attitudes of some scientists and producers who perpetuate the outdated notion that, somehow, anyone involved in the food production chain should be regarded as a protected species, free to operate as he or she sees fit with no constraints. Such people fail to realize that, in the eye of the public, they are in a position of trust. The indiscriminate use of drugs, deliberate pollution of the environment and lack of concern about welfare are all problems which cause people to reconsider their automatic acceptance of the meat-eating habit. A recent informal survey of students in one faculty of a university revealed that one-third did not normally eat meat. Many support this view on what they perceive to be moral grounds.

11.2. The Future Possibilities for Technical Advance

11.2.1. Nutrition

The application of many ideas advanced in earlier chapters would, in many circumstances, greatly improve the efficiency of performance of growing animals.

Genetic engineering is a concept which may well be more acceptable for crop plants than it is for animals. New strains of familiar crops could be produced which are much more closely aligned to the needs of simple-stomached animals. For example, the protein content and protein quality of wheat could be changed to be much more similar to that required in a complete diet. Completely new crops may become available and existing crops may be modified not only in terms of their composition but also in relation to the areas in which they may be grown. For example, it is quite possible to think of soybean strains which can grow at much higher latitudes than the varieties currently available.

Genetic engineering of bacteria may also affect the ease with which certain nutrients such as the amino acids threonine and lysine can be produced by fermentation, thus making them very much cheaper. It is also conceivable that genetic engineering of bacteria might produce some advantages in the bacterial population of the gut of ruminant animals.

11.2.2. Technology and growth

The way in which the body controls the types of tissue it produces is now much better understood. It has been shown that the regulation of speed of growth and the proportion of fat to lean can be profoundly affected by such substances as the specific somatotrophin for the species (see chapter 4). These substances can now be produced in quantity by recombinant DNA technology. Trials in Europe and the USA have already shown that this material can transform the perspectives that we have for what is possible in terms of rate of growth and the partition between fatty tissues and lean. The big technical problem is how to provide growing animals with these substances so that one may avoid the labour and hazard to the animal of daily injections. The big ethical issue is whether the public will accept the scientific fact that it produces lean meat without any risk of detrimental effect to the humans who eat it, since these substances do not survive without degradation during cooking and since they are in any case far removed chemically from human somatotrophin and are totally inactive in humans.

Other substances shown to have a profoundly beneficial effect on the rate of muscle growth, and so on the leanness of the meat, include the analogues of adrenaline, the so-called β-agonists. These have been produced by many pharmaceutical companies and have the potential advantage that they are destroyed very rapidly in the tissues and leave no effective residue in the meat.

Again the ethical question of acceptability must be considered. It is important to see these developments in perspective. All animals control their own growth in one way or another and the indigenous factors they use are present in all meat. For example, the testicles and ovaries of the animal produce quite natural but very potent growth substances in meat which mankind has been consuming over the millennia without apparent ill-effect. Many of the aspects of the argument are really more about the role of technology in animal production than about whether there is any risk to the consumer. Over the years many so-called growth-promoting substances have been withdrawn from use. Among these have been the sex steroids and their analogues, several modifying agents of the bacterial flora of the gut of non-ruminants such as arsenilic acid, and many of the common antibiotics. Many other similar substances have been placed on a restricted list and may only be used under strict veterinary supervision. The reasons for their removal have usually been on the grounds of generalized contamination of the environment or because of suspected widespread abuse and failure to conform to the conditions of the licence.

11.2.3. Health of those engaged in animal production

A further factor which is increasingly being taken into account is the need of the operator to work in a healthy, dust-free environment. Cynics may regard this as an unnecessary elaboration, since protection could be given by masks. This, however, is symptomatic of an uncaring attitude since few would opt to pursue the whole of their working life encumbered by a mask unless it was absolutely necessary. Nor should it be presumed that the environment which humans find distressing is alright for animals. It is probably not, and is almost certainly responsible for suboptimal production and the exacerbation of respiratory disease. Two or more environments, one suitable for humans, can easily be designed into buildings, and if high quality staff are to be retained, it is essential that their working environment is made satisfactory and free of health hazard.

11.2.4. Breeding

There are many exciting possible developments in breeding. First there are the combinations of specially bred lines to give good reproductive performance on the female side, but with good meat characteristics in the slaughter generation. The best sire lines combine good distribution of meat with good meat quality.

The possibilities for genetic engineering or highly selective breeding are illustrated by the potential of certain of the Chinese breeds of pigs. If, for example, their greater prolifacy and sexual precocity could be transferred to the advanced white breeds, without too great a cost in terms of growth and efficiency, then there would be an enormous increase in potential.

Genetic engineering might also be used in other ways. For example, it could be used to enhance the effects of natural growth substances, such as somato-trophin, or perhaps improve the immune system of animals to make it even more effective and so reduce losses due to poor performance. The possibilities are endless, but all are subject to the procedures and the products being acceptable to the consumer.

Some studies in Australia have concentrated on the possibilities of producing animals which are capable of synthesizing limiting amino acids by introducing the genetically engineered pathways which are present in much simpler types of animal. This includes the synthesis of the sulphur amino acids in sheep and lysine in pigs.

11.2.5. Meat processing and the image of meat

Probably the most critical area for the future of meat as a marketable commodity is the role played by the processing industry. Every fat carcass can be converted into lean joints by the processor if the appropriate technology is applied. Although fat may be an embarrassment to the processor, it could be turned into an asset. Fat has a value in its own right, even if it does not continue in the human food chain. At the worst, it can be recycled through the animal feed chain where it has a considerable value as a high energy constituent of diets, particularly for the younger growing animal and for the lactating sow. Many retailing chains have demonstrated their requirements by rejecting the products of some processors, and concentrating on obtaining the product they want from any country in the world which can meet their standards. In the world of supermarket and hypermarket chains, the new 'royalty' of the production chain are the buyers. They have virtually absolute discretion over what is bought and sold, and their views must be considered not only by the processors but also by the producers.

It may seem extraordinary to those who understand the science, that meat labelled as 'naturally produced' has any kind of consumer preference. One could argue that the only natural meat is that obtained from a wild animal shot in the forest and riddled with all the natural diseases. However 'natural' is romantically whatever the buyer chooses to define as natural, and there are undoubtedly considerable market opportunities for those who are prepared to go along with their view. Again, the right of choice must be conceded to the buyer, even when such a choice involves an element of gimmick and charade. If by following this type of lead, meat can be reinstated in the purchase preferences of those who have otherwise lost confidence in the product, then this is all to good. Again it must be stressed that the producer who has accepted the terms must not break faith, for in so doing he or she may, when found out, alienate many others of the consuming public.

11.3. Conclusions

The future of meat production depends on all components of the production chain acting harmoniously together with a common objective, namely producing attractive, wholesome, cheap meat for the consuming public. To achieve this, it is necessary for all sectors to accept some discipline and to do their utmost to improve communication between all the sectors. The starting point in all this is to be absolutely sure what the consumer wants and what influences his or her choice. If the consumer has the choice to eat meat or starve, then he or she will eat meat. If there is a surfeit of choice, then the meat has to be something exceptional within this range of choice. Image has become very important. The application of science to producing better livestock must be sensitive to consumer requirements and must also ensure that proper and balanced information is supplied about all aspects of meat consumption. This must include the positive and negative aspects, for in the long run humans have a right to decide, in the light of all the evidence, whether meat production and consumption is a proper activity for a civilized society.

Index

Note: Page numbers in *italic* refer to figures and/or tables